Sea Level Fluctuations

Sea Level Fluctuations

Editor

Grigory Ivanovich Dolgikh

MDPI • Basel • Beijing • Wuhan • Barcelona • Belgrade • Manchester • Tokyo • Cluj • Tianjin

Editor
Grigory Ivanovich Dolgikh
V.I. Il'ichev Pacific Oceanological Institute
Far Eastern Branch Russian Academy of Sciences
Russia

Editorial Office
MDPI
St. Alban-Anlage 66
4052 Basel, Switzerland

This is a reprint of articles from the Special Issue published online in the open access journal *Journal of Marine Science and Engineering* (ISSN 2077-1312) (available at: https://www.mdpi.com/journal/jmse/special_issues/Grigory_sea_level_fluctuations).

For citation purposes, cite each article independently as indicated on the article page online and as indicated below:

LastName, A.A.; LastName, B.B.; LastName, C.C. Article Title. *Journal Name* **Year**, *Volume Number*, Page Range.

ISBN 978-3-0365-4293-5 (Hbk)
ISBN 978-3-0365-4294-2 (PDF)

Cover image courtesy of Dr. Grigory Ivanovich Dolgikh.

Contents

About the Editor

Grigory Ivanovich Dolgikh

Grigory Ivanovich Dolgikh (born 1954) is a Russian specialist in the field of ocean physics and laser interferometry. He is an Academician of the Russian Academy of Sciences (2016). He is Deputy Chairman of the Far Eastern Branch of the Russian Academy of Sciences (since 2008). Since December 2021, he has been the director of the Pacific Oceanological Institute of the Far East Branch of the Russian Academy of Sciences. He is a Doctor of Physical and Mathematical Sciences, Professor. He conducts scientific research in the field of studying the physics of the emergence, development, and transformation of geospheric processes in the "atmosphere-hydrosphere-lithosphere" system. He has conducted work on the experimental study of the regularities of generation, the dynamics of marine wave processes in the infrasonic range, and their transformation into seismoacoustic oscillations of the earth's crust at the "hydrosphere-lithosphere" boundary. He is also the creator of the theoretical and experimental foundations for the use of laser interference methods for studying the ocean.

Preface to "Sea Level Fluctuations"

Sea level changes and atmospheric pressure fluctuations, superimposed on these changes, are of tremendous significance, and should be studied when researching various non-uniformly scaled processes.

When studying various non-uniformly scaled processes in the sea/ocean that cause fluctuations in its level, it is necessary to identify their primary source, which can be located in any of the geospheres: atmosphere, hydrosphere, and lithosphere.

The identification of primary sources, studying patterns of dynamics and the transformation of wave and non-wave processes at the interface of the geospheres is one of the main tasks of modern science. The complexity of this identification is associated with ambiguous interpretation of the results of observations, obtained in only one geosphere.

Development of this research direction will allow us to integrate several directions into the comprehensive whole, in which the causes of sea level fluctuations, from the hydroacoustic range to secular fluctuations, are studied, making it possible to establish the patterns of their occurrence, dynamics, and transformation, with assessment of their impact on the earth's biosphere.

Grigory Ivanovich Dolgikh
Editor

Journal of
Marine Science and Engineering

Editorial

Sea Level Fluctuations

Grigory Ivanovich Dolgikh

V.I. Il'ichev Pacific Oceanological Institute, Far Eastern Branch Russian Academy of Sciences (FEB RAS), 690041 Vladivostok, Russia; dolgikh@poi.dvo.ru

Citation: Dolgikh, G.I. Sea Level Fluctuations. *J. Mar. Sci. Eng.* **2022**, *10*, 330. https://doi.org/10.3390/jmse10030330

Received: 2 February 2022
Accepted: 21 February 2022
Published: 25 February 2022

Publisher's Note: MDPI stays neutral with regard to jurisdictional claims in published maps and institutional affiliations.

1. Introduction

We do not consider sea level change due to global warming, but only sea level fluctuations in our time scale. However, we understand that sea level changes and atmospheric pressure fluctuations, superimposed on these changes, are of tremendous significance, and should be studied when researching various non-uniformly scaled processes.

When studying various non-uniformly scaled processes in the sea/ocean that cause fluctuations in its level, it is necessary to identify their primary source, which can be located in any of the geospheres. The identification of primary sources, studying patterns of dynamics and the transformation of wave and non-wave processes at the interface of the geospheres, is one of the main tasks of modern science. The complexity of this identification is associated with ambiguous interpretation of the results of observations, obtained in only one geosphere.

Development of this direction will allow us to integrate several directions into the comprehensive whole, in which the causes of sea level fluctuations, from the hydroacoustic range to secular fluctuations, are studied, making it possible to establish the patterns of their occurrence, dynamics and transformation, with assessment of their impact on the earth's biosphere.

2. Some Results of the Sea Fluctuation Research

It is undeniable that the main cause of sea level change is gravity and infragravity sea waves at work [1], which occur in various regions of the world ocean; some of them are regional in nature. This paper considers several mechanisms of generation and propagation of waves with different periods during typhoon movement. The relationship between the variations of the main periods of gravity sea waves with dispersion and the Doppler effect, and the variations in wind speed and direction was studied. A thorough study of the dynamics of the main parameters of gravity sea waves in paper [2] revealed new patterns of surface progressive gravity waves and their transformation into primary microseisms, when the waves move on the shelf with decreasing depth. The non-isochronal behavior of progressive waves was established, which manifests itself in a decrease in the gravity waves' periods due to the conversion of a part of their energy into the energy of primary microseisms. A new method for studying modulation effects, based on regression analysis and general period change functions, was presented in work [3]. Using this method, it was shown that when wind waves are modulated by tides, waves with a large period and amplitude are concentrated at high points of the tide. However, when extraneous wave processes occur, such as seiches, the modulation of wave amplitude may have an extremum in the low point of the tide, i.e., modulation of the wave period and its amplitude will be in antiphase. These gravity sea waves interacting with the bottom transform into Rayleigh surface waves, moving along the water–bottom boundary. These waves are also known as surface Scholte waves. The general properties of these waves' propagation are actively used in solving some problems of geoacoustic parameter inversion [4].

Changes in the ocean level by various wave and non-wave processes, the action of typhoons and cyclones over the water area, and the presence of various currents directly affect the propagation of sound in the ocean [5,6]. On the other hand, solving the inverse problem, we can make hydroacoustics serve as a tool for studying various sea processes. For example, in paper [7], a mesoscale vortex in the Pacific Ocean and its dynamics are studied using hydroacoustic and oceanic methods.

Abnormal changes in ocean level can lead to devastating environmental consequences. Thus, the anomalous ocean level variability during the summer months of 2020 affected the routine behavior and food resource formation of microalgae at work [8]. The strongest temperature anomalies of the aquatic environment in the entire history of observations, combined with specific conditions of coastal water circulation, became extremely favorable for the seasonal development of phytoplankton, which ultimately led to an intense red tide and had a negative impact on the coastal waters of the Kamchatka Peninsula (even leading to the mass death of hydrobionts).

The studies of the processes leading to changes in the ocean level are carried out by contact and remote methods. Remote methods primarily include aerospace methods. A method for obtaining spatial spectra of slopes and heights of surface waves from aerospace optical images was developed, taking into account the nonlinear modulation of the brightness field at work by the slopes of the sea surface [9]. This method can be used for aerospace monitoring of sea areas, to identify and study the variability of sea waves spectra associated with abnormal phenomena and processes, including those of anthropogenic origin. The remote methods of a more economical nature undoubtedly include the technology of surface and underwater video monitoring [10]. This paper describes the technology for express analysis of images and videos recorded by coastal video surveillance systems. Its main feature is the ability to measure or evaluate in real time the parameters of sea waves, sea level fluctuations, underwater current fluctuations, etc.

When considering marine processes that lead to changes in sea level, one cannot disregard tsunamis, which bring significant troubles to humankind. Paper [11] presents a new method for short-term tsunami forecasting. General regularities of distribution of deformation anomalies associated with seafloor motions and leading to tsunamis were established in three seismically active zones of the earth [11]. The established regularities, the foremost of which is the general law of divergence, confirm the indisputable fact that these deformation anomalies, the behavior of which is described by the sine-Gordon equation, are related to the tsunami generation process. The heights of the tsunamis, however, are not directly proportional to the magnitude of the seabed displacements.

In conclusion, we can state that a comprehensive analysis of the papers, published in this thematic issue, is extremely useful for understanding the processes and phenomena of the interacting geospheres of the earth, its biosphere, and man. These papers are completely united by the common theme of the "Study of the fundamental foundations of the emergence, development, transformation and interaction of hydroacoustic, hydrophysical and geophysical fields in the World Ocean".

Funding: The work was carried out with financial support from the Russian Federation represented by the Ministry of Science and Higher Education of the Russian Federation, Agreement No. AAAA-A20-120021990003-3.

Institutional Review Board Statement: Not applicable.

Informed Consent Statement: Not applicable.

Data Availability Statement: I do not own the data.

Conflicts of Interest: The author declares no conflict of interest.

References

1. Dolgikh, G.; Budrin, S.; Dolgikh, S. Fluctuations of the Sea Level, Caused by Gravitational and Infra–Gravitational Sea Waves. *J. Mar. Sci. Eng.* **2020**, *8*, 796. [CrossRef]
2. Dolgikh, G.I.; Gromasheva, O.S.; Dolgikh, S.G.; Plotnikov, A.A. Dynamics and Transformation of Sea Surface Gravity Waves at the Shelf of Decreasing Depth. *J. Mar. Sci. Eng.* **2021**, *9*, 861. [CrossRef]
3. Dolgikh, G.I.; Budrin, S.S. Method of Studying Modulation Effects of Wind and Swell Waves on Tidal and Seiche Oscillations. *J. Mar. Sci. Eng.* **2021**, *9*, 926. [CrossRef]
4. Dong, Y.; Piao, S.; Gong, L.; Zheng, G.; Iqbal, K.; Zhang, S.; Wang, X. Scholte Wave Dispersion Modeling and Subsequent Application in Seabed Shear-Wave Velocity Profile Inversion. *J. Mar. Sci. Eng.* **2021**, *9*, 840. [CrossRef]
5. Liu, J.; Piao, S.; Zhang, M.; Zhang, S.; Guo, J.; Gong, L. Characteristics of Three-Dimensional Sound Propagation in Western North Pacific Fronts. *J. Mar. Sci. Eng.* **2021**, *9*, 1035. [CrossRef]
6. Khan, S.; Song, Y.; Huang, J.; Piao, S. Analysis of Underwater Acoustic Propagation under the Influence of Mesoscale Ocean Vortices. *J. Mar. Sci. Eng.* **2021**, *9*, 799. [CrossRef]
7. Liu, J.; Piao, S.; Gong, L.; Zhang, M.; Guo, Y.; Zhang, S. The Effect of Mesoscale Eddy on the Characteristic of Sound Propagation. *J. Mar. Sci. Eng.* **2021**, *9*, 787. [CrossRef]
8. Bondur, V.; Zamshin, V.; Chvertkova, O.; Matrosova, E.; Khodaeva, V. Detection and Analysis of the Causes of Intensive Harmful Algal Bloom in Kamchatka Based on Satellite Data. *J. Mar. Sci. Eng.* **2021**, *9*, 1092. [CrossRef]
9. Bondur, V.; Murynin, A. The Approach for Studying Variability of Sea Wave Spectra in a Wide Range of Wavelengths from High-Resolution Satellite Optical Imagery. *J. Mar. Sci. Eng.* **2021**, *9*, 823. [CrossRef]
10. Fischenko, V.K.; Goncharova, A.A.; Dolgikh, G.I.; Zimin, P.S.; Subote, A.E.; Klescheva, N.A.; Golik, A.V. Express Image and Video Analysis Technology QAVIS: Application in System for Video Monitoring of Peter the Great Bay (Sea of Japan/East Sea). *J. Mar. Sci. Eng.* **2021**, *9*, 1073. [CrossRef]
11. Dolgikh, G.; Dolgikh, S. Deformation Anomalies Accompanying Tsunami Origination. *J. Mar. Sci. Eng.* **2021**, *9*, 1144. [CrossRef]

Journal of
Marine Science and Engineering

Article

Fluctuations of the Sea Level, Caused by Gravitational and Infra–Gravitational Sea Waves

Grigory Dolgikh *, Sergey Budrin and Stanislav Dolgikh

V.I. Il'ichev Pacific Oceanological Institute FEB RAS, 690041 Vladivostok, Russia; ss_budrin@mail.ru (S.B.);
sdolgikh@poi.dvo.ru (S.D.)
* Correspondence: dolgikh@poi.dvo.ru

Received: 4 September 2020; Accepted: 10 October 2020; Published: 13 October 2020

Abstract: In the article we analyzed the results of processing experimental data of the range of surface gravity sea wind waves (2–20 s) and the range of infra-gravitational sea waves (30 s–10 min), obtained on the laser meter of hydrosphere pressure variations. The laser meter of hydrosphere pressure variations was installed for a long time on the bottom at different points of the Sea of Japan shelf. This paper presents the results of the analysis of swell waves caused by the KOMPASU typhoon, which passed over the Sea of Japan on 2–3 September 2010. Several mechanisms of the generation and propagation of waves with different periods during the typhoon movement are considered. In the course of the analysis, we studied the connection between variations of the main periods of gravitational sea waves with the dispersion and the Doppler effect, variations of speed and direction of the wind in a typhoon zone. The nonlinearity of the process of wave period change caused by dispersion is estimated. In the combined analysis of variations of hydrosphere pressure in the ranges of gravitational and infra-gravitational sea waves, we studied their energy relationships and determined regional infra-gravitational sea waves, which make a significant contribution to the energy of the infra-gravitational range.

Keywords: gravitational waves; pressure variations; period variation; laser meter of hydrosphere pressure variations; infragravity waves; gravity wind waves; laser strainmeter; typhoon

---

1. Introduction

Many results of the theory of sea surface wind waves can be considered the classics of hydrodynamics, the starting point of which is the Archimedes' work "On floating bodies" [1]. In the works of Newton, Pascal, and Euler, this principle was developed, with receipt of analytical equations of hydrodynamics, which are the basis for describing sea gravity wind waves. All basic equations of hydrodynamics are inherently nonlinear; the main way to solve them was to find particular solutions for certain initial conditions. The next stage in developing the theory of waves started with the creation of a new method for solving nonlinear partial differential equations. In 1967, physicists J. Green, C. Gardner, and M. Kruskal [2], using the method of the inverse scattering problem that they created, showed that the Korteweg-de Vries equation has solutions absolutely for all initial conditions.

The same year, T. Benjamin and J. Feyer [3] by theoretical calculations managed to show that because of the instability of a periodic wave in deep water, the waves break into the groups. V.E. Zakharov obtained the equation, describing this process, in 1968. This equation describes the formation of groups of several dozen waves, while the average wave in the enveloping curve of the entire group is the largest, but if there are too many waves in the group, then this group will split into several ones. Basically, after that, theoretical studies were reduced to numerical solutions and modeling of nonlinear wave fields. In the early 1980s, there was rapid progress in experimental and numerical research, in this respect, we should mention the works of M. Longuet-Higgins [4–6], K. Hasselman [7,8], and E. Caponi [9].

Currently, in the world there are several most common models used to calculate and predict wind surface waves: spectral-parametric wave models of the 2nd generation AARI-PD2, discrete models of the 3rd generation—WAM, WAVEWATCH, and SWAN, which take into account nonlinear effects and are designed specifically for calculating wind waves in shallow water areas [10–12].

Further research developed with varying success, but many problems have not been solved yet: mainly, the problems associated with matching the obtained experimental data and the results of model-theoretical studies. We can mention some of them in this work: (1) spatiotemporal development of the wave process during nonlinear motion of sea gravity surface wind waves initiation sources (typhoons, cyclones, etc.); (2) variations in the periods of gravity waves not associated with the dispersion process; (3) role of gravity waves in the initiation of sea infragravity waves and their regional peculiarities; and (4) initiation of rogue waves and the role of gravity waves in this process. In this paper, special attention will be paid to solving the above problems, basing on obtained experimental data, the results of their processing and analysis.

2. Experimental Data

The experimental data analyzed in this paper were obtained using a laser meter of hydrosphere pressure variations, created on the basis of an equal-arm Michelson interferometer with a frequency-stabilized helium-neon laser [13]. Instruments of this type carry out measurements all year round in the permanent mode at their installation point; the registration is interrupted only for short periods of time for adjustment and troubleshooting. The appearance and optical scheme of the laser meter of hydrosphere pressure variations are shown in Figure 1.

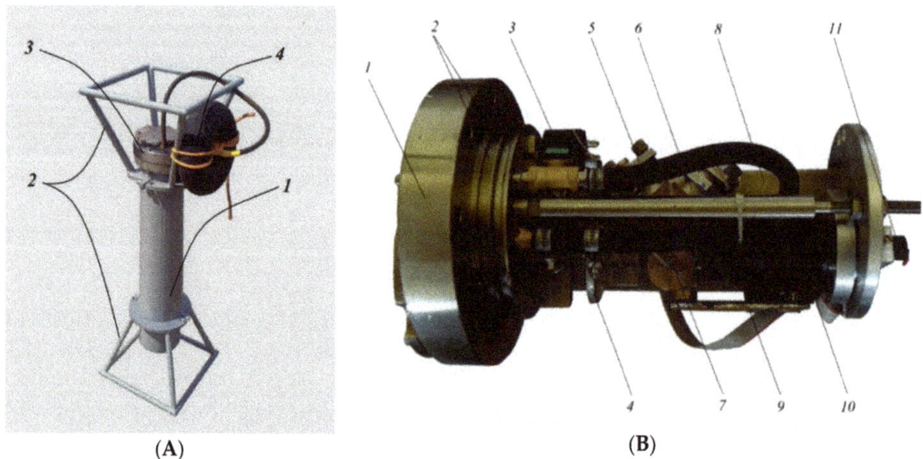

Figure 1. External view (**A**) and optical scheme of a laser meter of hydrosphere pressure variations (**B**).

The laser meter of hydrosphere pressure variations (Figure 1A) is a cylindrical housing (1), which is mounted in the protective grating (2). One side of the housing is hermetically sealed and has a cable port. The other side is sealed with a removable cover (3). Outside the instrument, there is an elastic air-filled container (4). Its outlet is connected with a tube to a compensation chamber, located in the removable cover. The sensitive element of the laser meter of hydrosphere pressure variations is the round membrane fixed in its removable cover in such a way that one side of it is in contact with water, and the other side faces the inside of the instrument and is a part of the Michelson interferometer.

All optical elements of the laser meter of hydrosphere pressure variations (Figure 1B) are rigidly fixed on the optical base (8). The laser beam (11) comes through the collimator (10) to the plane-parallel dividing plate (7), which splits it into two beams—measuring and reference. The first (measuring) beam is directed to the lens (4), then, through the optical window—to the mirror-coated membrane located in the cover (1). After reflecting from the membrane, the beam again comes to the lens and then to the dividing plate (7), from which it is reflected to the photodiode of the resonance amplifier (9). The second (reference) beam, after the dividing plate, passes through the system of control mirrors (5) and (6), mounted on piezoceramic bases. Then, like the measuring beam, it comes to the photodiode of the resonance amplifier. By means of these two beams, the interference pattern is adjusted; its change corresponds to the variations of hydrosphere pressure affecting the membrane.

It is designed to record hydrosphere pressure variations in the frequency range from 0 (conventionally) to 1000 Hz, with an accuracy of 0.1 mPa at sea depths of up to 500 m using a specially designed compensation chamber. The laser meter of hydrosphere pressure variations was installed on the bottom at various points of the Sea of Japan shelf and measurements were taken with a duration from several days to several months. The obtained experimental data were transferred in real time to the laboratory room located at Schultz Cape, the Sea of Japan, Primorsky Region of the Russian Federation, and, after preliminary processing, were loaded in the experimental database. Subsequently, the experimental data were processed according to the set tasks. In the low-frequency sound range (20–1000 Hz), the propagation regularities of hydroacoustic signals, generated, among other things, by low-frequency hydroacoustic emitters, were investigated [14]. In the higher frequency range (1–20 Hz), the formation peculiarities of the microseisms in the "voice of the sea" range were studied [15]. The scope of our interests in the lower frequency range included studies aimed at investigation of the free oscillations of the Sea of Japan and some of its bays, recession-setup phenomena, tidal dynamics, and the relationship of oscillations and waves of the infrasonic range of the hydrosphere with oscillations and waves of the corresponding period of the atmosphere and lithosphere. Of interest are also processes in the range of gravity sea wind waves (2–20 s), in the range of infragravity sea waves (30 s–10 min), and nonlinear effects leading to the formation of surface and internal soliton-like disturbances of large amplitude. It is these processes, described in the previous sentence, that we will pay attention to in this paper. Along with tsunamis, these processes lead to significant variations in sea level, in some cases causing serious damage to human activities.

When processing the experimental data of the gravity sea wind waves range, we found that the temporal change in the periods of wind waves, emerging from the typhoon action zone, so-called swell waves, did not occur linearly as a consequence of the dispersion law. For further analysis, 16 record fragments with a characteristic change in the period of surface waves were selected. The fragments sampling rate was 1000 Hz. When processing the data, high frequencies were filtered by the Hamming window, the filter order was 3000, the cutoff frequency was 1 Hz, followed by 1000 times decimation. Thus, the extreme frequency in the spectral analysis was 500 mHz (2 s). To get rid of the low-frequency background in the signal, low frequency filtering was also carried out by the Hamming window, the filter length was 1500, the cutoff frequency was 50 mHz (20 s). Thus, a frequency range, corresponding to periods of surface wind waves, was selected from the entire signal.

After processing and filtering the signal, the signal spectrograms were constructed for all selected fragments. Then, with a step of 60 samples (1 min), the frequency maxima of the region, corresponding to wind waves, were singled out. In order to visually represent the period change nature, on the obtained spectral maxima, by means of regression analysis, the functions were constructed, which were the polynomials of the 6th degree, since they adequately described the nature of the changes. Figures 2 and 3 show the spectrograms of some record fragments and the corresponding regression curves.

As we can see from Figures 2 and 3, changes in the period of wind waves in all cases had a different character. In the case of the regression curves, presented in Figure 2, there were areas of abrupt period change and areas in which the period practically did not change, while Figure 3 shows a fragment, in which the period decreased monotonically. Most likely, this effect was associated with the dispersion of wind waves and changes in the wind conditions in the part of the water area, where these waves were generated [16]. For more detailed description of the reasons for the wind waves period change, let us study this process while analyzing the data obtained during the movement of the KOMPASU typhoon in the Sea of Japan.

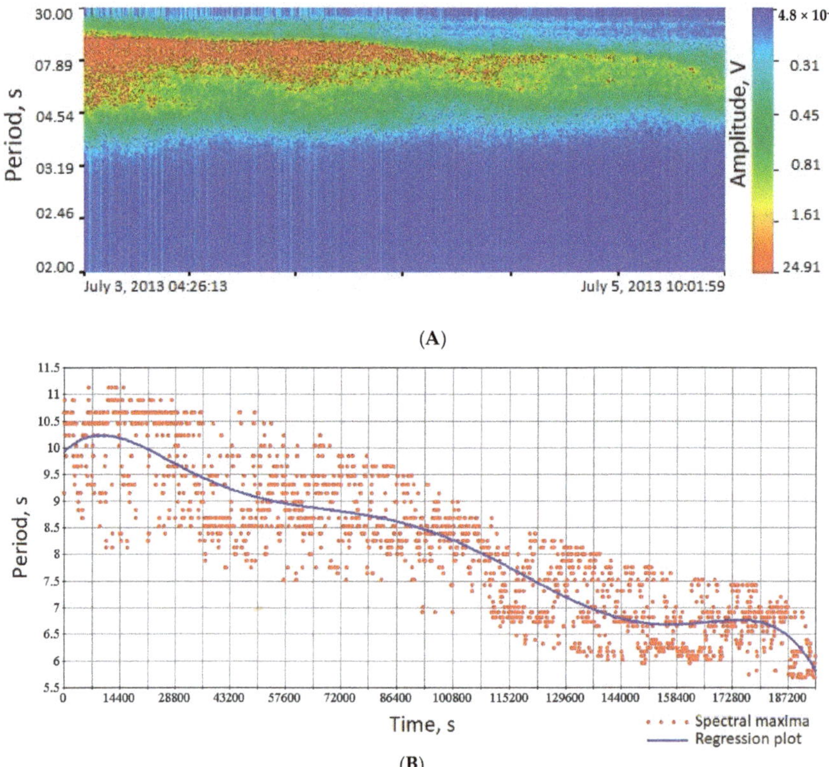

Figure 2. (**A**) Swell wave spectrogram recorded by a laser meter of hydrosphere pressure variations in the period from 3 July to 5 July 2013 and (**B**) regression plot, constructed on the spectral maxima singled out from the signal spectrogram recorded in the period from 3 July to 5 July 2013.

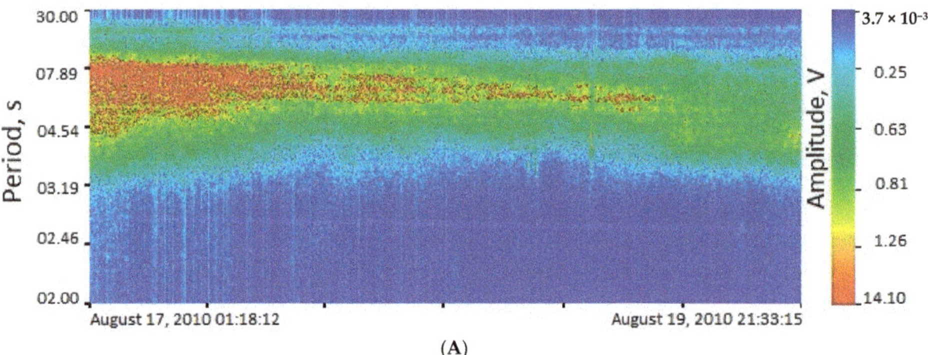

Figure 3. *Cont.*

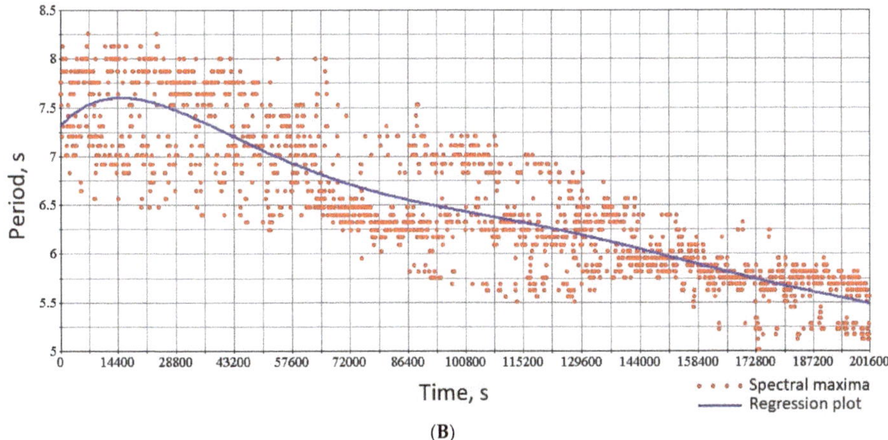

Figure 3. (**A**) Swell wave spectrogram recorded by a laser meter of hydrosphere pressure variations in the period from 17 August to 19 August 2010 and (**B**) regression plot, constructed on the spectral maxima singled out from the signal spectrogram recorded in the period from 17 August to 19 August 2010.

3. KOMPASU Typhoon Swell Waves Study

On 2 September 2010, typhoon KOMPASU entered the Sea of Japan, at 10:00 a.m. it passed over the territory of the Korean Peninsula; its trajectory is shown in Figure 4.

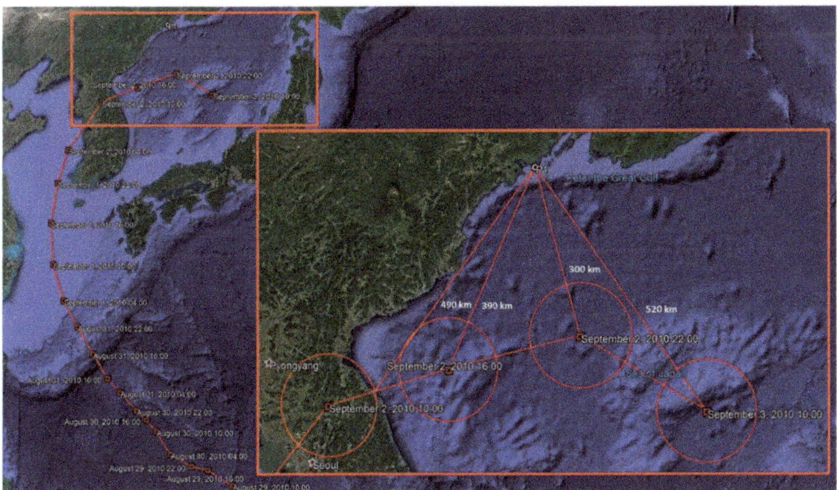

Figure 4. Typhoon KOMPASU trajectory. Point P1—"Vityaz" Bay, the points of the typhoon passage are indicated by the date and time (local time), red circles indicate areas with winds exceeding 18 m/s, the area radius is 90 km. Distances to "Vityaz" Bay are noted for three points of the typhoon distribution.

The dates and time are shown in Figure 4 at each point, where the typhoon passed. The typhoon began to enter the Sea of Japan on 2 September 2010 at 10:00, the last point of the typhoon trajectory—on 3 September 2010 at 04:00 (local time).

Figure 5 shows a spectrogram of the record of the laser meter of hydrosphere pressure variations located in Vityaz Bay during the typhoon passage and regression plot, constructed on the spectral maxima singled out from the signal spectrogram.

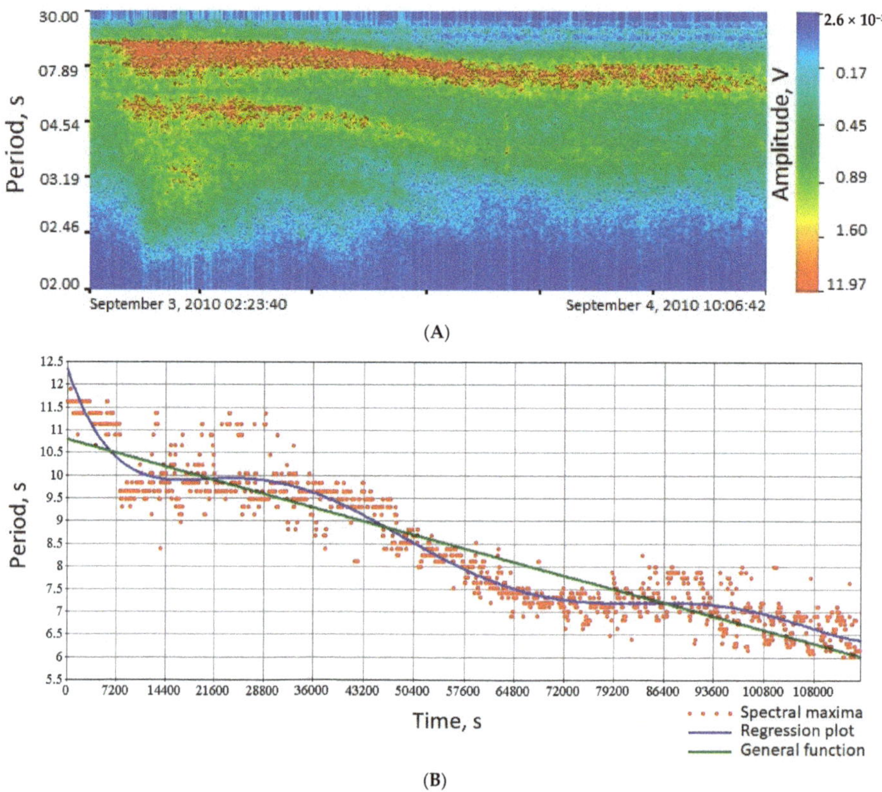

(A)

(B)

Figure 5. (**A**) Spectrogram of the record of the laser meter of hydrosphere pressure variations located in Vityaz Bay during the KOMPASU typhoon passage and (**B**) regression plot, constructed on the spectral maxima singled out from the signal spectrogram.

At the initial stage of Figure 2A, the period of the incoming swell was about 11.4 s. As we can see from the spectrogram, the lowest frequency process was nothing more than swell waves coming from the typhoon, which had a fairly wide range from 11.4 to 6 s. In this case, higher frequency processes were of no interest to us and they were local wind waves. Let us compare the general period variation function [17], constructed from the spectral maxima, with a more complex regression type that can better describe the effects associated with the typhoon passage. To estimate the polynomial regression and the general function of the period change, we used two criteria for the regression analysis evaluation: the coefficient of determination R^2 (equal to 1 in the ideal case) and the standard deviation S (equal to 0 in the ideal case). The values of these quantities are given in Table 1.

Table 1. Regression analysis evaluation criteria.

	R^2	S
Polynomial regression	0.926	0.4
Period function	0.865	0.54
Difference	0.061	0.14

As we can see from Table 1, the difference between two regressions was not significant, but was still present. Now let us take a closer look at the effects of period change, associated with the propagation of this typhoon. Let us calculate the time of swell propagation from the supposed source. Let us use the formula for the propagation velocity in the deep sea approximation $c = gT/2\pi$. Entering the period 11.4 s into the express (A) spectrogram of the record of the laser meter of hydrosphere pressure variations and (B) regression plot, constructed on the spectral maxima singled out from the signal spectrogram. Taking into account that we need the group propagation velocity, and not the phase one, i.e., $c_{gr} = c_{ph}/2$, we estimated that the propagation speed in this case will be 8.98 m/s. Considering the distance of 390 km from the 2nd typhoon point in Peter the Great Bay, we found that the swell waves should come to Vityaz Bay in 43,870 s or 12.18 h. If we take into account that the typhoon came to the point on 2 September 2010 at 16:00, then, according to calculations, the swell waves should come to the place of registration on 3 September 2010 at about 04:00. Figure 6 shows a spectrogram with a mark of the main group of swell waves arrival to Vityaz Bay. Considering the estimated time of arrival, the coincidence is very good, but as we can see, before this time there is also swell with a period of 11.4 s on the spectrogram. We could explain this arrival by the fact that the swell waves began to form earlier than the calculated point, and, taking into account the time of 2 h, this point was approximately 60 km down the typhoon trajectory.

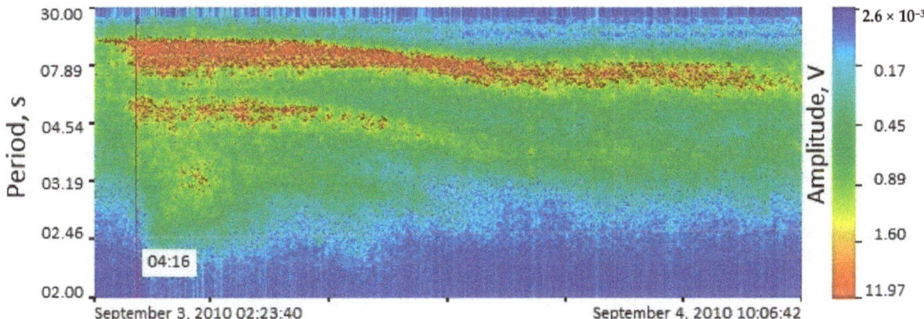

Figure 6. Time of swell waves arrival to the registration point.

Let us consider points 3 and 4 (final) of the typhoon movement, the distance from each of them to Vityaz Bay is 300 and 520 km. Considering the swell propagation speed, we found that the swell waves should propagate to the registration point for 9 and 16 h (Figure 7). In theory, all this time the typhoon wind conditions should feed the swell waves with energy, and the wave period should not decrease during this period.

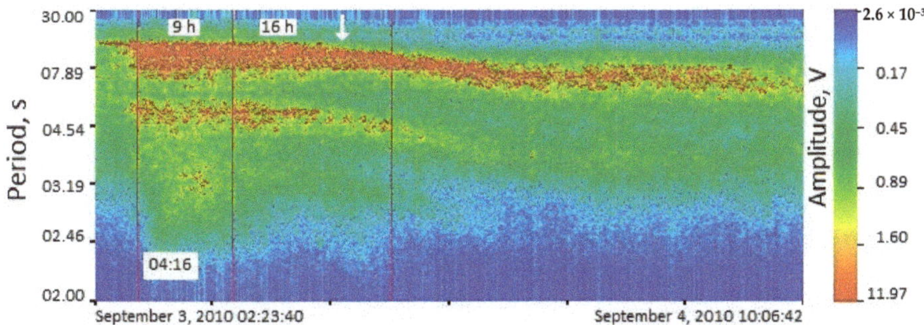

Figure 7. Segments of swell waves arrival time from points 3 and 4 of the typhoon trajectory.

Let us look at Figure 7. It is assumed that at the selected time intervals, the swell waves with a period of 11.4 s should be fed, and the period of the waves should not change. However, as we can see from the 16-h fragment, the wave period decreased (shown by the white arrow). This effect could have several explanations: (1) change in the wind conditions. Indeed, according to the typhoon data, when passing points 2 and 4, the typhoon wind speed changed from 23 to 15 m/s, which could affect the period changes. (2) Doppler effect. If we look at the typhoon trajectory, we can clearly see that from point 2 to point 3 the typhoon moved almost parallel to the section line of Vityaz Bay, and at point 4 it moved from it at a certain angle, which in turn could also influence the changes in the wave period.

Let us consider the influence of the typhoon movement direction, its speed and the angle of its movement direction to the receiver. The starting point for calculations was the point where the typhoon enters the Sea of Japan, i.e., a point located in the middle of the section between point 1 and 2 of the typhoon passage. The typhoon was at this point on 2 September 2010 at 13:00; we took this point as 0, the time from this point we counted in seconds. Tables 2 and 3 show the parameters of the typhoon movement and wind speed during its movement.

Table 2. Typhoon parameters.

Time, s	Speed, m/s	Angle, Degrees
0	9.76	140
10,800	9.81	130
32,400	11.2	90
54,000	11.9	30

Table 3. Wind data.

Time, s	Wind Speed, m/s
0	23
10,800	20
32,400	18
54,000	15

We also wrote information on wind speed into Table 3.

Let us calculate the changes in the wave period by the formula:

$$T(t) = T_0\left(1 + \frac{V(t) * \cos(\theta(t))}{c}\right) \tag{1}$$

where: T_0 is initial value of the wave period, $V(t)$ is the typhoon movement speed, and c is wind wave speed. Figure 8 shows the calculated graph of the possible period change due to the Doppler effect and data on the wind speed change.

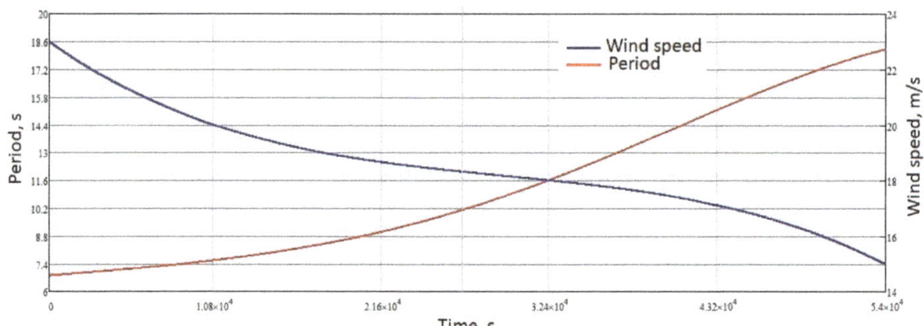

Figure 8. Graph of the swell period change on the typhoon speed and the angle of direction to the receiver (left scale, red curve), change in the typhoon wind speed (right scale, blue curve).

According to the graph of the swell period change, presented in Figure 8, we compiled a table of the swell waves periods on the time of the typhoon movement, every 3 h, starting from the starting point in the water area (2 September 2010 at 13:00), taken as 0, we also indicate the distance to the receiver, propagation time, time at a point, and time of arrival to the receiver.

We can see from Table 4, that waves with periods of 9 and 11.5 s during the typhoon movement should arrive simultaneously, and ones with periods of 7.5 and 15 s. What is interesting is that 11.4 and 9 s were the main periods on the spectrogram, if we display the data from Table 4 on it, it will look like this (see Figure 9).

Table 4. Estimated data.

Time on Graph, s	Period, s	Propagation Time, h	Time at a Point	Arrival Time
0	6.8	24	2 September 2010 13:00	3 September 2010 13:00
10,800	7.5	18	2 September 2010 16:00	3 September 2010 10:00
21,600	9	12	2 September 2010 19:00	3 September 2010 07:00
32,400	11.5	9	2 September 2010 22:00	3 September 2010 07:00
43,200	15	9	3 September 2010 01:00	3 September 2010 10:00
54,000	18	10	3 September 2010 04:00	3 September 2010 14:00

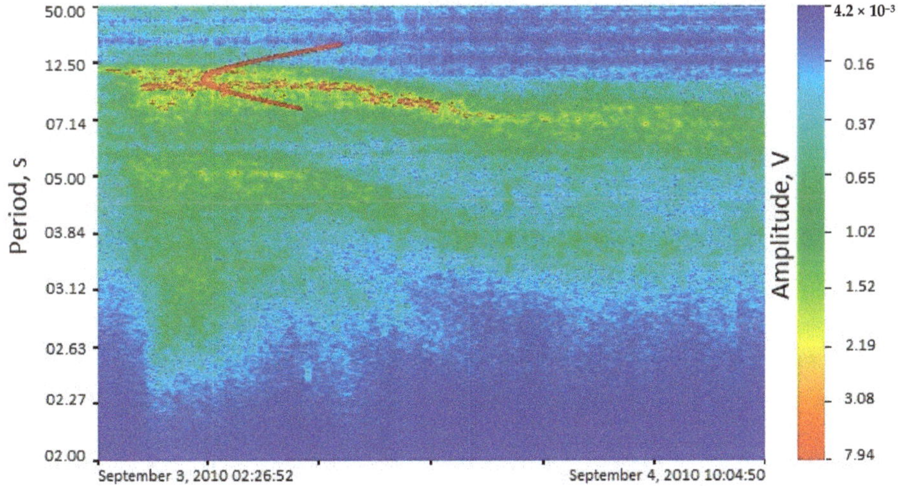

Figure 9. Calculated Doppler effect.

However, the time difference between the arrival of the main front of the swell waves in relation to the arrival of the 9 and 11.5 s waves, theoretically generated by the typhoon movement, was 3 h. The question arises whether this effect can occur not in the very center of the typhoon movement, but in its vicinity. Let us estimate the calculation error in kilometers, i.e., the distances that the waves of these periods would pass in 3 h. Error estimates are given in Table 5.

Table 5. Error estimates.

Period, s	Propagation Speed, m/s	Distance, km
9	7	75
11.5	8.9	96

Initially, it was said that the typhoon had a radius of 90 km; based on the data in Table 5, we could assume that simultaneous arrival of the swell front with periods of 9 and 11.5 s caused by the Doppler effect for this typhoon was possible only in one case: when this effect was formed by the leading front of the typhoon or near it. There is a question: if this effect really exists, then, as we already mentioned, when moving to the last point, the typhoon should initiate waves with a period of 18 s, and there were none of them on the spectrogram. We can explain this by a decrease in wind speed, which is shown in Figure 7. Thus, the waves with a period of 18 s will have very small amplitude and be invisible against the background of other processes. In support of this assumption, we should consider the wave spectrum in the interval from the moment of the first arrival of the wave packet to the beginning of the decrease in the swell period, shown in Figure 10.

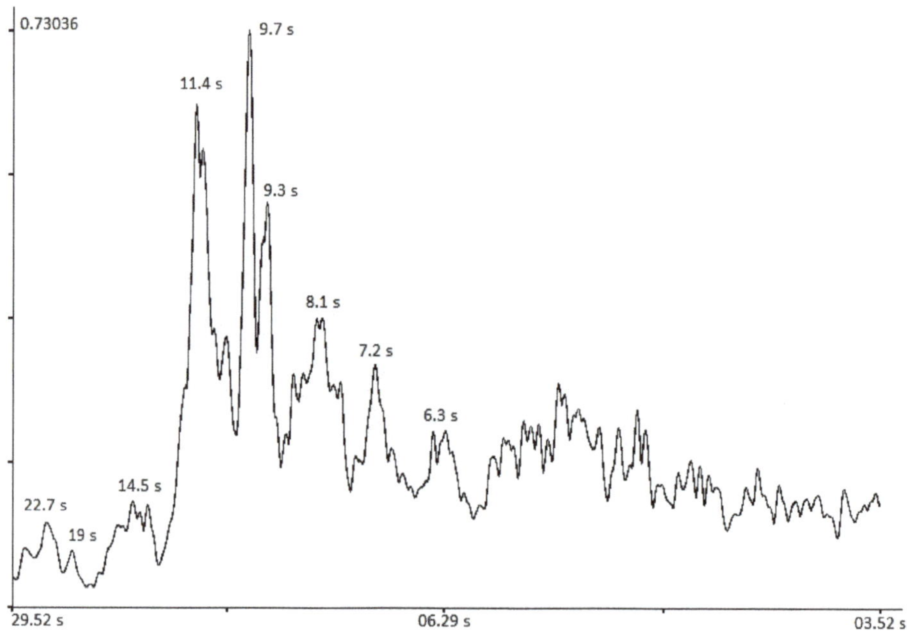

Figure 10. Swell waves spectrum.

To begin with, we should say that the wave condition is a complex and non-stationary process, the period is constantly changing within small limits and it is impossible to talk about any exact figures, only within certain limits. As we can see from Figure 9, the periods of the main harmonics approximately coincided with the calculated periods from Table 4, which may testify in favor of the assumption that a number of harmonics of swell waves are formed during the movement of a typhoon leading front. We can also see the harmonic of the period 19 s on the spectrum, which, in turn, can confirm the assumption about the small amplitude of these waves due to the wind decrease, and their formation when the typhoon moved in the direction from the source.

Of course, there is also a simple explanation for the simultaneous arrival of waves with periods of 9 and 11.5 to the receiving point. This will happen if 9 s waves originated at a point, located between the first and second points of the typhoon trajectory, indicated in Figure 4, at 490 km from the receiving point. The typhoon was at the indicated point on 2 September 2010 at 13:00. Then the waves of the specified period will reach the receiving point on 3 September 2010 at 6:00. The waves with a period of 11.5 s should be generated in the second point of the typhoon trajectory, at the distance of 390 km from the receiver, and will come to the receiving point on 3 September 2010 at 4:00. The difference of 2 h can also be leveled with the typhoon radius of 90 km, since the error for the period of 9 s will be 50 km. There is an even simpler explanation, which is in the concept itself of the swell waves group propagation speed. When a typhoon affects the water surface, the waves of a sufficiently small period begin to appear, which propagate with a phase speed; over time, the waves period increases and, reaching a maximum during propagation, begins to capture waves of a close period; they, in turn, begin to capture waves of a period close to them. As a result, the entire wave packet propagates at the speed of the maximum possible period, and at a speed 2 times lower than the phase one, i.e., with group speed. However, not all periods can capture each other, which is why such a "comb" of harmonics appeared in the spectrum. We can easily show the formation of wave groups by making calculations by a simple equation.

$$y(t) = h \times \cos\left(\frac{\omega_1 - \omega_2}{2}\right)\sin\left(\frac{\omega_1 + \omega_2}{2}\right) \tag{2}$$

Figure 11 shows an example of the formation of waves with periods of 11.4, 9 and 7 s, the main period of which was 11.4 s, and propagating with the group speed.

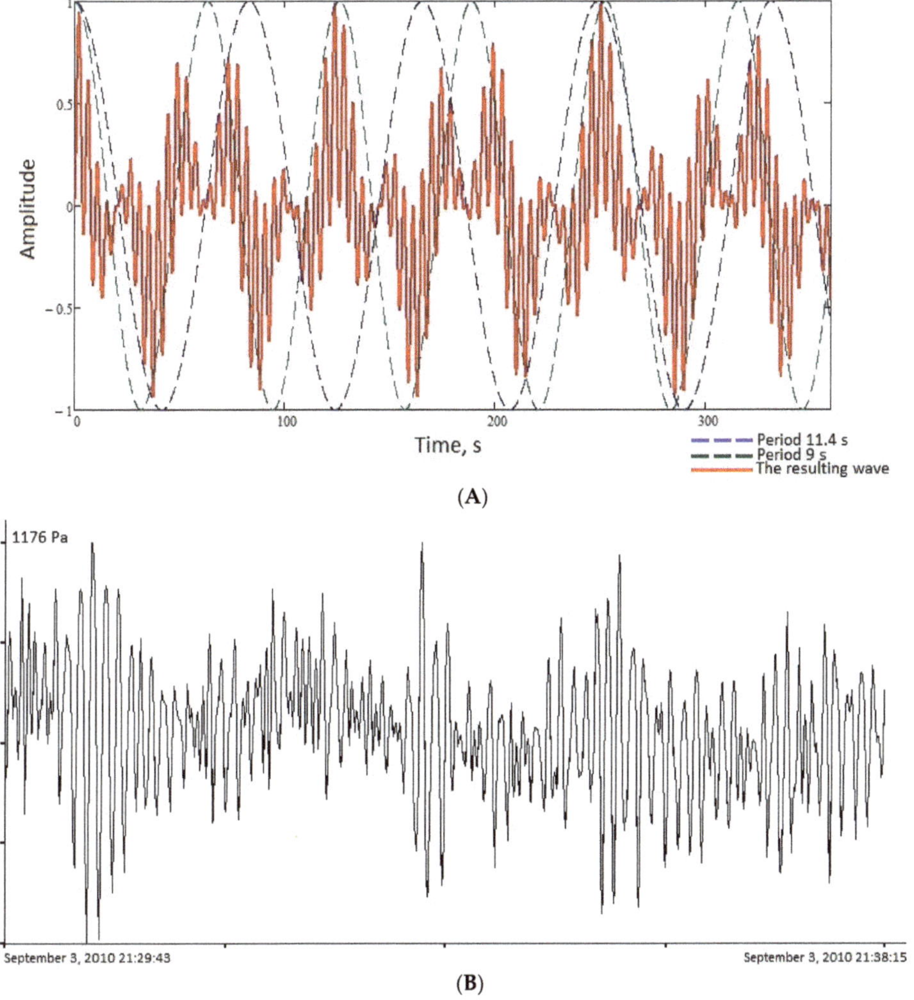

Figure 11. (**A**) Group of waves formed with a period of 11.4 s, which captured a wave with a period of 9 s, which, in turn, captured a wave with a period of 7 s (11.4 s enveloping curve—blue dotted line, 9 s enveloping curve—green dotted line, and red line is the resulting wave, consisting of three wave periods). (**B**) Record fragment of the laser meter of hydrosphere pressure variations of the typhoon arrival area, shown for comparison.

Now let us consider the question of how, from a fragment with a changing period, can we determine the distance to the place of origin of the wave process. From the spectrogram, shown in Figure 6, we can see that about 13 h passed from the beginning of the period decrease to its final value, and to determine the distance to the source, we could use the simple formula $S = c \cdot \Delta t$, where Δt is the time interval in seconds and c is the group propagation speed. In this case, in the formula for the group speed, we should substitute the largest period, from which the decrease begins, the final period in this case did not matter, we can show this using the general function of period change. According to [18], we wrote down the general formula for period change as:

$$\overline{T}(t) = K_{10} \times \frac{\Delta T}{\Delta t} \times t + T_0 \tag{3}$$

Substitute this expression into the formula for finding the distance

$$S = \frac{g(K_{10} \times \frac{\Delta T}{\Delta t} \times t + T_0)\Delta t}{4\pi} \tag{4}$$

After transformations we get

$$S = \frac{g(K_{10} \times \Delta T \times t + T_0 \Delta t)}{4\pi} \tag{5}$$

Taking into account the value of the coefficient K_{10}, equal to $K_{10} = -2.753 * 10^{-4}$ then, with $t = 1$ the term $K_{10} * \Delta T * t$ can be neglected, and then we get the following expression:

$$S = \frac{g \times T_0 \times \Delta t}{4\pi} \tag{6}$$

This is nothing other than the group speed multiplied by the time interval. Thus, knowing that the period began to change from 11.4 s and its complete change occurred in about 13 h (46,800 s), we get that the source should be located at the distance of 416 km, which practically corresponded to point 2 of the typhoon passage. Thus, we could claim that the linear change in the swell period was nothing other than the usual dispersion of propagation, in which waves with a large period, having a high speed, came first, and with a smaller one, arrived last.

4. Shelf Infragravity Sea Waves

There are still various points of view on the nature of oscillations and waves in the range of periods from 20–30 s to 8–10 min, recorded in the sea, each of which finds a balanced confirmation in the observed experimental data. The considered range of periods corresponds to the so-called "Infragravitational noise of the Earth", the origin of which may be associated with various processes in all geospheres, any of which is suitable for explaining the appearance of oscillations and waves of this periods range. In this paper, we will pay attention to the mechanism of occurrence of the considered periods range oscillations related to infragravity sea waves, arising on the shelf as a result of the action of sea gravity waves, i.e., so-called wind surface waves or swell waves. When interpreting the obtained results, we used the experimental data from the laser meter of hydrosphere pressure variations, installed on the shelf of the Sea of Japan at the depth of 27 m to the south from Shultz Cape (see Figure 12).

Hourly data files from the laser meter of hydrosphere pressure variations were filtered by a low frequency Hamming filter, followed by decimation of up to 2 s, then the decimated experimental data were filtered by a high-frequency Hamming filter with a cutoff frequency of 0.002 Hz (8.3 min) and length of 2500. In the process of spectral processing of the filtered hour files, significant peaks were singled out in the range of surface sea wind waves (periods from 1 to 20 s) and in the range of infragravity sea waves (periods from 20 s to 8 min). Additionally, the relative energy was determined from the harmonics values of all spectral components in the given period ranges (1–20 s and 20 s–8 min).

Thus, the values of the total relative energy were obtained in the range of sea surface wind waves (U_{sw}) and sea infragravity waves (U_{iw}). In total, 267 files of one-hour duration each were processed, i.e., duration of the continuous series of observations was a little over 11 days. In the considered observation interval, the values of the periods and amplitudes of surface sea waves (swell and wind sea waves) changed significantly, which allowed us to determine the relationship between infragravity sea waves and surface sea waves. Figure 13 shows the graphs of changes in the relative energy of surface sea waves (1–20 s) and the relative energy of infragravity sea waves (20 s–8 min). As we can see from this figure, the relative energy of infragravity sea waves generally repeats the curves of surface sea waves.

Figure 12. Map of locations of the laser meter of hydrosphere pressure variations (LMHPVs) and the laser strainmeter (LS).

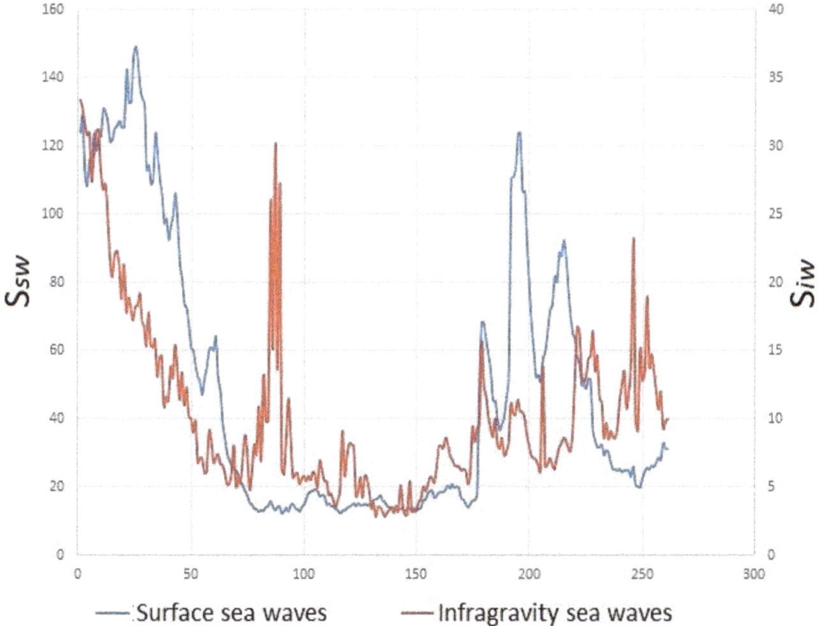

Figure 13. Change in the relative energy of surface sea waves (S_{sw}) and infragravity sea waves (S_{iw}).

Some abrupt increase in the relative energy of infragravity sea waves in comparison with the relative energy of surface sea waves (abrupt surges in files 87–91) was associated with the presence of solitary waves in the records of the laser meter of hydrosphere pressure variations, which, upon spectral processing, significantly increased the relative energy in the infragravity range and did not increase the relative energy in the sea wave range.

5. Conclusions

After the studies in the range of gravity sea waves, the following conclusions could be drawn: (1) When comparing the polynomial regression and the general function of the period change, with the help the analysis we conducted of the change in the swell waves periods during the passage of the typhoon through the Sea of Japan, we determined that the difference in the quality of describing this process was insignificant.

Thus, to describe wave processes of this type, the general period change function could be used instead of high-order polynomial regressions. (2) The assumption about the generation of waves of a certain period when the typhoon moved at a certain speed at an angle to the receiver looked convincing, if we assume that these periods are formed by the leading front of the typhoon, while the wind condition in the typhoon, which can affect the amplitude of these wave periods, plays an important role. This effect could also explain the variety of periods of swell waves, created by the typhoon. (3) Separate peaks of certain periods in the typhoon spectrum and simultaneous arrival to the registration point could be explained by the group propagation of swell waves, when waves with the largest period captured the waves with a period close to their own, and the entire group propagated at a speed of the largest period, which was two times less than the phase speed. (4) Using the formula for the group speed of wave propagation in deep water conditions, knowing the duration of the interval of period change from maximum to minimum, we could quite accurately determine the distance to the source, from which these waves came. The linear effect of the period decrease was associated exclusively with dispersion during their propagation. We established that change in the total energy of the sea infragravity waves (20 s–8 min) correlated with change in the total energy of sea gravity waves (2–20 s), which testified in favor of the theory of generation of sea infragravity waves by gravity sea waves. We discovered powerful solitary perturbations of the sea surface of considerable duration, exceeding the amplitude of local tides. According to some indications, they can be related to rogue waves, but with a big difference in the duration of the process. Further research in this direction is extremely necessary to establish the physics of their origin, development, and transformation.

Author Contributions: G.D.—problem statement, experimental work experimental data processing, guidance in writing an article. S.D.—processing of experimental data, calculations, creation of graphic material dedicated to infragravity sea waves. S.B.—processing of experimental data, calculations, creation of graphic material dedicated to Swell Waves. All authors have read and agreed to the published version of the manuscript.

Funding: This work was carried out with partial financial support of the topic AAAA-A20-120021990003-3 "Research of fundamental foundations of the origin, development, transformation and interaction of hydroacoustic, hydrophysical and geophysical fields of the World Ocean".

Acknowledgments: We would like to express our deep gratitude to all employees of the Physics of Geospheres laboratory.

Conflicts of Interest: The authors declare no conflict of interest.

References

1.   Newton, I. *The Principia: Mathematical Principles of Natural Philosophy*; Cohen, I.B.; Whitman, A., Translators; University of California Press: Berkeley, CA, USA, 1999.

2.   Gardner, C.S.; Greene, J.M.; Kruskal, M.D.; Miura, R.M. Method for solving the Korteweg-de Vries equation. *Phys. Rev. Lett.* **1967**, *19*, 1095–1097. [CrossRef]

3.   Benjamin, T.B.; Feir, J.E. The disintegration of wave train on deep water. *J. Fluid Mech.* **1967**, *27*, 417–430. [CrossRef]

J. Mar. Sci. Eng. **2020**, *8*, 796

4. Longuet-Higgins, M.S.; Cleaver, R.P. Crest instabilities of gravity waves. Part 1. The almost-highest wave. *J. Fluid Mech.* **1994**, *258*, 115–129. [CrossRef]

5. Lounge-Higgins, M.S. The instabilities of gravity waves of infinite amplitude in deep sea. II. Subharmonics. *Proc. R. Soc. Lond. Ser. A* **1978**, *360*, 489–505.

6. Lounge-Higgins, M.S.; Smith, N.D. An experiment on third order resonant wave interactions. *J. Fluid Mech.* **1966**, *25*, 417–436. [CrossRef]

7. Hasselmann, K. On the nonlinear energy transfer in gravity-wave spectrum. General theory. *J. Fluid Mech.* **1962**, *12*, 481–500. [CrossRef]

8. Hasselmann, K. Weak-interaction theory of ocean waves. In *Basic Development in Fluid Dynamics*; Academic Press: New York, NY, USA, 1968; pp. 117–182.

9. Caponi, E.A.; Saffman, P.G.; Yuen, H.C. Instability and confinedchaos in a nonlinear dispersion wave system. *Phys. Fluids* **1982**, *25*, 2159–2166. [CrossRef]

10. Tolman, H.L.; Balasubramaniyan, B.; Burroughs, L.D.; Chalikov, D.V.; Chao, Y.Y.; Chen, H.S.; Gerald, V.M. Development and implementation of wind generated ocean surface models at NCEP. *Weath. Forecast.* **2002**, *17*, 311–333. [CrossRef]

11. Tolman, H.L. Numerics in wind wave models. In Proceedings of the ECMWF Workshop on Ocean Wave Forecasting, Reading, UK, 2–4 July 2001; pp. 5–14.

12. Donelan, M.A. A nonlinear dissipation function due to wave breaking. In Proceedings of the ECMWF Workshop on Ocean Wave Forecasting, Reading, UK, 2–4 July 2001; pp. 87–94.

13. Dolgikh, S.G.; Budrin, S.S.; Plotnikov, A.A. Laser meter for hydrosphere pressure variations with a mechanical temperature compensation system. *Oceanology* **2017**, *57*, 600–604. [CrossRef]

14. Dolgikh, G.I.; Piao, S.; Budrin, S.S.; Song, Y.; Dolgikh, S.G.; Chupin, V.A.; Yakovenko, S.V.; Dong, Y.; Wang, X. Study of low-frequency hydroacoustic waves behavior at the shelf of decreasing depth. *Appl. Sci.* **2020**, *10*, 3183. [CrossRef]

15. Dolgikh, G.; Chupin, V.; Gusev, E. Microseisms of the "Voice of the Sea". *IEEE Geosci. Remote Sens. Lett.* **2020**, *15*, 750–754. [CrossRef]

16. Dolgikh, G.I.; Budrin, S.S.; Dolgikh, S.G.; Ovcharenko, V.V.; Plotnikov, A.A.; Chupin, V.A.; Shvets, V.A.; Yakovenko, S.V. Dynamics of wind waves during propagation over a shelf of decreasing depth. *Dokl. Earth Sci.* **2012**, *447*, 1322–1326. [CrossRef]

17. Budrin, S.S.; Dolgikh, G.I.; Dolgikh, S.G.; Yaroshchuk, E.I. Studying the variability of the wind wave period. *Russ. Meteorol. Hydrol.* **2014**, *39*, 47–52. [CrossRef]

18. Dolgikh, G.I.; Budrin, S.S. Some Regularities in the Dynamics of the Periods of Sea Wind Waves. *Dokl. Earth Sci.* **2016**, *468*, 536–539. [CrossRef]

Publisher's Note: MDPI stays neutral with regard to jurisdictional claims in published maps and institutional affiliations.

Journal of
Marine Science and Engineering

Article

Dynamics and Transformation of Sea Surface Gravity Waves at the Shelf of Decreasing Depth

Grigory I. Dolgikh *, Olga S. Gromasheva, Stanislav G. Dolgikh and Alexander A. Plotnikov *

V.I. Il'ichev Pacific Oceanological Institute, Far Eastern Branch Russian Academy of Sciences, 690041 Vladivostok, Russia; gromasheva@poi.dvo.ru (O.S.G.); sdolgikh@poi.dvo.ru (S.G.D.)
* Correspondence: dolgikh@poi.dvo.ru (G.I.D.); lotos_toi@mail.ru (A.A.P.)

Abstract: This paper reviews the results of the processing of synchronized data on hydrosphere pressure variations and the Earth's crust deformation in the microseismic range (5–15 s), obtained over the course of numerous experiments, using a coastal laser strainmeter and laser meters of hydrosphere pressure variations installed in various points of the Sea of Japan shelf. Interpreting the results, we have discovered new regularities in the dynamics of surface progressive gravity waves, and their transformation into primary microseisms, when waves move at the shelf of decreasing depth. For example, we found non-isochronous behavior of progressive waves, which manifests itself in a decrease in the periods of gravity waves due to the transformation of a part of their energy into the energy of primary microseisms. Furthermore, when processing the synchronous fragments of the records, made by laser strainmeters and laser meters of hydrosphere pressure variations, we identified approximate zones of the most effective transformation of the energy of gravity progressive waves into the energy of primary microseisms, which start from the depth of less than a half-wavelength and stretch to the surf zone.

Keywords: wind waves; progressive waves; standing waves; primary microseisms; secondary microseisms; wave dynamics; wave transformation

Citation: Dolgikh, G.I.; Gromasheva, O.S.; Dolgikh, S.G.; Plotnikov, A.A. Dynamics and Transformation of Sea Surface Gravity Waves at the Shelf of Decreasing Depth. *J. Mar. Sci. Eng.* **2021**, *9*, 861. https://doi.org/10.3390/jmse9080861

Academic Editor: Francesca De Serio

Received: 30 June 2021
Accepted: 9 August 2021
Published: 10 August 2021

Publisher's Note: MDPI stays neutral with regard to jurisdictional claims in published maps and institutional affiliations.

1. Introduction

Many works are devoted to the process of the emergence and development of sea progressive and standing surface gravity waves. We do not discuss it in this paper, but let us only note that the main parameters of emerging sea gravity waves depend on wind speed and the time of its activity over a particular water area, acceleration magnitudes, geometric characteristics of the water areas, natural conditions that generate head waves with the period, close to the period of an incident sea gravity wave. Interesting are the issues associated with dynamics of surface gravity waves during their movement at the shelf of decreasing depth and their transformation into primary microseisms. When moving at the shelf, these waves, especially at depths of less than a half-length of a sea gravity wave, are subject to various changes, associated primarily with the impact of nonlinear and dissipative processes. Modern theoretical and experimental studies show that nonlinear processes are dominant, both in deep and shallow water.

Of particular interest is the issue of the transformation of the spectra of gravity waves as they propagate at the shelf of decreasing depth. In this case, the main maximum can shift to both the high-frequency and low-frequency spectral regions. For example, [1] presents the results of experimental and model studies of the evolution of frequency spectra for groups of dispersive focusing waves in a two-dimensional wave reservoir. It shows that there is a nonlinear energy transfer to low-frequency spectral components, which cannot be detected easily. The energy dissipation in the spectral maximum region is ascertained, which increases the energy in the higher-frequency range, which depends on the Benjamin–Feir index. It is demonstrated that the energy increase in the low-frequency range occurs,

21

at least partly, due to nonlinear energy transfer before a wave breaks, and that breaking of a wave cannot necessarily increase the energy in this range.

As stated in [2], the waves covering a long distance over a very gradually sloping bottom undergo more nonlinear evolution than the waves coming over a steeply sloping bottom, so the peculiarities of the waves' nonlinearity depend on propagation distance and water depth. In particular, additional frequency peaks can be formed due to spectrum broadening as a result of backward energy transfer from higher frequencies [3] and from lower (infragravity) frequencies to the main peak [4]. The mechanisms of energy transfer between spectral peaks of sea gravity waves in the frequency range of 0.05–0.5 Hz are considered. These energy transfers occur among the three phase-coupled frequencies and are called nonlinear triadic interactions. The summarized interactions transfer energy from the main spectral peak to the spectral peaks many times, which is often referred to as higher harmonics. The development of higher harmonics is associated with a skewed shape of sea waves during shallowing and an asymmetric saw-like shape during breakdown very close to the coast [5]. Simultaneously with transformation of the sea wave shape, the energy is transferred to lower frequencies of infragravity (0.005–0.05 Hz) due to difference interactions, forming bound long waves.

There are interesting results in [6–8]. It was demonstrated that as nonlinearity increases, the crest distribution deviates more and more from its linear analog. The general nature of this deviation is propagation of the density mass towards larger and smaller crests, which is consistent with vertically asymmetric properties of nonlinear waves, which are known to have shallower troughs and sharper and larger crests than their linear analog. Nonlinearity increases and peak period decreases due to the destructive interference of the second-order free harmonics. In [8], it is stated that variability of wave periods for waves of a given height is inversely proportional to the wave height. However, at the same time, even the influence of nonlinearities on the wave periods has not been fully studied. This requires, first of all, complex experimental field studies. The mechanism of transformation of the progressive sea surface wave energy into the energy of primary microseisms is also associated with the nonlinear transformation of sea gravity waves as they move at the shelf of decreasing depth. At the same time, the issue of the bottom sloping, required for effective transformation of this energy, has not been fully investigated. Hasselman [9] considered a constant sloping seabed and calculated the seismic response for frequencies from 0.05 to 0.1 Hz, which is known as the main microseismic band. The pressure pattern at the seabed is characterized by a wide power spectrum in the wavenumber range, and thus is equivalent to the vertical point force, by which seismic response can be estimated, e.g., [9,10]. This theory was successfully applied to slowly changing bottom slopes at seismic noise frequencies below 0.03 Hz [11]. For primary microseisms with a frequency of $0.05 < f < 0.1$ Hz, Hasselman [9] used the lowest slope of 3%. Recently, Ardhuin et al. [11] showed that the primary microseismic signal, measured at the French seismic station Saint Sauveur en Rue (the geoscope network), requires an average ocean bottom slope of 6% at depths of about 20 m, while the average slopes on the French continental shelves are about 0.1% or less. It is difficult to explain such a large difference by additional effects not included in these models. Another reason why a constant or slowly changing bottom slope is not a satisfactory model for primary microseisms is that the horizontal component of the equivalent point force, equal to the vertical force multiplied by the bottom slope, is too weak to explain the observed kinetic wave energy [12–14].

In the presented work, when processing and analyzing the experimental data, we will consider nonlinear mechanisms of the transformation of the energy of surface gravity waves and their transformation into primary microseisms when moving at the shelf of decreasing depth.

2. Experimental Studies

The paper analyzes the results of the processing of the experimental data on hydrosphere pressure variations and the Earth's crust deformations, obtained over the course of experimental work using laser meters of hydrosphere pressure variations and laser strainmeters. The laser strainmeters with measuring arms of 52.5 and 17.5 m are stationary, installed at Schulz Cape of Peter the Great Bay, the Sea of Japan (Russia), at the depth of 3–5 m in underground hydro- and thermally insulated rooms at coordinates 42.58° N and 131.157° E [14]. The laser strainmeters are located at about 70 m above the sea level. The optical part of each laser strainmeter is built based on the use of a modified Michelson interferometer of an unequal-arm type, using frequency-stabilized helium-neon lasers with long-term stability from 10–9 to 10–12 as a light source. The main interference unit of the Michelson interferometer of each laser strainmeter, together with the digital record-ing system, is located in a separate hydro- and thermally insulated room. The auxiliary interference unit of the Michelson interferometer, consisting of a corner reflector with an alignment system, is located in another hydro- and thermally insulated room. Between the main interference unit and the auxiliary interference unit, a laser beam propagates through a sealed pipeline, consisting of stainless-steel pipes, hermetically sealed at the ends by optical windows. Figure 1 shows photographs of the main interference unit of the 52.5 m laser strainmeter (Figure 1a) and an underground vacuum pipeline (Figure 1b). The measuring arms of laser strainmeters are practically orthogonal to each other and relatively positioned at the angle of 92°. The laser strainmeter with a measuring arm of 52.5 m is oriented at the angle of 18° relative to the meridian line, and the strainmeter with a measuring arm of 17.5 m is oriented at the angle of 110° relative to the same line. The main interference unit of the 52.5-meter laser strainmeter is installed on a massive concrete monolith with height of about 3 m, which is rigidly connected in the lower part to the hard rocks of Shultz Cape, consisting of compressed loam. The corner reflector of the 52.5-meter laser strainmeter is rigidly attached to a 1-meter-high concrete monolith, which is connected to a granite cliff. The described laser strainmeters have the following technical characteristics: an operating frequency range from 0 (conditionally) to 100 Hz, and measuring accuracy of the displacement of the Earth's crust plates 0.03 nm, which can be improved through appropriate technical measures.

Figure 1. Shows photographs of the main interference unit of 52.5 m laser strainmeter (**a**) and an underground vacuum pipeline (**b**).

The laser meters of hydrosphere pressure variations [15] in their improved version were described in [16]. For several years, we installed them at various shelf points, both inside and outside Vityaz Bay, in the seaward part off Shultz Cape. At present, these laser meters of hydrosphere pressure variations have the following technical characteristics: an operating frequency range from 0 (conditionally) to 1000 Hz, a measuring accuracy of hydrosphere pressure variations 0.24 mPa, and operating depths to 50 m. Figure 2 shows the external view of the instrument, created based on the use of a modified Michelson interferometer of the homodyne type and a frequency-stabilized helium-neon laser, which ensures stability of the radiation frequency in the ninth decimal place. It is enclosed in a cylindrical stainless-steel housing, which is fastened in a protective grating designed to protect the instrument in severe operating conditions (rocky or slimy bottom). One side has a hole for cable entry. The other side is sealed with a lid. In addition to the protective grating, an elastic air-filled container is located outside the instrument. Its outlet is connected by a tube to a compensation chamber located in the removable cover. The housing contains a Michelson interferometer, the compensation chamber, an electromagnetic valve, and a digital recording system. The sensitive element of the supersensitive detector is the round stainless-steel membrane, which is fixed at the end face of the device. On the outside, the membrane interacts with water. A mirror is fixed on the thin pin inner side of the membrane, which is a part of the "cat's eye" system, consisting of a biconvex lens with the appropriate focal length and this mirror. The mirror with the lens is included in the structure of the measuring arm of the interferometer. The mirror, rigidly fixed on a thin pin in the center of the membrane, shifts along the interferometer axis under the influence of the hydrosphere pressure variations. A change in the length of the measuring arm leads to the change in the intensity of the interference pattern, recorded by the digital registration system. The output signal of the supersensitive detector, after preprocessing by the digital registration system, is the hydrosphere pressure variations. It is possible to obtain the limiting technical characteristics of these systems by reducing photoelectronic equipment noise and the compensation of temperature noise, more accurately equalizing the length difference between the measuring and reference arms of the interferometer, which correspond to the following design parameters: an operating frequency range from 0 (conditionally) to 10,000 Hz, and a measuring accuracy of hydrosphere pressure variations 1.8 μPa.

Figure 2. Photo of the laser meter of hydrosphere pressure variations records. (1)—stainless-steel housing, (2)—protective grill, (3)—air container, (4)—laser meter of hydrosphere pressure in the small scale.

Studying the dynamics and transformation of sea surface gravity waves at the shelf of decreasing depth, we will focus on the results of processing the experimental data, obtained when the laser meters of hydrosphere pressure variations were located at points 1, 2, and 3 (see Figure 3). Points 1 and 2 were in Vityaz Bay, the Sea of Japan, and point 3in the seaward part of the experimental site. At points 1 and 2 at the depths of 11.8 and 4.5 m, two laser meters of hydrosphere pressure variations were located at the same time, along the line perpendicular to the coastline in the instruments' location area (perpendicular to the front of the surface gravity wave propagating from the deep-water part of Vityaz Bay towards the coastline). The distance between the systems was 96 m. At point 3, at the depth of 27 m, the laser meter of hydrosphere pressure variations was operating in the summer–autumn period of the year.

Figure 3. Map scheme of the experimental works. Points 1, 2, and 3 show the positions of laser meters of hydrosphere pressure variations, point 4 shows the position of laser strainmeters.

Additionally, we used the following equipment in the measuring complex: a TRIMBLE 5700 GPS, a meteorological station, low-frequency hydroacoustic emitters of an electromagnetic type, hydroacoustic receiving systems, hydrological profilers, and boats. The experimental data, obtained from all measuring systems, were transferred in real-time mode to the laboratory room, where, after filtration and decimation pre-processing, it was recorded on solid media with subsequent placement in the previously created experimental database.

3. Processing and Analysis of the Obtained Results

As appropriate, the results of the experimental studies, loaded into the experimental database, were processed further. Considering that the sampling frequency of the recorded data files was, in general, 1000 Hz, they were preliminarily filtered with a low-pass Hamming filter to the frequency of 1 Hz and further averaged to the sampling frequency of 2 Hz. The performed procedure allowed us to avoid the effect of power-consuming, high-frequency components overlapping the considered range of surface gravity sea waves (1–0.04 Hz). Depending on the planned processing results, various methods of experimental data processing were used: spectral methods (periodogram method, i.e., Fast Fourier Transform; the maximum likelihood method), averaging over different time windows with a suitable number of averagings, filtering with a high-frequency Hamming filter, statistical estimation methods, dynamic spectrogram.

Experimental studies of ocean wave processes in the range of surface gravity waves were carried out in the Sea of Japan. We know that periods of wind waves, and free wind waves–swell waves, depend on wind characteristics (speed, direction, duration), and also on the characteristics of the basin, over which a cyclone/typhoon is acting (geometrical dimensions of the sea, acceleration value, depth). For the Sea of Japan, the largest periods of wind waves are about 12–14 s, and background progressive surface gravity waves have periods of about 5–6 s. Therefore, these progressive waves, when interacting with the bottom, should generate primary microseisms of the corresponding periods (from 12–14 to 5–6 s). Standing surface waves should generate secondary microseisms with periods two times shorter than those of interacting progressive gravity waves. Further, we will not pay attention to these interactions and the formation of secondary microseisms in our work. Let us dwell on the dynamic changes of progressive gravity sea waves, as well as their transformation into primary microseisms. Figure 4 shows a dynamic spectrogram of a record of the laser meter of hydrosphere pressure variations in the range of sea progressive gravity waves. The data was obtained when registering swell waves, coming from the central areas of the Sea of Japan, where the speed of wind was high. The behavior of gravity sea waves shown in the figure is typical for all records of laser meters of hydrosphere pressure variations. As we can see from the figure, besides the linear component of the decrease in the period of wind waves (swell), there are also nonlinear components, associated with both a decrease and increase in the periods of wind waves.

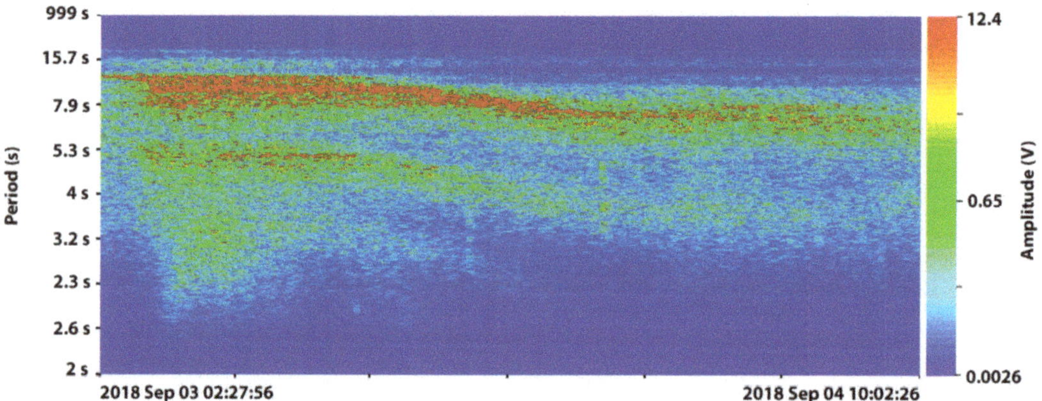

Figure 4. Dynamic spectrogram of a record of laser meter of hydrosphere pressure variations.

The linear decrease in wind waves' periods can be explained by dispersion, according to which free wind waves of large periods have high speeds. Therefore, the waves of large periods should be the first to reach the observer. For deep water, this ratio has the form:

$$c^2 = \frac{g\lambda}{2\pi},\tag{1}$$

where c is speed, g is gravitational acceleration, λ is a wavelength, $\lambda = gT^2/2\pi$, and T is a wave period. When a wave moves in shallow water, the situation will not change much, since

$$c^2 = gh,\tag{2}$$

where h is sea depth in the wave propagation place. The following question is interesting: what mechanisms/processes cause the nonlinear change in the waves' periods over time? We will disregard a possible slight increase in the periods during the initial phase, which is associated with a slight increase in the periods of swell after it leaves the wind zone due to attenuation of the high-frequency components and the energy transfer to the lower-frequency components. Let us consider the change in the periods of surface gravity waves when they propagate in the remaining deep-water part and at the shelf of decreasing depth. The change in the periods of surface gravity waves as they move through the rest of deep water can be partially associated with the Doppler Effect [17]. Analysis of the data of many months convinces us that the increase and decrease in the periods of wind waves, along with the mechanisms of developing waves and dispersion, are associated with the degree of change in the magnitude and direction of the speed of typhoons or other pressure depressions. That is, the change in the periods of wind waves is associated with the Doppler Effect. In accordance with the Doppler Effect, we write the expression:

$$f = f_0(1 \pm u/c) \tag{3}$$

where: f is frequency of the received signal, f_0 is frequency of the emitted signal, u is the speed of source movement (typhoon, etc.), c is wind waves' speed, and "+" is used when approaching, "-" when moving away. Here, are some calculations. Let $c = 20$ m/s and $u = 36$ km/h, then $\Delta f = \pm 0.5 f_0$. At 72 km/h, $\Delta f = \pm 1.0 f_0$. Changes in wave periods, associated with the Doppler Effect, were studied in detail in [18], using data from the laser strainmeter, which recorded primary microseisms, caused by gravity sea waves generated by a distant typhoon moving in the Pacific Ocean.

These changes in the swell periods can be associated with a typhoon moving at high speeds. However, how can we associate the observed variations in the periods of surface gravity waves in the case of an almost stationary location of a powerful cyclone as the "generator" of wind waves? Especially when surface gravity waves move at the shelf of decreasing depth?

Let us consider the results of processing synchronous records of two laser meters of hydrosphere pressure variations, located at points 1 and 2 (see Figure 3). The distance between the systems is 96 m, and they are installed at depths of 11.8 (point 2 in Figure 3) and 4.5 m (point 1 in Figure 3). In this case, the average bottom slope is about 7.6%. We will pay attention to the dynamics of wind waves with periods from 5 to 8 s. The length of one processed dataset is 8192 s. Considering that a wind wave covers the distance between two systems in no more than 10 s, we can state that during the length of one dataset of 8192 s, the data from these two systems is synchronous. It follows from the analysis of all processed data that the periods and energy of surface gravity waves decrease as they move from the position of the laser meter of hydrosphere pressure variations, located at the depth of 11.8 m (point 2 in Figure 3), to the position of another laser meter of hydrosphere pressure variations, located at the depth of 4.5 m (point 1 in Figure 3). When moving in this area, the wind wave transfers part of its energy to the upper layer of the sea Earth's crust. This process is observed not only at significant levels, but also at small amplitudes of wind waves, which, with some approximation, can be considered nonlinear. When wind waves propagate in the studied area, the average percentage of changes in the periods of wind waves in the range of 5–6 s is about 6%, and the average percentage of changes in the periods of wind waves in the range of 6–7 s is about 7%. It is clear that this change in the periods of wind waves is associated not only with the length of a wind wave but also with its amplitude. Figure 5 shows, as a typical example of the dynamics of wind waves during their movement at the shelf of decreasing depth, the spectra of synchronous fragments of the record of laser meters of hydrosphere pressure variations, installed at points 1 and 2 (Figure 3). We can see that the main maximum of wind waves with a period of 6.45 s transforms into the maximum with a period of 6.00 s.

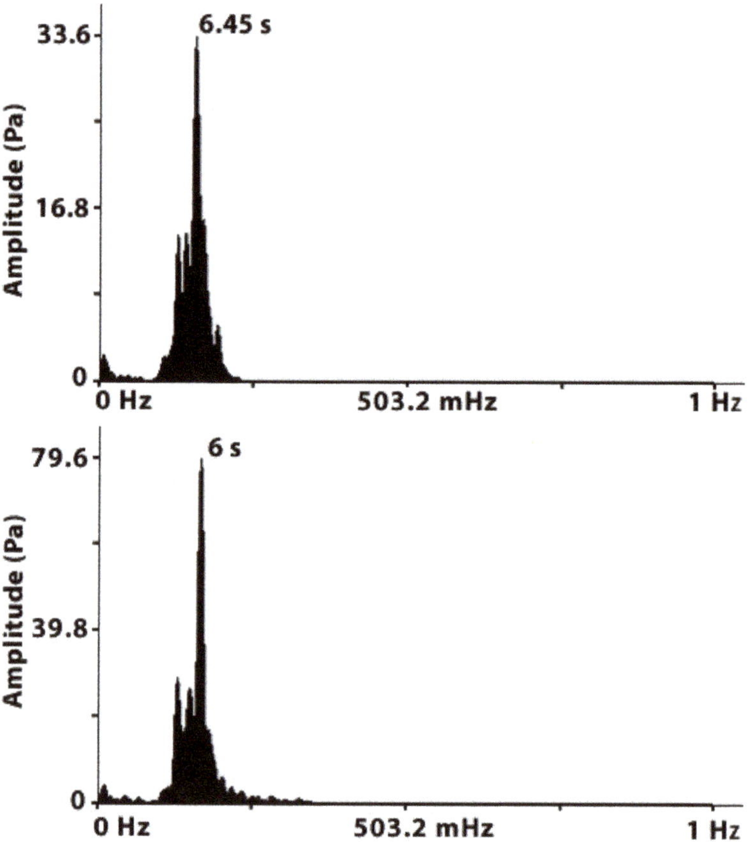

Figure 5. Spectra of synchronous fragments of the record of laser meters of hydrosphere pressure variations installed in point 1 (**top**) and point 2 (**bottom**).

Let us look at other synchronous experimental data, obtained from the above systems. Taking into account the fact that the speed of wind waves in shallow water approximately equals 6–10 m/s, the pairs of record fragments of the laser meters of hydrosphere pressure variations were taken with a short time shift (about 10 s). Additionally, 10 synchronous record fragments were processed. The results of processing are listed in Table 1. The wavelength and wave speed for these periods of wind waves and depths are taken from [18]. According to the formula below, taken from [18], relative amplitudes of sea gravity waves on the water surface were calculated from relative pressure amplitudes, measured by the instruments installed on the bottom:

$$P = \frac{g\rho a}{ch(kh)} \tag{4}$$

where g is gravitational acceleration, ρ is water density, a is a wave amplitude, $k = 2\pi/\lambda$ is the wave number, and h is sea depth.

Table 1. Experimental data.

№№	Period, s	Sea Depth, m	Wave-Length, m	Speed, m/s	Relative Pressure at the Bottom	Relative Wave Amplitude
1	4.90	11.8	36.26	7.4	414	0.160
	4.47	4.5	25.03	5.6	874	0.150
2	5.65	11.8	45.80	8.1	276	0.072
	5.15	4.5	30.40	5.9	440	0.065
3	5.50	11.8	44.00	8.0	285	0.079
	5.36	4.5	31.60	5.9	550	0.078
4	5.82	11.8	48.30	8.3	388	0.094
	5.58	4.5	33.50	6.0	640	0.088
5	6.17	11.8	52.40	8.5	561	0.120
	4.79	4.5	27.80	5.8	663	0.100
6	5.07	11.8	38.5	7.6	815	0.320
	4.71	4.5	26.85	5.7	1928	0.310
7	4.87	11.8	36.04	7.4	498	0.198
	4.58	4.5	26.11	5.7	1111	0.183
8	6.90	11.8	62.1	9.0	1449	0.261
	6.50	4.5	40.3	6.2	1798	0.226
9	6.4	11.8	55.68	8.7	2010	0.407
	6.2	4.5	37.82	6.1	2436	0.315
10	6.4	11.8	55.68	8.7	1865	0.278
	6.2	4.5	37.82	6.1	1996	0.258

As we can see from Table 1, there are no special dependences between changes in the periods of wind waves and changes in the amplitudes of wind waves, except one: almost always, the decrease in the period of wind waves is associated with the decrease in the amplitudes of wind waves. We can note that for periods shorter than 6.2 s, there is an almost (on average) linear relationship between the percentage change of the periods of wind waves and the percentage change of the amplitudes of wind waves. For periods longer than 6.2 s, there is a nonlinear relationship—in percentage terms, the change in the amplitudes of wind waves is significantly greater than the change in the periods of wind waves. Let us consider the reasons for these changes. Calculations are carried out in accordance with the theory described in [18]. Yet, only linear cases are considered there, which is not correct. However, unfortunately, no one can analytically solve nonlinear equations in which all the forces acting on the water are taken into account. Therefore, we will use what we have, and will try to describe a possible reason for the period change, and to quantify this change. In addition, we will verify the obtained estimates using other field data. The change in the periods of wind waves when they move at the sloping shelf can be associated with the average value of integral pressure along the entire path of wave propagation (96 m). It is clear that the more water particles, moving under the action of the wind wave near the bottom, press on the bottom, the greater part of the wave energy will go to the bottom. Therefore, the energy carried into the bottom will be proportional to the total pressure exerted on the bottom by the sea wind wave in the studied area. Let us calculate the total pressure for the pairs of waves listed in Table 1. For each element of the pair, this total pressure can be estimated by the formula:

$$P_{sum} = \int_{h_2}^{h_1} \frac{g\rho a}{ch(kh)}\,dh \tag{5}$$

Expression (5) can be presented as follows:

$$P_{sum} = \frac{g\rho a\lambda}{\pi}\left(arctg\left(e^{\frac{2\pi h_1}{\lambda}}\right) - arctg\left(e^{\frac{2\pi h_2}{\lambda}}\right)\right) \tag{6}$$

At $a = 1$ for a sea wind wave with a wavelength of 48 m, we have $P_{tot} = 44.34$ kPa, and for a sea wind wave with a wavelength of 66 m, we have $P_{tot} = 54.17$ kPa. Thus, with continuous impact on the sea bottom of a total pressure of 7.39 kPa, in the first case, the wave period changes (decreases) by 1%, and for the second case, with continuous impact on the sea bottom of a total pressure of 7.74 kPa, the wave period changes (decreases) by 1%. That is, taking into account possible approximation errors, we can assume that with continuous impact of the sea wind wave of any period on the bottom at a total pressure of 7.56 kPa, its period changes (decreases) by 1%.

Next, let us consider another case of the dynamics of the sea wind wave period when the laser meter of hydrosphere pressure variations was installed at the depth of 27 m. The change in the wave period from 10.4 to 6.9 s occurred within 32.5 h. In accordance with [18], from the dispersion relation for the group velocity of wind and swell waves, we have:

$$c_g = \frac{gT}{4\pi} \tag{7}$$

Then, the distance to the place of generation of these waves can be calculated by the formula:

$$L = \frac{T_1}{T_0 - T_1} \times \frac{gT_0}{4\pi}\Delta t \tag{8}$$

The place of generation should be approximately at the distance of 1871 km, i.e., outside the Sea of Japan. For the second case, when the wave period changes from 9.4 to 6.9 s with a time interval between them of 16.4 h, according to the dispersion Equation (7), we find that the source of wind waves should be located at a distance of about 1200 km. That is, in the first and second cases, the sources of wind waves should be located outside the Sea of Japan/the East Sea, which is impossible, because this sea is practically closed to wind waves from the open part of the Pacific Ocean. Of course, only the linear change in the period of wind waves, associated only with the dispersion relation, is taken into account here. The Doppler Effect and decrease in the period of wind waves due to friction of a wave (water particles, with a loss of energy of the latter) when moving at the shelf of decreasing depth are not taken into account.

With regard to our Expression (6) and the obtained estimated parameters of the change in the periods of wind waves due to friction of the wave particles on the sea bottom, let us estimate the possible change in the periods in the cases described in the previous paragraph, when the waves propagate from the depths $h = \lambda/2$ to $h = 27$ m. We will assume that a wind wave with a period of 10.4 s has a wavelength (deep sea) of 169 m, a wind wave with a period of 6.9 s has a wavelength (deep sea) of 74 m, and a wind wave with a period of 9.4 s has a wavelength (deep sea) 138 m. Then, the total pressure acting on the bottom, with a wave amplitude equal to 1, for the wave with a period of 10.4 s equals 184.6 kPa. On the condition that when the total pressure on the bottom is equal to 7.56 kPa, the wave period will decrease by 1%, we obtain that at 184.6 kPa, the wave period should decrease by 24%, i.e., the initial period of the wave that "enters" the shelf at the sea depth of 169 m will equal 13.7 s, not 10.4 s. For a wave with a period of 6.9 m, through the same calculations, we see that when the wave "enters" the shelf at the sea depth of 74 m, its period should be 7.1 s, not 6.9 s. In the same way, for a wave with a period of 9.4 s, we see that when a wind wave "enters" the shelf at the sea depth of 138 m, its period should be not 9.4 s, but 11.2 s. Further, in accordance with (8), we calculate the possible distance to the source of wind waves for new, calculated period values. For the dynamics of wind waves with periods from 13.7 to 7.1 s within 32.5 h, this source should be located at a distance of 1340 km from the registration site (previously, it was 1871 km). For the dynamics of wind waves with periods from 11.2 to 7.1 s within 16.4 h, this source should be located at a distance of 890 km from the registration site (previously, it was 1200 km).

Now, let us discuss peculiarities of using Formula (6) for calculations. In all cases, we assumed that the wavelength at the registration point would be equal to the wavelength in "deep" water. In fact, this is not quite correct. Thus, we should modify our calculations, considering that the wavelength for the registration point (sea depth 27 m) in Expression (6) will be shorter. Second: in our calculations, we took wave amplitude as equal to 1 everywhere. This is not quite correct. For waves with periods of 13.7 and 7.1 s, they will be different at the registration point, even if we assume that they have left the pressure depression zone with equal amplitudes. During propagation even in the "deep" sea, the amplitude of the wave with a period of 13.7 s will decrease less than the amplitude of the wave with period of 7.1 s due to dissipative effects. Unfortunately, we cannot calculate the wavelengths at the sea depth of 27 m using the available theoretical formulas, since they were all obtained with a rough linear approximation and upon the condition that the wave period never changes, which is completely wrong. The only hope is to carry out a large-scale complex experiment to register spatio-temporal dynamics of wind waves' parameters from the places of their origin to the places of breaking. In addition, taking into account the fact that the dispersion Equation (7) was obtained under the conditions of linear approximation, without taking into account the summands of higher orders and, moreover, without taking into account the influence of many forces, except for gravitational, we can state that it is correct only with some rough approximation. We can obtain the value of this approximation only by analyzing such experimental data, which, unfortunately, nobody in the world has received yet. In view of the above, we cannot accurately estimate, using the above formulas, how the period of wind waves changes with time at the registration point. We will try to obtain these estimates, further analyzing the obtained experimental data.

Let us now consider some other effects of nonlinear behavior of surface gravity waves. Figure 6 shows a dynamic spectrogram of the laser meter of hydrosphere pressure variations record, which clearly shows low-frequency modulation of sea gravity progressive waves by a 12-hour tide. In some works, for example [2,4], the nonlinear transformation of the periods of surface gravity waves is associated with inter-harmonic transfer. That is, energy transfer is only possible from harmonic to harmonic, i.e., abruptly, not smoothly. Let us study this transformation, using experimental data. Let us take a dataset, a certain processing result of which is shown in Figure 6. Let us carry out spectral processing of successive fragments of the laser meter of hydrosphere pressure variations record. We take these fragments with a noticeable overlap, when the beginning of the next record fragment is shifted a little from the beginning of the previous fragment. In this case, the concentration of the maximum energy should change smoothly between frequencies from one fragment to another or jump from one harmonic to another. As follows from Figure 6, and from the results of additional processing, we do not observe an abrupt change in the period of wind waves. This change in the period of wind waves occurs smoothly.

In conclusion, let us consider some regularities in the transformation of the energy of sea surface gravity waves into the energy of microdeformations of the upper layer of the sea Earth's crust, i.e., into primary microseisms. Studying these processes, we will use the experimental data of synchronous records of the laser meter of hydrosphere pressure variations, installed in point 3 in Figure 3 at the depth of 27 m and the 52.5-meter laser strainmeter, installed in point 4 in Figure 3. We will be interested in the issues associated with zones of transformation of wind waves (swell waves) into primary microseisms, and we will also try to find some regularities that confirm or refute the above results. Please note that we consider as primary those microseisms which resulted from transformation of surface wind progressive waves. For the Sea of Japan, the periods of these waves are in the range of 5 to 15 s (approximately). Of course, during the propagation of progressive swell waves with periods of about 14 s, under special conditions, standing waves with a period of 7 s can occur, which will excite secondary microseisms with the same period. However, it is not important. Important is that the primary microseisms are generated by progressive sea waves. In the course of the same processing of synchronous experimental data, we noted that for all fragments, the spectra of microseismic range in the records of the laser strainmeter are wider than the spectra of the same frequency range in the records of the laser meter of hydrosphere pressure variations. As an example, Figure 7 shows the spectra of the synchronous fragments of these devices' records.

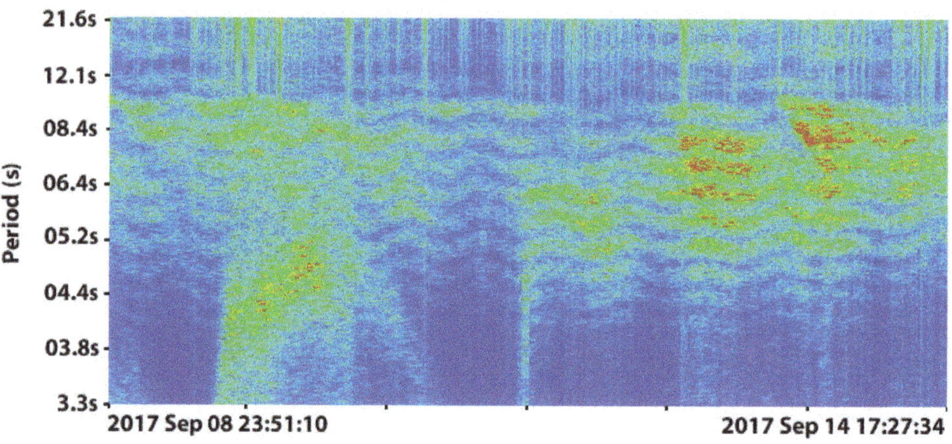

Figure 6. Dynamic spectrogram of a record of laser meter of hydrosphere pressure variations.

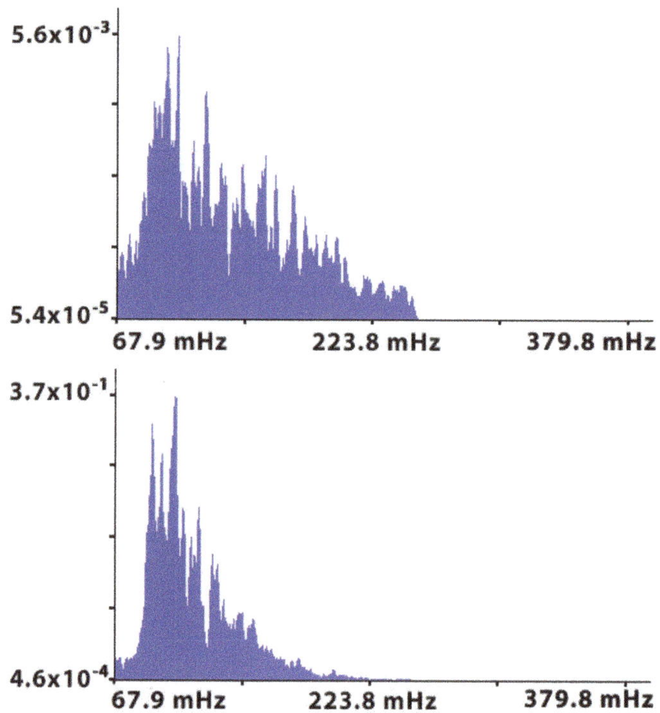

Figure 7. Spectra of synchronous fragments of laser strainmeter (**top**) and laser meter of hydrosphere pressure variations (**bottom**) records. Relative amplitudes are shown on the ordinate axis.

As we can see in Figure 5, the periods of waves and the periods of microseisms are, with some approximation, equal, and their maximum energy is concentrated in the range of 9–11 s. Let us consider the regularities of transformation of swell waves into primary microseisms with different periods. On the synchronous records of the instruments, within some time intervals, we identified oscillations with periods from 8 to 12 s, and within other time intervals, oscillations with periods from 5 to 7 s. Further, we will consider dynamic spectrograms and spectra of records of the laser strainmeter and the laser meter of hydrosphere pressure variations of different duration.

On dynamic spectrograms and spectra of records of the laser strainmeter and the laser meter of hydrosphere pressure variations, oscillations with periods from 9 to 12 s are identified (see Figure 8). On the dynamic spectrogram of the laser strainmeter (Figure 8a), the amplitude of oscillations in the range from 9 to 11.5 s is several times higher than the amplitude of background oscillations, which we can see in the spectrum below. Within the same time interval, oscillations with similar periods were identified in the records of the laser meter of hydrosphere pressure variations. On the dynamic spectrogram of the laser meter of hydrosphere pressure variations (Figure 8c), the amplitude of oscillations in the periods ranging from 9 to 12 s is several times higher than the amplitude of background oscillations. On the spectrum of the instrument record, shown below, in this range of periods, the amplitude of oscillations is also several times higher than the amplitude of the background oscillations. During further processing and analysis, we found a large number of fragments of the laser strainmeter and the laser meter of hydrosphere pressure variations records, where oscillation periods of microseisms coincide with periods of wind waves. This effect was observed on the instruments' records over several days of the experiment.

In the second fragment of the laser strainmeter and the laser meter of hydrosphere pressure variations records, we identified microseisms and wind waves with periods of about 6 s. When analyzing dynamic spectrograms and spectra of the laser strainmeter and the laser meter of hydrosphere pressure variations records, we found microseisms with periods from 4.5 to 6 s, while periods of wind waves ranged from 5.5 to 7 s. The dynamic spectrogram (Figure 9a) shows microseisms with periods from 5 to 6 s. On the spectrum of the laser strainmeter, shown below, in this range of periods, the amplitude of oscillations is also several times higher than the amplitude of background oscillations. On the dynamic spectrogram (Figure 9b), wind waves with periods from 6 to 7 s are identified, which is also confirmed by the record spectrum of the laser meter of hydrosphere pressure variations, where amplitudes of these oscillations are several times higher than amplitudes of background oscillations. Examining many other fragments of the laser strainmeter and the laser meter of hydrosphere pressure variations records, we found that microseisms with periods of about 6 s gradually transformed into microseisms with periods of about 5 s. Accordingly, the period of wind waves gradually decreased from 7 to 6 s.

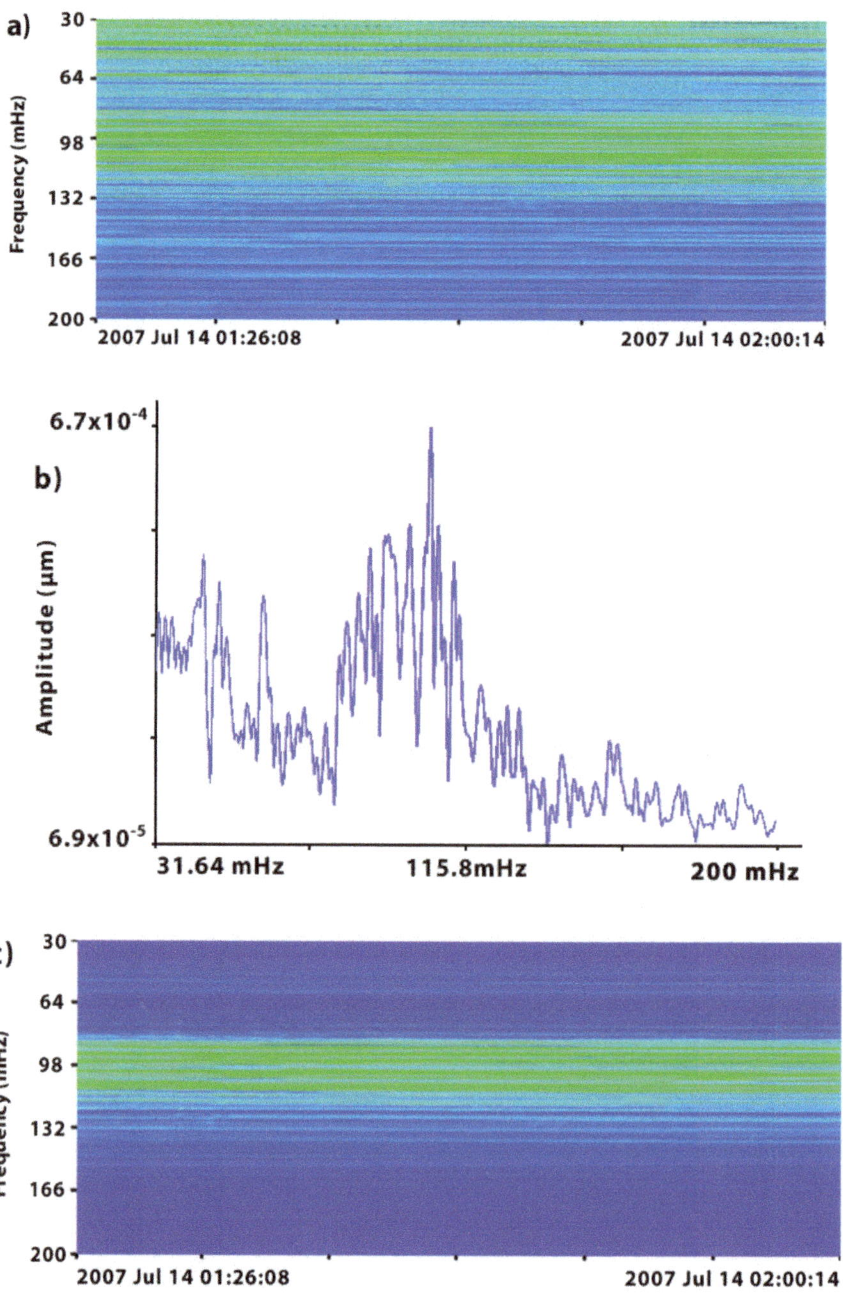

Figure 8. *Cont.*

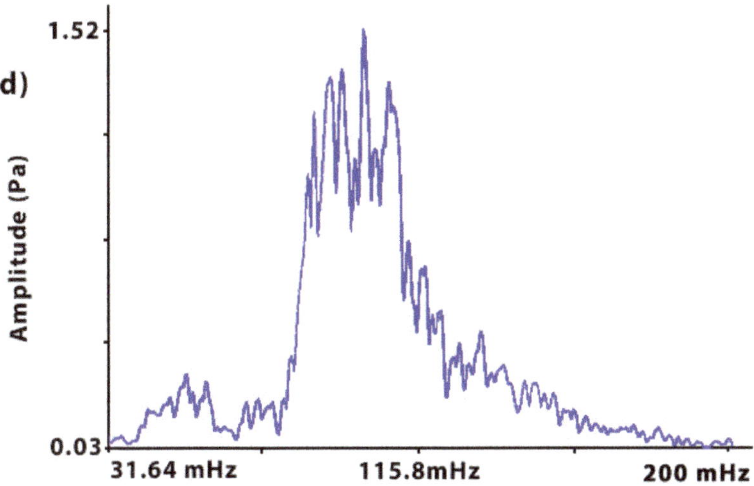

Figure 8. Dynamic spectrograms and spectra of laser strainmeter and laser meter of hydrosphere pressure variations records. (**a,b**)—Laser strainmeter, (**c,d**)—laser meter of hydrosphere pressure variations records.

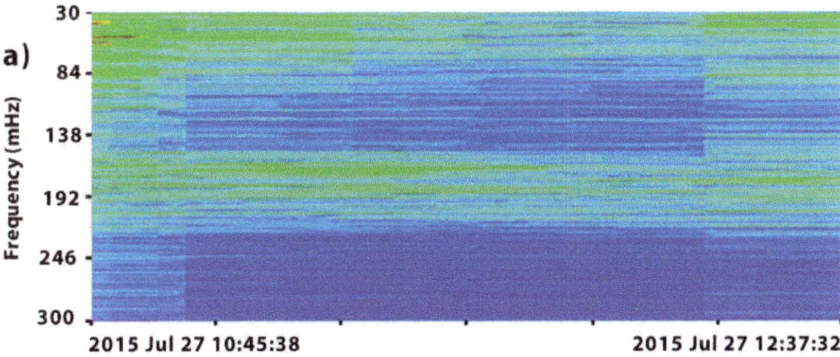

Figure 9. *Cont.*

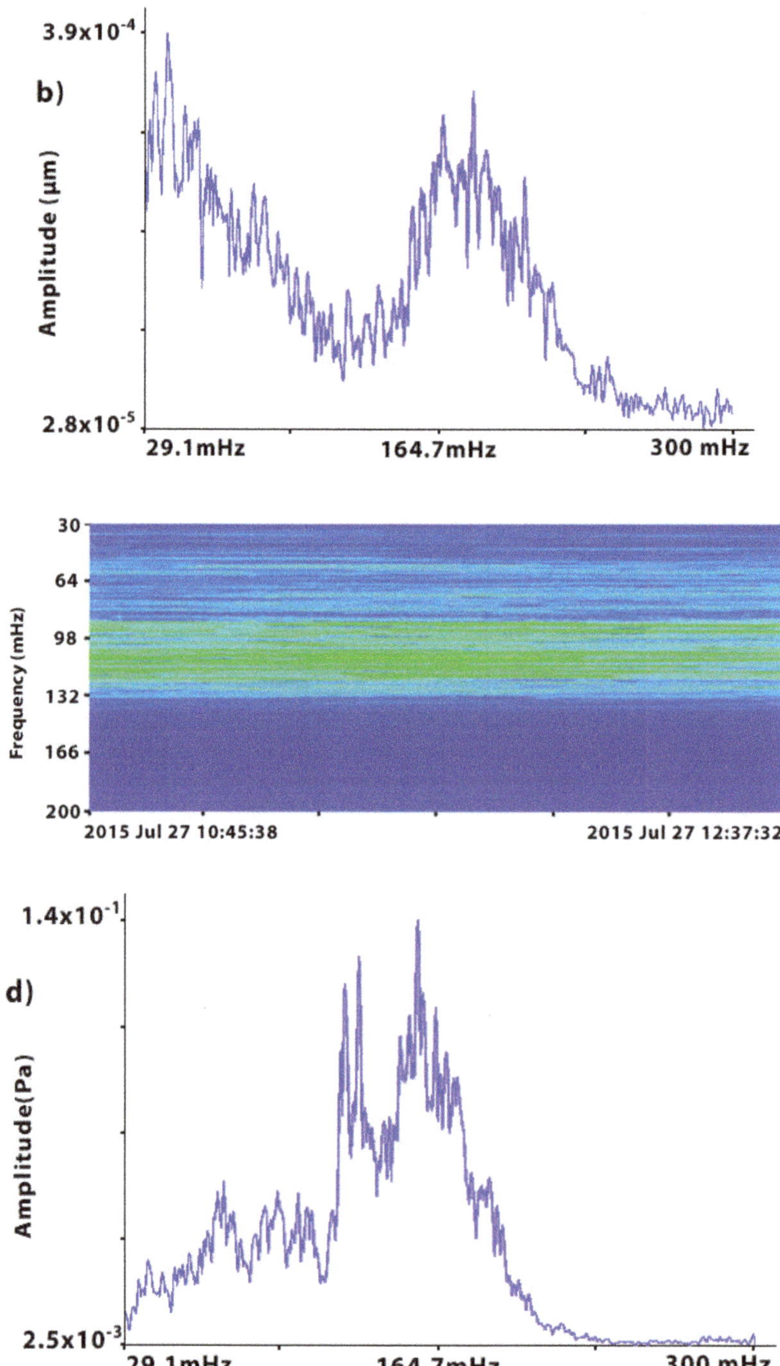

Figure 9. Dynamic spectrograms and spectra of synchronous records of laser strainmeter and laser meter of hydrosphere pressure variations. (**a,b**)—Laser strainmeter, (**c,d**)—laser meter of hydrosphere pressure variations records.

The study of microseisms, using the data of the laser strainmeter, and of wind waves, using the data of the laser meter of hydrosphere pressure variations, showed that when wind waves' periods were about 9 s, microseisms with the same periods were identified, and when there were wind waves with periods of about 6 s, the period of recorded microseisms was lower. It is related to the location of the measuring instruments. Wind waves move towards the coast through the area of the laser meter of hydrosphere pressure variations installation. The length of the wind wave with a period of about 9 s is several times greater than the depth of the instrument installation, and the energy transfer from the wind wave to the bottom begins to occur at a depth equal to or less than a half-wavelength. The length of the wind wave with a period of about 6 s is comparable with the depth of the instrument installation, and energy transfer occurs at a shallower depth. When a wind wave propagates at the shelf, it loses part of its energy when interacting with the bottom, and its energy is redistributed to a higher frequency range. This fact is confirmed by registration of microseisms with periods less than the periods of wind waves by the laser strainmeter. Moreover, we can assert that the zone of the most effective transformation of the wind waves energy into the energy of primary microseisms with periods of about 5–6 s is located at depths less than 27 m. Additionally, for progressive wind waves with periods of 9–10 s, such a zone is located near the depth of 27 m.

4. Conclusions

Over the course of processing synchronous experimental data from two spatially distanced laser meters of hydrosphere pressure variations, installed on the shelf of decreasing depth along a line perpendicular to the wave front, we found that a wind wave, propagating at the shelf of steadily decreasing depth, when interacting with the bottom, loses part of its energy (transfers it to the bottom), which leads to the transformation of the wave spectrum, and the energy in the spectrum is redistributed to a higher frequency region. The magnitudes of these changes depend not only on the lengths of wind waves and the degree of depth decrease, but also on their amplitudes. We have obtained an analytical equation for approximate estimation of the energy of a gravity wave, transformed into the energy of the upper layer of the Earth's crust. However, this equation does not take into account the dynamic change in the amplitude and length of a sea wave. Further theoretical studies, we hope, will allow us to formulate an analytical equation based on experimental data, which will take into account all the dynamic parameters of a sea gravity wave. We have established that the average percent of changes in the wave periods depends on the wavelength (the initial period of the incoming wave), when propagating through a specific area: the longer the period of a wave, the greater its change. This effect is associated with the amplitude of oscillations of water particles near the bottom, which is greater for waves with large periods. Because of this, total water pressure on the bottom is greater for waves with a longer period (wavelength).

To clarify our results on the percentage change in the periods of gravity sea waves as they move along the shelf of decreasing depth, and to assess the relationship of this percentage change in the periods of gravity sea waves to the energy carried away into the bottom, additional multiscale experimental studies are required. To perform this, on the shelf of monotonically decreasing depth, one-type instruments should be placed perpendicular to the front of a gravity sea wave at different depths, at 5–7 points (preferably, laser meters of hydrosphere pressure variations). In addition, receiving systems must operate on land to record primary microseisms (in our case, laser strainmeters). Based on the obtained experimental data, we would be able to estimate: (1) the periods and amplitudes of the gravity sea waves of a particular train at each point, and (2) the speed of a gravity sea wave from point to point. Equation (5) should be transformed, taking into account different amplitudes and periods at each point, speed and length of waves from point to point, using the full dispersion ratio $c^2 = (g\lambda/2\pi)th(2\pi h/\lambda)$, to derive the analytical equation. The result should be an equation, similar to Equation (5), with a triple integral on the right-hand side. It is the solution of this equation that will allow us to estimate the dynamics of sea gravity waves as they move along the shelf of decreasing depth of various configurations. The results of this solution are also useful in the applied field to accurately forecast the parameters of sea waves in specific regions of different water areas.

Based on additional synchronous data from the coastal laser strainmeter, we have established that sea progressive gravity waves excite primary microseisms of the corresponding periods. We have experimentally ascertained that the zones of transformation of gravity waves into microseisms depend on the periods of surface gravity waves and are associated with their length.

Author Contributions: G.I.D.—Problem statement, discussion and writing the article. O.S.G.—Data processing. S.G.D.—Data processing, discussion and writing the article. A.A.P.—Data processing and discussion. All authors have read and agreed to the published version of the manuscript.

Funding: The work was carried out with the financial support of the project "Study of the fundamental foundations of the origin, development, transformation and interaction of hydroacoustic, hydrophysical and geophysical fields in the World Ocean".

Acknowledgments: We would like to express our deep gratitude to all employees of the Physics of Geospheres laboratory.

Conflicts of Interest: The authors declare no conflict of interest.

References

1. Tian, Z.; Perlin, M.; Choi, W. Frequency spectra evolution of two-dimensional focusing wave groups in finite water depth water. *J. Fluid Mech.* **2011**, *688*, 169–194. [CrossRef]
2. Elgar, S.; Guza, R.T. Nonlinear model predictions of bispectra of shoaling surface gravity waves. *J. Fluid Mech.* **1986**, *167*, 1–18. [CrossRef]
3. De Bakker, A.T.; Tissier, M.M.F.S.; Ruessink, B.G. Beach steepness effects on nonlinear infragravity wave interactions: A numerical study. *J. Geophys. Res. Ocean.* **2016**, *121*, 554–570. [CrossRef]
4. Elgar, S.; Guza, R.T. Observations of bispectra of shoaling surface gravity waves. *J. Fluid Mech.* **1985**, *161*, 425–448. [CrossRef]
5. Cho, Y.J. Joint Distribution of the Wave Crest and Its Associated Period for Nonlinear Random Waves of Finite Bandwidth. *J. Mar. Sci. Eng.* **2020**, *8*, 654. [CrossRef]
6. Guedes Soares, C.; Carvalho, A.N. Probability distribution of wave heights and periods in measured two-peaked spectra from the Portuguese coast. In Proceedings of the OMAE' 20th International Conference on Offshore Mechanics and Artic Engineering, Rio de Janeiro, Brazil, 3–8 June 2001.
7. Longuet-Higgins, M.S. On the Joint Distribution of Wave Periods and Amplitudes of Sea Waves. *J. Geophys. Res.* **1975**, *80*, 2688–2694. [CrossRef]
8. Hasselmann, K. A statistical analysis of the generation of microseisms. *Rewiews Geophys.* **1963**, *1*, 177–210. [CrossRef]
9. Gualtieri, L.; Stutzmann, E.; Capdeville, Y.; Ardhuin, F.; Schimmel, M.; Mangenev, A.; Morelli, A. Modelling secondary micsoseismic noise by normal mode summation. *Geophys. J. Int.* **2013**, *193*, 1732–1745. [CrossRef]
10. Ardhuin, F.; Gualtieri, L.; Stutzmann, E. How ocean waves rock the Earth: Two mechanisms explain seismic noise with periods 3 to 300 s. *Geophys. Res. Lett.* **2015**, *42*, 765–772. [CrossRef]
11. Friedrich, A.; Krager, F.; Klinge, K. Ocean-generated microseismic noise located with the Grafenberg array. *J. Seismol.* **1998**, *2*, 47–64. [CrossRef]
12. Juretzek, C.; Hadziioannou, C. Where do ocean microseisms come from? A study of Love-to-Rayleigh wave ratios. *J. Geophys. Res.* **2016**, *121*, 6741–6756. [CrossRef]
13. Nishida, K.; Kawakatsu, H.; Fukao, Y.; Obara, K. Background Love and Rayleigh waves simultaneously generated at the Pacific ocean floors. *Geophys. Res. Lett.* **2008**, *35*, L16307. [CrossRef]
14. Dolgikh, G.I.; Kovalev, S.N.; Koren', I.A.; Ovcharenko, V.V. A Two-Coordinate laser strainmeter. Izvestiya. *Phys. Solid Earth* **1988**, *34*, 946–950.
15. Grigoriy, D.; Stanislav, D.; Sergey, K.; Vladimir, C.; Vyacheslav, S.; Sergey, Y. Super-low-frequency laser instrument for measuring hydrosphere pressure variations. *J. Mar. Sci. Technol.* **2009**, *14*, 436–442.
16. Dolgikh, G.; Budrin, S.; Dolgikh, S.; Plotnikov, A. Supersensitive detector of hydrosphere pressure variations. *Sensors* **2020**, *20*, 6998. [CrossRef] [PubMed]
17. Dolgikh, G.I.; Mukomel, D.V. Dependence of microseism variation periods upon the cyclone propagation velocity and direction. *Dokl. Earth Sci.* **2004**, *394*, 141–144.
18. Bowden, K.F. *Physical Oceanography of Coastal Waters*; Halsted Press: New York, NY, USA, 1983; 302p.

Journal of
Marine Science and Engineering

Article

Method of Studying Modulation Effects of Wind and Swell Waves on Tidal and Seiche Oscillations

Grigory Ivanovich Dolgikh [1,2,*] **and Sergey Sergeevich Budrin** [1,2,*]

[1]   V.I. Il'ichev Pacific Oceanological Institute, Far Eastern Branch Russian Academy of Sciences, 690041 Vladivostok, Russia

[2]   Institute for Scientific Research of Aerospace Monitoring "AEROCOSMOS", 105064 Moscow, Russia

*   Correspondence: dolgikh@poi.dvo.ru (G.I.D.); ss_budrin@mail.ru (S.S.B.)

Abstract: This paper describes a method for identifying modulation effects caused by the interaction of waves with different frequencies based on regression analysis. We present examples of its application on experimental data obtained using high-precision laser interference instruments. Using this method, we illustrate and describe the nonlinearity of the change in the period of wind waves that are associated with wave processes of lower frequencies—12- and 24-h tides and seiches. Based on data analysis, we present several basic types of modulation that are characteristic of the interaction of wind and swell waves on seiche oscillations, with the help of which we can explain some peculiarities of change in the process spectrum of these waves.

Keywords: wind waves; swell; tides; seiches; remote probing; space monitoring; nonlinearity; modulation

Citation: Dolgikh, G.I.; Budrin, S.S. Method of Studying Modulation Effects of Wind and Swell Waves on Tidal and Seiche Oscillations. *J. Mar. Sci. Eng.* **2021**, *9*, 926. https:// doi.org/10.3390/jmse9090926

Academic Editor: Lev Shemer

Received: 19 June 2021
Accepted: 22 August 2021
Published: 26 August 2021

1. Introduction

The phenomenon of modulation of short-period waves on long waves is currently widely used in the field of non-contact methods for sea surface monitoring. These processes are mainly investigated during space monitoring by means of analyzing optical [1,2] and radar images [3,4] received from the satellites to restore the structure of the rough sea surface. A two-scale model of the sea surface was used in analysis of radar images. The effect of short waves was taken into account in the framework of the Bragg scattering mechanism, and the effect of the large-scale component is taken into account by changing the slope of the surface. As a result, the small-scale wave component turned out to be responsible for the backscattering of radar signals, and the large-scale component was responsible for spatial modulation of the scattered signals [5,6].

Of particular interest during remote probing of seas and oceans in these types of research is the study of internal waves (IW) [7], current fields [8,9], and anthropogenic impacts on the aquatic environment [10]. Internal waves propagating in the ocean appear on the sea surface due to horizontal components of orbital velocities near the surface, which lead to variations in the characteristics of short wind waves. Thus, on the sea surface, IWs appear in the form of stripes and spots with increased (tidal rip) and decreased (slick) intensity of short gravity waves [11–13]. Modulation effects can occur in a wide range of wind waves. The modulation of wind waves can be described and explained within the kinematic mechanism framework [14], but there are also other theories [15,16]. For example, the authors of [17] propose a model where an increment modulation effect caused by the variation of the wind speed field over the water surface on which there is a field of currents generated by internal waves (IW) acts as the mechanism of this model. The modulation effect of wind increments is being actively discussed nowadays [18].

There are many works and studies devoted to modulation of infragravity (IG) waves. For example, [19] presents the results of studying short IG length changes, while longer waves move on the surface. In [20], the changes in wavelengths and amplitudes of a short-period wave process are carefully calculated with regard to the nonlinear interactions

39

between two wave processes. The change in the energy of short waves, in this case, corresponds to the effect of longer waves against the radiation stress of short waves, which was previously neglected. The radiation wave stress should be understood as an overflow of the impulse caused by wind waves, or, more exactly, by their nonlinearity. In [21], on the basis of field data and numerical models of sea surface gravity waves, it is shown that in the coastal zone during nonlinear wave-wave interaction, energy is transferred from low-frequency long waves back to higher-frequency motions. This explains the tidal modulation of the energy of IG waves, observed in the records of near-bottom pressure on the southern shelf of Sakhalin Island. Similar results were obtained in [22], where the strong tidal modulation of infragravity (200 to 20 s period) waves observed on the southern California shelf is shown to be the result of nonlinear transfers of energy from these low-frequency long waves to higher-frequency motions.

What is interesting is question of the evolution of the wind wave spectrum, which has many mechanisms. For example, [23,24] describes the mechanism of change in the wave spectrum, while waves are going out to a shallow; in this case, due to nonlinear interactions with the bottom, the frequency of the wave process decreases. In [25], numerical and experimental studies of the process of nonlinear energy transfer between two main spectral maxima conditioned by the mechanism of dispersive focusing indicate that nonlinear energy transfer plays a greater role than linear superposition. There is also a large number of works devoted to the evolution of the wave spectrum associated with the modulation effect caused by the interaction of wind waves and swell waves [26,27]. In this paper, we will try to show how the effects of the modulation of wind waves at long waves, such as tides and seiches, can affect changes in the wave spectrum.

We should also note the participation of modulation effects in the emergence of rogue waves. For example, in [28], it is stated that rogue waves (extreme waves) naturally originate as a result of evolution of spectrally narrow packets of gravity waves. We can say that rogue waves are a nonlinear stage of modulation instability. In [29], the Euler equation was solved for liquid with a free surface in deep water. Periodic and boundary conditions were created in the form of a Stokes wave, which was slightly modulated by a low frequency (10^{-5}). At the same time, such a wave is unstable and modulation should increase with time, thereby generating an extremely high wave.

This paper presents a method for studying the effects of modulation of wind and swell waves on tidal and seiche oscillations. This method was developed and tested on hydrophysical data obtained from high-precision, modern instruments based on laser interference methods [30]. The paper also presents the results of using this method for processing and analyzing experimental data, the results of which several main types of modulation of wind waves on seiche oscillations were identified and described. The possibility of introducing this method for processing hydrophysical data received in real time mode is being considered.

All of the experimental data presented in this work were obtained using laser meters of hydrosphere pressure variations [31]. These devices were installed on the bottom for a period from several days to several months, transmitting data in real time to laboratory rooms located at Shultz Cape, the Sea of Japan, in the Primorsky Territory of the Russian Federation. There are also mobile versions of these devices, with which we conduct measurements of tidal and seiche oscillations in various harbors of the Posiet Bay and the Peter the Great Gulf of the Sea of Japan. Although most of the data presented were obtained in Vityaz Bay, the Sea of Japan, we also analyzed and compared the results obtained in other bays. Thus, the presented results will be valid for other closed water areas of the World Ocean.

2. Wave Modulation Research Method

As we know, the main change in the period of swell waves during propagation from the point of generation occurs due to dispersion during propagation; this change in the period is linear and has a decreasing character. However, if we consider large time-scale phenomena, such as typhoons, which originate hundreds of kilometers from the place of registration, the process of swell wave propagation can take several days. Additionally, in addition to dispersion, waves can be influenced by both large-scale phenomena, such as tides, and local phenomena, such as seiche oscillations that occur in closed sea areas and, as a result of these phenomena, the process of changing the period becomes nonlinear. At the same time, in order to study the emerging nonlinear processes, it is first necessary to separate the process of changing the period associated with dispersion from other processes that affect the variations in the period of wind waves and swell waves.

Earlier, from the fragments of the record, which contained swell waves created by passing typhoons, we derived the general function of period change [32,33]. With high accuracy, this function describes the dispersion of waves as they propagate from the source to the receiving point; the general view of the function is presented below.

$$\overline{T}(t) = K_{10} \cdot \frac{\Delta T}{\Delta t} \cdot t + T_0 , \tag{1}$$

where $K_{10} = -2753 \cdot 10^{-4}$, ΔT is the total change in the period in the investigated section, Δt is the total duration of the section, and T_0 is the initial period of wave for $t = 0$.

Thus, as applied to the above problem, we only need to identify the nonlinear part of the period change process. To do this, we will use regression analysis.

Based on the selected values of the spectral maxima, the surface wave signal, the coefficients of the polynomial regression are calculated followed by its construction. The coefficients are calculated from the system of equations presented below.

$$\begin{cases} b_0 + b_1 t + b_2 t^2 + \cdots + b_k t^k = T \\ b_0 t + b_1 t^2 + b_2 t^3 + \cdots + b_k t^{k+1} = Tt \\ b_0 t^2 + b_1 t^3 + b_2 t^4 + \cdots + b_k t^{k+2} = Tt^2 \\ \cdots\cdots\cdots\cdots\cdots\cdots\cdots\cdots\cdots\cdots\cdots\cdots\cdots \\ b_0 t^k + b_1 t^{k+1} + b_2 t^{k+2} + \cdots + b_k x^{2k} = Tt^k \end{cases} \tag{2}$$

where T—values of spectral maxima at time t, $b_0 \ldots b_k$—regression coefficients.

This system can be represented in matrix form as $AB = C$, where

$$A = \begin{pmatrix} 1 & t & t^2 & \cdots & t^k \\ t & t^2 & t^3 & \cdots & t^{k+1} \\ t^2 & t^3 & t^4 & \cdots & t^{k+2} \\ \vdots & \vdots & \vdots & \ddots & \vdots \\ t^k & t^{k+1} & t^{k+2} & \cdots & t^{2k} \end{pmatrix}, \; B = \begin{pmatrix} b_0 \\ b_1 \\ b_2 \\ \vdots \\ b_k \end{pmatrix}, \; C = \begin{pmatrix} T \\ Tt \\ Tt^2 \\ \vdots \\ Tt^k \end{pmatrix} \tag{3}$$

We can write the general regression equation as follows:

$$T(t) = b_0 + tb_1 + t^2 b_2 + \cdots + t^{2k} b_k \tag{4}$$

To estimate the polynomial regression and the general function of the period change, we will use two criteria for evaluating the regression analysis: the coefficient of determination R^2 (equal to 1 in the ideal case) and the standard deviation S (equal to 0 in the ideal case). However, as practice and the theory of regression analysis show, to describe processes with the n number of extrema, it is sufficient to use a polynomial of degree $n + 1$. Thus, to describe the effects of modulation of wind waves on tidal oscillations, which can have 4–5 extrema in a few days, polynomials of 5–6 degrees will be sufficient for the description. As the practice of the conducted research shows, this approach to the choice of the degree of polynomial regression is also the best for parameters of the coefficient of determination and standard deviation. Thus, the use of polynomial degrees greater than $n + 1$ is not practical.

Subtracting the period change function from the obtained expression, we eliminate the constant component of the graph, which is the dispersion that occurs during propagation of wind and swell waves. Thus, we obtain the absolute values of nonlinearities arising during propagation. The general expression for the function describing the absolute values of nonlinearities can be written as:

$$T(t) = \left(b_0 + tb_1 + t^2 b_2 + \cdots + t^{2k} b_k\right) - \left(K_{10} \cdot \frac{\Delta T}{\Delta t} \cdot t + T_0\right) \tag{5}$$

3. Modulation of Swell Waves on Tides

Figure 1 shows an example of processing experimental data using this processing method, highlighting the nonlinearity of the period change.

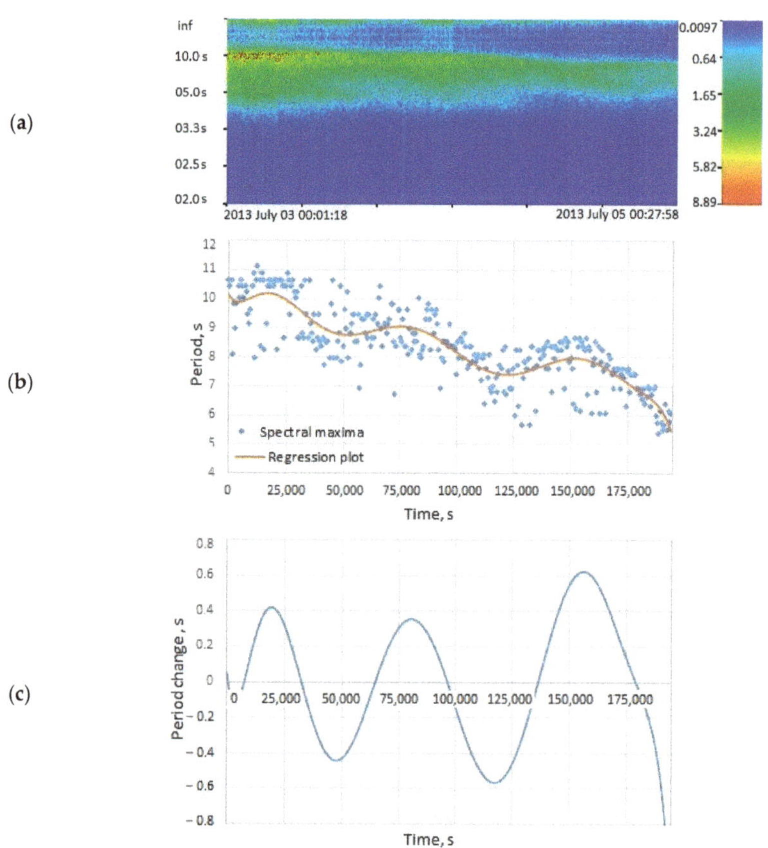

Figure 1. (a) Spectrogram of the wind wave signal obtained in the period from 3 to 5 July 2013 in Vityaz Bay, the Sea of Japan; (b) spectral maxima and the regression graph constructed according to (4); (c) absolute values of the change in the period of the swell waves identified according to (5).

As we can see from Figure 1c, the graph contains three pronounced maxima: the time interval between them is 12 h, which may indicate the modulation of swell waves by tidal fluctuations, which is also proved by other analyzed fragments. Thus, we can conclude that the nonlinearity of the period change is associated with the effect of the modulation of swell waves by low-frequency wave processes, in this case, tidal oscillations. Let us consider one more case of the modulation of wind waves by tidal oscillations of 24 h, shown in Figure 2.

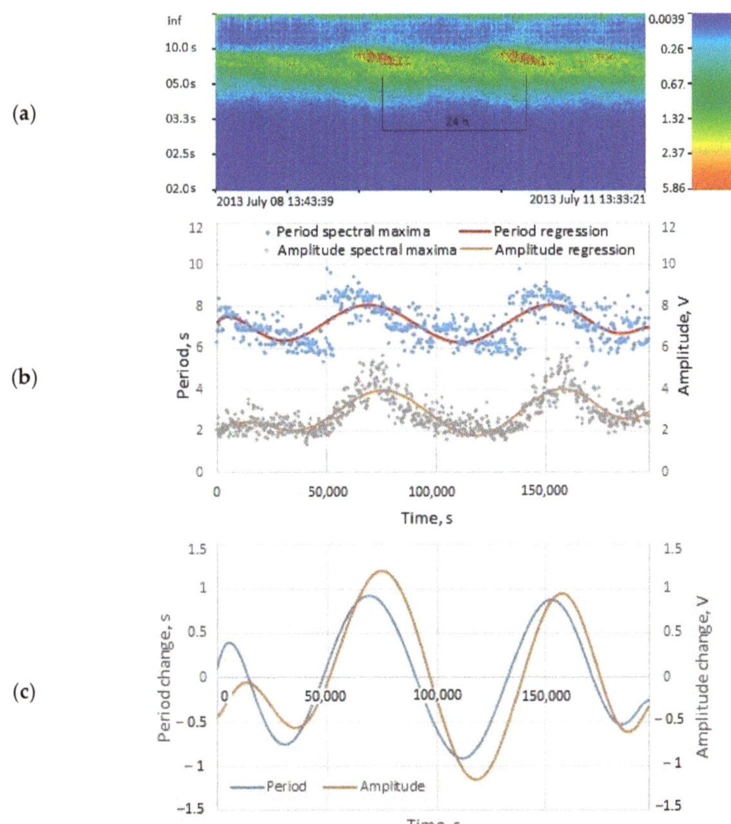

Figure 2. (**a**) Spectrogram of the wind wave signal obtained in the period from 8 to 11 July 2013 in Vityaz Bay, the Sea of Japan; (**b**) spectral maxima and regression graphs; (**c**) absolute changes of period and amplitude of wind waves on tidal oscillations.

In Figure 2c, there is an explicit modulation of both the wind wave period and the amplitude. As a result of modulation at the points of maximum tide values, waves with a large period and amplitude are concentrated, which corresponds to the general idea of this process. However, not everything is as clear at first glance as it seems. When processing the data, we found a number of fragments that do not correspond to generally accepted concepts. One of these fragments is shown in Figure 3.

As we can see in Figure 3a, the record contains two pronounced tides within a period of 12 h, but in Figure 3b, the modulations of the period and the amplitude of the waves are in antiphase, i.e., despite the fact that the waves within aa large period are at the maximum tide point, waves within a smaller period have a greater amplitude. At the same time, all similar cases have one common feature: in all of the fragments, there are seiche oscillations; in this case, these are oscillations are within period of 18 min, which is characteristic of the place where the measurements were completed (Figure 4).

In connection with the above, we can assume that these ambiguous cases may arise due to complex modulation processes associated with the "submodulation" of several wave phenomena.

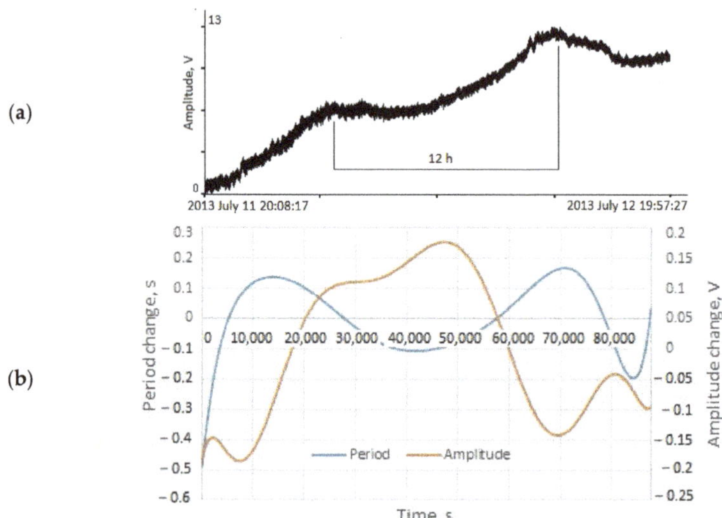

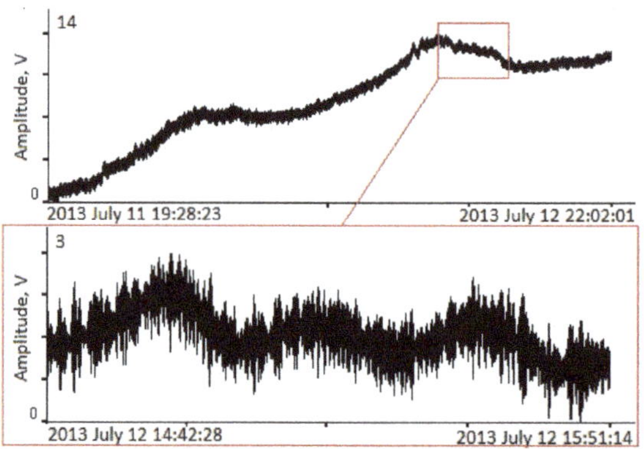

Figure 3. (**a**) Record of tidal fluctuations with a period of 12 h obtained in the period from 11 to 12 July 2013 in Vityaz Bay, the Sea of Japan; (**b**) change in the period and amplitude of waves associated with modulation on tidal oscillations and calculated by the following section of the record.

Figure 4. Seiche oscillations within a period of 18 min against the background of a 12 h tide.

4. Influence of Atmosphere on Modulation of Wind Waves by Tidal Oscillations

An additionally important issue in this topic is the influence of variations in atmospheric pressure and wind regime on the process of the modulation of wind waves by tidal oscillations. Let us consider this issue on the example of a processed recording fragment with tidal oscillations with a period of 12 h and provide meteorological data received from the meteorological station for the same period of time (Figure 5).

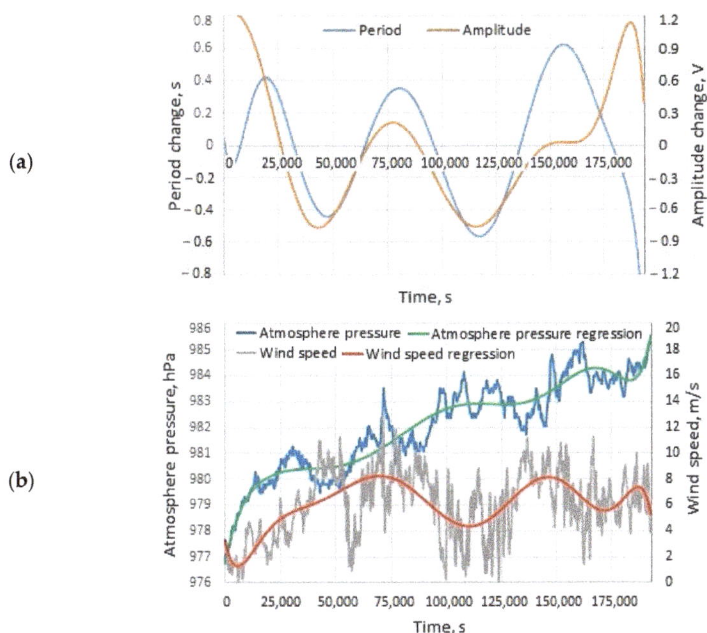

Figure 5. (**a**) Absolute values of wave periods and amplitude variations calculated from the site of the wind wave records obtained during the period from 3 July to 5 July 2013 in Vityaz Bay, the Sea of Japan; (**b**) data on atmospheric pressure and wind speed and regression graphs built on this data for the same period of time.

We can see from the graphs of the modulation of the amplitude and the period of the waves (Figure 5a) that they practically coincide in phase. This means that, as in the previously considered cases, at the maximum tidal points, there will be waves with a longer period and greater amplitude. The atmospheric pressure graph (Figure 5b) shows three peaks corresponding to atmospheric tides; the maxima of these tides are shifted several hours to the right of the maximum values of the swell modulation, which indicates that these oscillations occur after the sea tide. On the same graph, you can see that the graph of the wind speed variations is practically in antiphase with pressure variations and has a minor delay of several hours. Thus, we can conclude that diurnal pressure variations occur after sea tide. Variations in wind speed occur due to fluctuations in atmospheric pressure since air masses begin to move from the area of increasing pressure to areas with lower pressure, which results in a change in wind direction. An abrupt decrease in the wave amplitude at the second maximum (Figure 2a) at the max wind speed may be due to the fact that the wind direction was opposite to the direction of the swell wave propagation. At the same time, at the third maximum of the period change graph, we see an increase in the modulation effect, which may indicate that waves with periods of 4 s and below are more susceptible to the modulation effect than waves with higher periods. When processing and analyzing more than five more data fragments with modulations of wind waves and swell waves on tidal oscillations present on them and the meteorological data received over the same period of time, there were no impacts on the modulation effect on the part of the variations in the atmospheric pressure and wind speed.

5. Modulation of Wind Waves on Seiche Oscillations

Analyzing more than 30 fragments of the records on which seiche oscillations were explicitly presented simultaneously with strong wind waves, several characteristic types of modulations, "two-tone" and "four-tone", were identified. These types of modulations are shown in Figure 6.

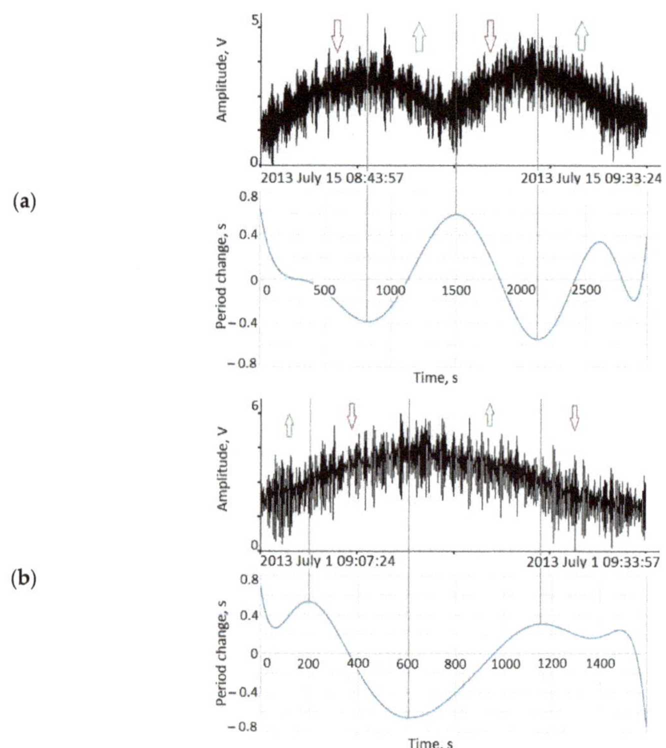

Figure 6. (a) A recording fragment with two seiche oscillations of 18 min and a change in the period of wind waves ("two-tone" modulation); (b) a recording fragment with one seiche vibration of 20 min and a change in the period of wind waves ("four-tone" modulation).

As we can see in Figure 6a, in one seiche oscillation with period of 18 min, the wind wave period changes two times, which means that on opposite seiche fronts, the processes of compression (red arrows) and extension (green arrows) occur, while in Figure 6b, in one oscillation with period of 20 min, the wind waves at the front have one compression–extension cycle, which, in turn, generates modulation.

We can describe this type of modulation using a simple equation.

$$\Delta T(t) = \sin\left[\left(\frac{2\pi t}{T_{\text{ww}}} + \alpha_1\right) + m \cdot \sin\left(\frac{2\pi t}{T_{mo}} + \alpha_2\right)\right] \tag{6}$$

where T_{ww} is the period of the wind waves, T_{mo} is the period of modulating oscillation, m is the modulation index.

Figure 7 shows the spectrum of wind waves within a period of 4.8 s modulated by seiche oscillations within a period of 18 min constructed on experimental data. Figure 6 also shows the spectrum calculated using Expression (5), with the same wave parameters and modulation index $m = 4$.

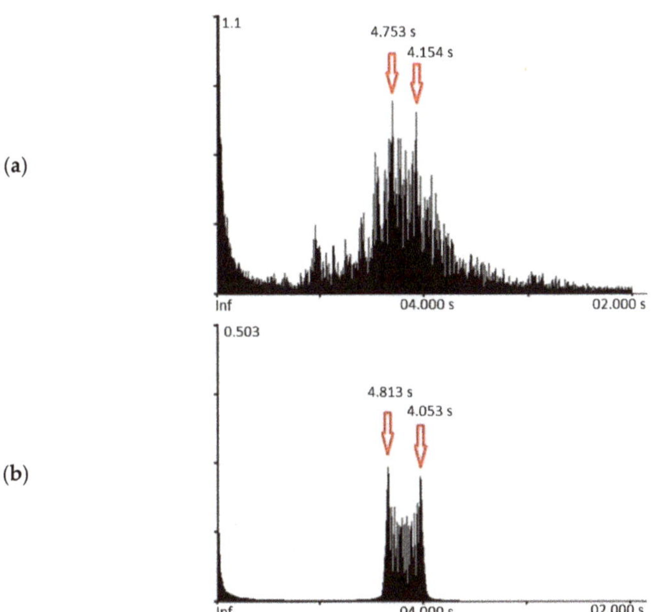

Figure 7. (**a**) Wave spectrum constructed on experimental data; (**b**) wave spectrum built on Expression (6).

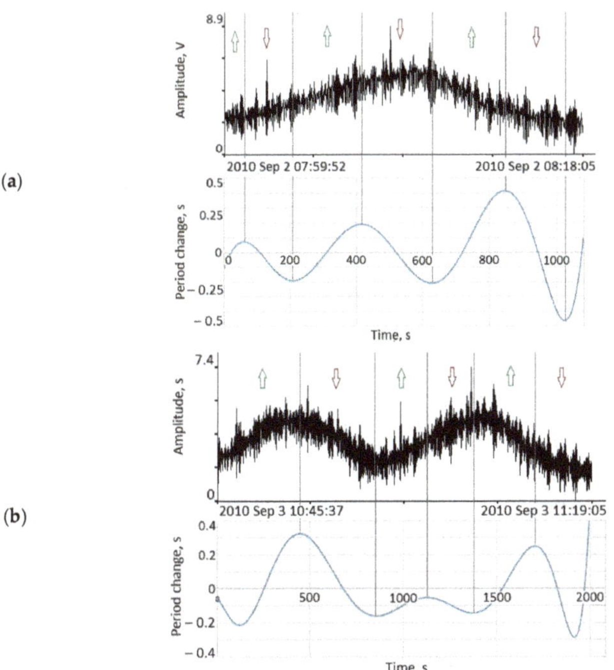

Figure 8. (**a**) Example of a combined modulation on one seiche oscillation; (**b**) example of a combined modulation on two seiche oscillations.

In the spectra shown in Figure 7, as in other similar cases, there are two characteristic maxima. The first is the main period of wind waves, while the second is responsible for the modulation process of wind waves on seiche oscillations. In this case, the main spectral maximum always remains in its place, and the position of the second (modulation) maximum can vary depending on the modulation index. Thus, when analyzing the spectra of experimental data based on two characteristic maxima, we can confidently speak about the presence of seiche oscillations and their modulation of wind waves.

We have now looked at the most common modulation ideal cases. However, we understand that in nature, not everything is so unambiguous and, of course, there are cases of combined modulations. As such, for example, on one (Figure 8a) and on two seiche oscillations, both "two-tone" and "four-tone" modulation can occur (Figure 8b). Such combined modulations, as we can assume, arise in cases of different steepness of the front of the modulating oscillation, which is caused by the seiche asymmetry. As a result, on the flatter part, wind waves and swell waves can be modulated several times.

These types of modulations can also be described using Expression (6); however, the modulation index, in this case, will be equal to the number of extrema per the number of studied modulating oscillations. To show this, let us calculate the spectrum of the signal using Expression (6) in Figure 8a and compare it with the spectrum constructed based on the experimental data. The initial data are as follows: $T_{ww} = 8.8$ s; $T_{mo} = 24$ s; $m = 6$. The calculation results are shown in Figure 9.

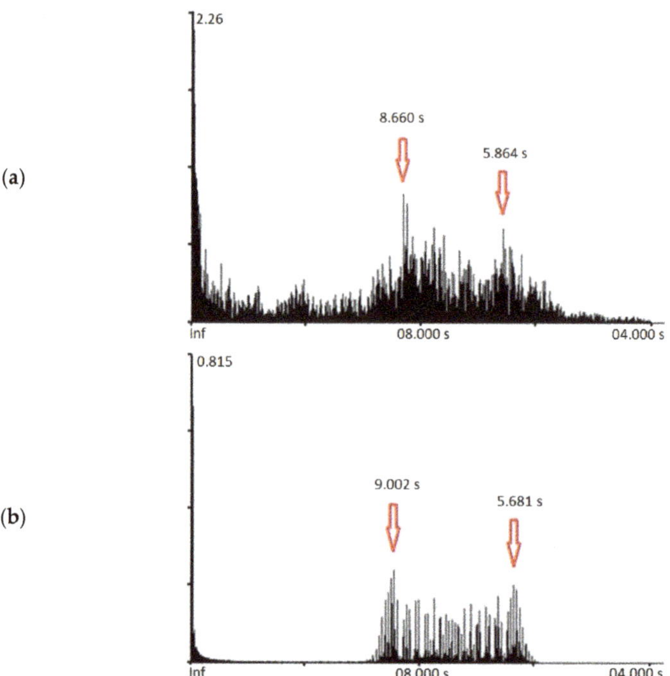

Figure 9. (**a**) Wave spectrum constructed on experimental data; (**b**) wave spectrum built on Expression (6).

As we mentioned above, the width of the modulated oscillations spectrum depends directly on the value of the modulation index, which is well demonstrated in Figure 9. The form of the spectra is almost identical although it has minor deviations in the numerical values of the maxima.

However, as we all perfectly understand, the change in the period is not only associated with the processes of modulation of wind waves but are also associated with dispersion during wave propagation. In order to account for this variance, we substitute period change Function (1) into the expression describing Modulation (6). As a result, we obtain an equation describing both the modulation process and the dispersion of wind waves during propagation.

$$T(t) = \sin\left[\left(\frac{2\pi \cdot t \cdot \Delta t}{K_{10} \cdot t \cdot \Delta t + T_0 \cdot \Delta T}\right) + m \cdot \sin\left(\frac{2\pi \cdot t}{T_{mo}} + \alpha\right)\right] \tag{7}$$

Figure 10 shows the spectrum of wind waves within a period of 5.2 s obtained based on experimental data; in the studied fragment, there is a pronounced dispersion of waves $\Delta T = 0.4$ s, and seiches within a period of 18 min are present. The figure also shows the spectrum calculated based on Expression (7), according to the initial data indicated above.

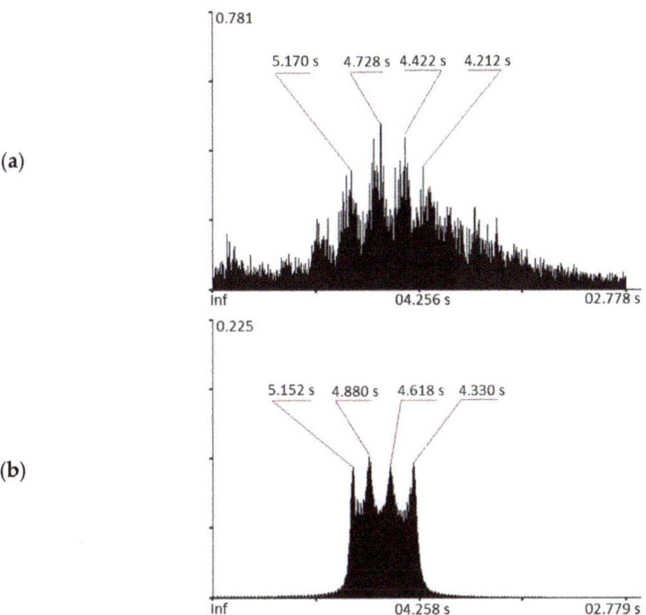

Figure 10. (**a**) Wave spectrum constructed on experimental data; (**b**) wave spectrum built on Expression (7).

As you can see in Figure 10, there are now four characteristic maxima in the spectra. The first two are responsible for the modulation of wind waves on seiches, and the other two are responsible for wave dispersion. In this case, the spectrum width does not depend on the modulation index. The modulation index is responsible for the amplitude of spectrum maxima, and the change in the wave period due to dispersion during propagation is responsible for the spectrum width.

6. Conclusions

The method of studying modulation effects presented in this work, which is based on regression analysis and the general functions of the period change, has shown good results when applied to studies of the modulation of wind and swell waves on tidal oscillations. Using this method, it was shown that, in general, when wind waves are modulated by tides, waves with large a period and amplitude are concentrated in the upper points of the tide. However, when extraneous wave processes, such as seiches, occur, the modulation of wave amplitude can have an extremum in the lower tide point, i.e., the modulation of wave period and its amplitude will be in antiphase.

Studying the modulation of wind waves on seiches by the above method, we identified several main types of modulation: "two-tone" and "four-tone". These types of modulation are well described using the common frequency modulation, Equation (6). When comparing the spectra of the experimental data and the spectrum calculated using the frequency modulation formula, we identified two spectral maxima, the first of which is responsible for the main wave period, and the second of which is responsible for the modulation process. From these two characteristic maxima, we can speak with good confidence about the presence of seiche oscillations and their modulation of wind waves, while the width of the spectrum depends on the modulation index, i.e., the number of wave modulations per one period of the seiche oscillation.

The obtained expression for frequency modulation accounting for Dispersion (7) well describes these phenomena. When comparing the spectra of the experimental data and the spectra calculated using this expression, four characteristic spectral maxima were identified. The first two are responsible for modulation on seiches; the rest are responsible for dispersion during propagation. In this case, the modulation index affects the amplitude of the maxima, and the change in period is responsible for the width of the spectra. In the presence of these maxima in the spectrum, we can speak not only about the presence of seiche oscillations, but also about the fact that the wave process modulated by them is not local in nature but most likely came from another point of the water area since its period varies linearly due to dispersion during propagation.

The efficiency of the considered method for studying the interaction of wind waves with wave processes of lower frequency—twelve-hour and round-the-clock tides and seiches—can be increased by combining it with space monitoring methods based on the analysis of optical and radar images, especially in closed water areas and in the shelf zone of the World Ocean.

Author Contributions: G.I.D. problem statement, discussion, and writing of the article. S.S.B. data processing, discussion, and writing of the article. All authors have read and agreed to the published version of the manuscript.

Funding: The work was conducted with the financial support for the project by the Russian Federation represented by the Ministry of Science and Higher Education of the Russian Federation, Agreement No. 075-15-2020-776.

Acknowledgments: We would like to express our deep gratitude to all of the employees of the Physics of Geospheres laboratory.

Conflicts of Interest: The authors declare no conflict of interest.

References

1. Bondur, V.G.; Murynin, A.B. Methods for retrieval of sea wave spectra from aerospace image spectra. *Izv. Atmos. Ocean. Phys.* **2016**, *52*, 877–887. [CrossRef]
2. Bondur, V.G.; Vorobyev, V.E.; Murynin, A.B. Retrieving Sea Wave Spectra Based on High-Resolution Space Images under Different Conditions of Wave Generation. *Izv. Atmos. Ocean. Phys.* **2020**, *56*, 887–897. [CrossRef]
3. Bondur, V.G.; Grebenyuk, Y.V.; Morozov, E.G. Satellite recording and modeling of short internal waves in coastal zones of the ocean. *Dokl. Earth Sci.* **2008**, *418*, 191–195. [CrossRef]
4. Donelan, M.A.; Pierson, W.J. Radar scattering and equilibrium ranges in wind-generated waves with application to scatterometry. *J. Geophys. Res.* **1987**, *93*, 4871–5029. [CrossRef]
5. Hasselman, K.; Raney, R.K.; Plant, W.J.; Alpers, W.; Shuchman, R.A.; Lyzenga, D.R.; Rufenach, C.L.; Tucker, M.J. Theory of synthetic aperture radar imaging: A MARSEN view. *J. Geophys. Res.* **1985**, *90*, 4659–4686. [CrossRef]
6. Kropfli, R.A.; Ostrovski, L.A.; Stanton, T.P.; Skirta, E.A.; Keane, A.N.; Irisov, V. Relationships between strong internal waves in the coastal zone and there radar and radiometric signatures. *J. Geophys. Res.* **1999**, *104*, 3133–3148. [CrossRef]
7. Bondur, V.G.; Grebenyuk, Y.V.; Sabinin, K.D. The spectral characteristics and kinematics of short-period internal waves on the Hawaiian shelf. *Izv. Atmos. Ocean. Phys.* **2009**, *45*, 598–607. [CrossRef]
8. Bondur, V.G.; Sabinin, K.D.; Grebenyuk, Y.V. Characteristics of inertial oscillations according to the experimental measurements of currents on the Russian shelf of the Black Sea. *Izv. Atmos. Ocean. Phys.* **2017**, *53*, 120–126. [CrossRef]
9. Bondur, V.G.; Vorobjev, V.E.; Grebenjuk, Y.V.; Sabinin, K.D.; Serebryany, A.N. Study of fields of currents and pollution of the coastal waters on the Gelendzhik Shelf of the Black Sea with space data. *Izv. Atmos. Ocean. Phys.* **2013**, *49*, 886–896. [CrossRef]

10. Bondur, V.G. Satellite monitoring and mathematical modelling of deep runoff turbulent jets in coastal water areas. In *Waste Water-Evaluation and Management*; IntechOpen: Rijeka, Croatia, 2011; pp. 155–180. ISBN 978-953-307-233-3. Available online: https://www.intechopen.com/chapters/14574 (accessed on 25 August 2021). [CrossRef]
11. Ermakov, S.A.; Salashin, S.G. Modulation of gravitational-capillary waves in the field of an internal wave. *Izv. Acad. Sci. USSR Phys. Atmos. Ocean.* **1984**, *20*, 394–404.
12. Ermakov, S.A.; Pelinovsky, E.N. Variation of the spectrum of wind ripple on coastal waters under the action of internal waves. *Dyn. Atmos. Ocean.* **1984**, *8*, 95–100. [CrossRef]
13. Hughes, B.A. The effect of internal waves on surface wind waves: Theoretical analysis. *J. Geophys. Res.* **1987**, *83*, 455–465. [CrossRef]
14. Basovich, A.Y.; Bachanov, V.V.; Talanov, V.I. Transformation of wind-driven wave spectra by short internal wave trains. *Izv. Acad. Sci. USSR Atmos. Ocean. Phys.* **1987**, *23*, 520.
15. Basovich, A.Y.; Bakhanov, V.V.; Bravo-Zhivotovskii, D.M.; Gordeev, L.B.; Zhidko, Y.M.; Muyakshin, S.I. Correlation of variations in the spectral density of centimeter and decimeter surface waves in the field of an internal wave. *Dokl. Acad. Sci. USSR* **1988**, *298*, 967.
16. Ermakov, S.A.; Salashin, S.G. On the effect of strong modulation of capillary-gravitational ripples by internal waves. *Dokl. Acad. Sci. USSR* **1994**, *337*, 108–111.
17. Gorshkov, K.A.; Dolina, I.S.; Soustova, I.A.; Troitskaya, Y.I. Modulation of short wind waves in the presence of strong internal waves: The effect of growth-rate modulation. *Izv. Atmos. Ocean. Phys.* **2003**, *39*, 596–606.
18. Landahl, M.T.; Widnall, S.E.; Hultgen, L. An interaction mechanism between large and small scales for wind-generated water waves. In Proceedings of the Twelfth Symposium on Naval Hydrodynamics, National Academy of Sciences, Washington, DC, USA, 5–9 June 1979; p. 541.
19. Unna, P.J. Sea waves. *Nature* **1947**, *159*, 239–242. [CrossRef]
20. Longuet-Higgins, M.S.; Stewart, R.W. Changes in the form of short gravity waves on long waves and tidal currents. *J. Fluid Mech.* **1960**, *8*, 565–583. [CrossRef]
21. Kovalev, P.D.; Kovalev, D.P. Modulation of short infragravity waves by tide. *Fundam. Prikl. Gidrofiz.* **2018**, *11*, 21–27. [CrossRef]
22. Thomson, J.; Elgar, S.; Raubenheimer, B.; Herbers, T.H.C.; Guza, R.T. Tidal modulation of infragravity waves via nonlinear energy losses in the surfzone. *Geophys. Res. Lett.* **2006**, *33*, L05601. [CrossRef]
23. Saprykina, Y.; Shtremel, M.; Volvaiker, S.; Kuznetsov, S. Frequency Downshifting in Wave Spectra in Coastal Zone and Its Influence on Mudbank Formation. *J. Mar. Sci. Eng.* **2020**, *8*, 723. [CrossRef]
24. Saprykina, Y.V.; Kuznetsov, S.Y.; Andreeva, N.; Shtremel, M.N. Scenarios of nonlinear wave transformation in coastal zone. *Oceanology* **2013**, *53*, 422–431. [CrossRef]
25. Kuznetsov, S.Y.; Saprykina, Y.V. An experimental study of near shore evolution of wave groups. *Oceanology* **2002**, *42*, 336–343.
26. Charles, L.V.; Jim, T.; Hans, C.G.; Clarence, O.C. Impact of swell on the wind-sea and resulting modulation of stress. *Prog. Oceanogr.* **2019**, *178*, 102164. [CrossRef]
27. Chen, S.; Qiao, F.; Jiang, W.; Guo, J.; Dai, D. Impact of Surface Waves on Wind Stress under Low to Moderate Wind Conditions. *J. Phys. Oceanogr.* **2019**, *49*, 2017–2028. [CrossRef]
28. Zakharov, V.E.; Dyachenko, A.I.; Prokofiev, A.O. Freak waves: Peculiarities of numerical simulations. In *Extreme Ocean Waves*; Springer: Dordrecht, The Netherlands, 2016; pp. 1–29.
29. Zakharov, V.E.; Shamin, R.E. Probability of the occurrence of freak waves. *JETP Lett.* **2010**, *91*, 62–65. [CrossRef]
30. Dolgikh, G.I.; Budrin, S.S.; Dolgikh, S.G.; Plotnikov, A.A. Supersensitive Detector of Hydrosphere Pressure Variations. *Sensors* **2020**, *20*, 6998. [CrossRef] [PubMed]
31. Dolgikh, S.G.; Budrin, S.S.; Plotnikov, A.A. Laser meter for hydrosphere pressure variations with a mechanical temperature compensation system. *Oceanology* **2017**, *57*, 600–604. [CrossRef]
32. Dolgikh, G.I.; Dolgikh, S.G.; Budrin, S.S. Fluctuations of the sea level, caused by gravitational and infra–gravitational sea waves. *J. Mar. Sci. Eng.* **2020**, *8*, 796. [CrossRef]
33. Dolgikh, G.I.; Budrin, S.S. Some regularities in the dynamics of periods of wind waves. *Dokl. Earth Sci.* **2016**, *468*, 536–539. [CrossRef]

Journal of
Marine Science and Engineering

Article

Deformation Anomalies Accompanying Tsunami Origination

Grigory Dolgikh [1,2,*] **and Stanislav Dolgikh** [1,2,*]

1   V.I. Il'ichev Pacific Oceanological Institute, Far Eastern Branch Russian Academy of Sciences, 690041 Vladivostok, Russia
2   AEROCOSMOS Research Institute for Aerospace Monitoring, 105064 Moscow, Russia
*   Correspondence: dolgikh@poi.dvo.ru (G.D.); sdolgikh@poi.dvo.ru (S.D.)

Abstract: Basing on the analysis of data on variations of deformations in the Earth's crust, which were obtained with a laser strainmeter, we found that deformation anomalies (deformation jumps) occurred at the time of tsunami generation. Deformation jumps recorded by the laser strainmeter were apparently caused by bottom displacements, leading to tsunami formation. According to the data for the many recorded tsunamigenic earthquakes, we calculated the damping ratios of the identified deformation anomalies for three regions of the planet. We proved the obtained experimental results by applying the sine-Gordon equation, the one-kink and two-kink solutions of which allowed us to describe the observed deformation anomalies. We also formulated the direction of a theoretical deformation jump occurrence—a kink (bore)—during an underwater landslide causing a tsunami.

Keywords: earthquake; tsunami; laser strainmeter; deformation jump; sine-Gordon equation; kink; anti-kink; underwater landslide

1. Introduction

We know that tsunamis are some of the most dangerous and catastrophic phenomena on Earth, which cause significant damage to humanity. A typical example is the tsunami that hit the Indian Ocean on 26 December 2004, killing more than 283,000 people. This was caused by a powerful earthquake with a maximum magnitude of about 9.3 [1]. Tsunamis affect various regions of the planet, although this is especially true for Japan. Considering the extent to which Japan suffers from the impacts of earthquakes and tsunamis and its high level of scientific and technical development, we can expect that more advanced scientific and technical ideas aimed at predicting the occurrence and development of earthquakes and tsunamis will be concentrated in this region of the planet. While the short-term forecasting of earthquakes is far from being solved, the detection of the moments of tsunami origination seems to be quite solvable. The Japanese Islands and the adjoining water areas are "crammed" with various seismic stations, GPS receivers, bottom seismic stations, and high-precision sea and ocean level meters. Nevertheless, the events of 2011 "exposed" the short-term tsunami forecasting problems even more.

Presently, the traditional method of short-term tsunami forecasting is based on seismological information (earthquake magnitude, main shock time, and epicenter location) [2]. An earthquake magnitude that exceeds a predetermined threshold, which will be different for different tsunamigenic zones, usually results in a tsunami warning. This approach, based on the "magnitude and geographical principle", is simple; it helps reduce the number of missed tsunamis, but also gives false alarms. Most current early tsunami warning systems are based on seismic data.

For example, the Pacific Tsunami Warning Center (PTWC) uses seismic data in conjunction with long-period wave (W-phase) data for global tsunami warnings in the Pacific [3]. Another tsunami warning center is the Japan Meteorological Agency (JMA), which provides local tsunami warnings within up to 3 min after near-field earthquakes by analyzing seismic data [4], then updates the warnings using the forms of seismic wave and tsunami

Citation: Dolgikh, G.; Dolgikh, S. Deformation Anomalies Accompanying Tsunami Origination. *J. Mar. Sci. Eng.* **2021**, *9*, 1144. https://doi.org/10.3390/jmse9101144

Academic Editor: Dong-Sheng Jeng

Received: 20 September 2021
Accepted: 12 October 2021
Published: 18 October 2021

Publisher's Note: MDPI stays neutral with regard to jurisdictional claims in published maps and institutional affiliations.

data [5]. The W-phase appears in seismic recordings between P-waves and S-waves and can be used to estimate the seismic moment, epicenter location, and fault mechanism. The efficiency of W-phase inversion has already been demonstrated in many studies and is actively used by tsunami warning centers [6,7].

Recently, two independent approaches have been proposed to determine the energy of a tsunami source: one is based on Deep-Ocean Assessment and Reporting of Tsunamis (DART) data during tsunami propagation, and the other on ground-based coastal global positioning system (GPS) data during tsunami generation. The GPS approach takes into account the dynamic earthquake process, while the DART inversion approach provides an actual estimate of the tsunami energy and propagating tsunami waves. Both approaches lead to coordinated energy scales for the earlier-studied tsunamis. Inspired by these promising results, the authors [8] researched an approach to determine the tsunami source energy in real time by combining the two methods. At the first stage, the tsunami source is determined immediately after the earthquake using the global GPS network for early warnings in the near-field zone. Then, the tsunami energy is defined more precisely based on the nearest DART measurements to improve the accuracy of forecasts or cancel the alarm. The combination of these two real-time networks could offer an attractive opportunity for the early detection of tsunami threats to save more lives and for the early cancellation of tsunami warnings to avoid unnecessary false alarms. In the past decade, the number of open ocean sensors capable of analyzing information about a passing tsunami has steadily increased, especially thanks to the national cable networks and international efforts such as DART systems. The received information is analyzed in order to warn people about a tsunami. Most of the current warnings, which include tsunamis, are aimed at the mid-to-far zone regions. In [8], the main directions of research using DART and GPS systems were formulated; however, the failure of this system in the Indian Ocean, when a powerful tsunami was missed, plunged once optimistic scientists into despondency. For early tsunami warnings, various methods can be used, including methods based on space monitoring of earthquake-endangered areas [9–11].

We should note that tsunamis can be caused not only by earthquakes, but also by underwater landslides, volcanic activity, and simply by the collapse of mountain massifs into the sea [12].

For the first time period, a deformation anomaly was recorded during the registration of a tsunamigenic earthquake, the laser strainmeter of which showed the form of a deformation jump that occurred after the earthquake started [13]. Later, this result was generalized in [14] and the deformation methods used for determining the tsunamigenic nature of earthquakes were based on it. The potential of this method is associated with the fact that its development will allow the nature of the movement of the Earth's crust to be determined remotely at planetary distances, which sets in motion huge masses of water that degenerate into tsunamis during their development. It is clear that oscillations originating in the source of an earthquake do not cause a tsunami. These oscillations are associated with the parameters of continuity breaches, i.e., with their geometrical dimensions and elastic plate deformations. As a rule, these oscillations occur in the time range from the first minutes to the first ten seconds. These oscillations will never cause a tsunami. Only quick displacements of huge masses of the Earth's crust, which unfortunately are not recorded by any broadband seismographs, lead to a tsunami. In this paper, we will consider the peculiarities of the appearance of the deformation anomalies accompanying (attending) tsunamis, using several examples and with descriptions of the physical mechanisms of their occurrence and development. The development prospects in this direction are associated with the fact that the speed of such deformation anomalies is more than an order of magnitude higher than the speed of tsunami propagation, which is extremely important for warning services.

2. Recording Complex

The deformation anomaly, described in [13] and to a greater extent corresponding to the concept of a "deformation jump", showed a small amplitude equal to about 60 μm and was recorded by a laser strainmeter at a distance of about 5600 km from its location of origin. It is clear that such deformation anomalies cannot be recorded by any broadband seismograph (velocimeter, accelerometer, etc.), since these instruments are unable to register such disturbances. We could talk about the prospects of using GPS receivers to register such anomalies; however, at such distances from the locations of deformation anomalies, their registration by GPS receivers is impossible because the main samples of GPS receivers can only provide displacement registration accuracy of about 1 mm, which is much larger than the values of the registered anomalies. In the Conclusions section of this paper, we will discuss where these GPS receivers can be used. Currently, to register deformation anomalies associated with the process of tsunami generation, the most efficient instruments are laser strainmeters, which are capable of measuring microdisplacements of the Earth's crust in the frequency range of 0 (conditionally) to 1000 Hz with high accuracy (up to 1 pm) [15]. In our study, we will use the data on variations in the microdisplacements of the Earth's crust obtained from the horizontal laser strainmeter, which were created on the base of a Michelson interferometer with an unequal measuring arm length of 52.5 m and a "north–south" orientation. This instrument uses a frequency-stabilized helium–neon laser manufactured by Melles–Griott with long-term stability of 10^{-9} as a light source [16]. Recently, we modernized this laser strainmeter, providing it with a frequency-stabilized laser, stabilized along the iodine lines in the eleventh digit, and an improved recording system. After the modernization process, it could detect variations of microdisplacements in the Earth's crust in the frequency range of 0 (conditionally) to 1000 Hz, with accuracy of 0.03 nm. The laser strainmeter was installed in thermally insulated underground rooms at depths of about 3–5 m under the Earth's surface at 42°34.798′ N, 131°9.400′ E, at an altitude of about 60 m above sea level. Figure 1 shows a general view of the underground part of the central interference unit (left) and the underground beam guide (right). Experimental data are transmitted via a cable line to the recording computer, where after pre-processing the data files are formed, with a sampling frequency of 1 kHz and duration of 1 h. The laser strainmeter is a part of the seismoacoustic and hydrophysical complex, located in the south of the Primorsky Territory of Russia at the sea hydrophysical study site of POI FEB RAS "Shultz Cape" [17]. The main purpose of the complex is studying the nature of variations in microdeformations of the Earth's crust; fluctuations in atmospheric and hydrospheric pressure over wide-frequency and dynamic ranges; and the regularities of the emergence, development, and transformation of oscillations and waves of the sonic and infrasonic ranges. The measurements at the complex are carried out in continuous mode and all obtained data are input into the database (approximately 10 terabyte capacity), which is being steadily complemented. A precise time clock based on the Trimble 5700 GPS instrument is used for synchronization.

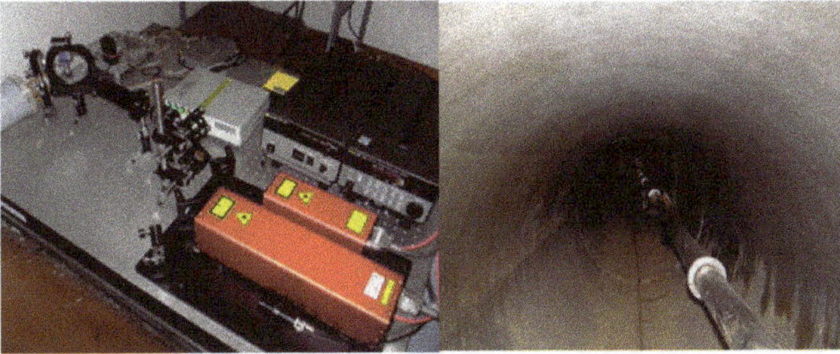

Figure 1. Modernized 52.5 m laser strainmeter. The central interference unit with a frequency-stabilized helium–neon laser, which is stabilized along the iodine lines (frequency stability 10^{-11}, (**left**)), as well as an underground beam guide (**right**).

3. Registration of Tsunamigenic Earthquakes Deformation Anomalies

The laser strainmeter, which has been operating since 2000, has recorded many tsunamigenic and non-tsunamigenic earthquakes. Let us consider the specific differences in the behaviors of tsunamigenic and non-tsunamigenic earthquakes using the two example earthquakes described below. The first powerful tsunamigenic earthquake occurred on 26 December 2004, while the second non-tsunamigenic earthquake occurred on 4 August 2000. The epicenter of the first tsunamigenic earthquake was located at 3.30° N, 95.87° E, about 160 km west of Sumatra, at a depth of 30 km below the sea level. The distance from the earthquake epicenter to the laser strainmeter location was approximately 5600 km. The laser strainmeter recording of a peculiar signal from a tsunamigenic earthquake is shown in Figure 2. The recording shows a powerful deformation anomaly that appeared a short time after the earthquake started, with an amplitude of about 59.3 μm. The amplitude of this anomaly was much greater than the amplitude of the daily tide, as observed at the instrument location. In Figure 2, the earthquake onset is marked with an arrow. The laser strainmeter recorded the tsunamigenic earthquake signal at 19 min 54 s after the earthquake started.

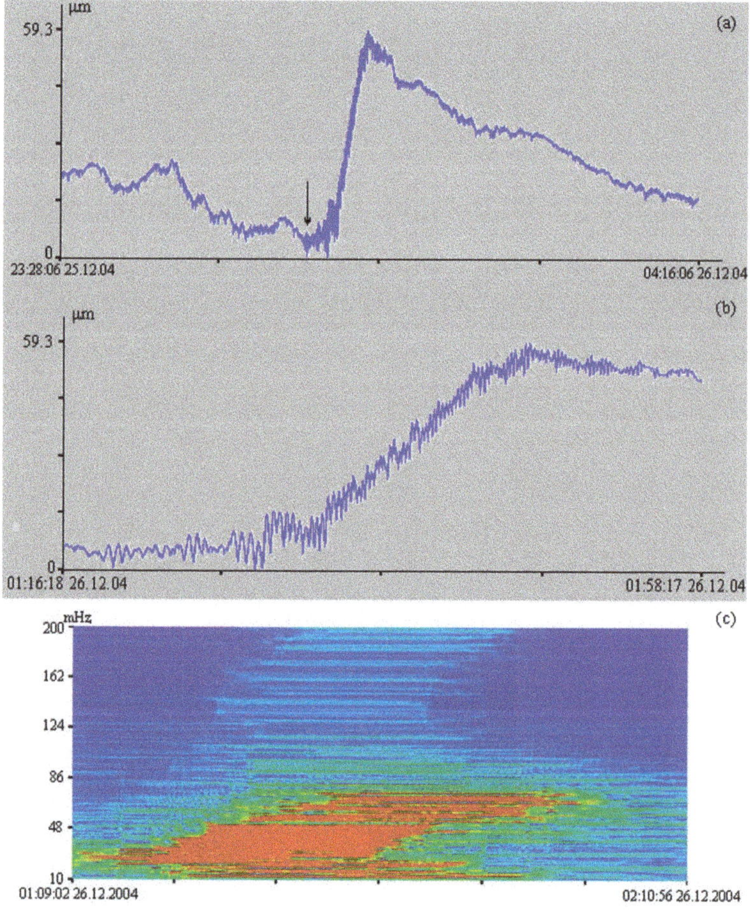

Figure 2. Fragment of the 52.5 m laser strainmeter recording from December 2004 (**a**), enlarged fragment of the laser strainmeter recording (**b**), and dynamic spectrogram of the laser strainmeter recording (**c**). Universal time is shown on the abscissa axis.

Analyzing the dynamic spectrogram shown in Figure 2c, we found that the periods of the main oscillations caused by the earthquake gradually decreased from 30 to 14 s. If we know the relation connecting the propagation speed of the elastic waves with the period of oscillations, the magnitude of the change in the period of the main oscillations, and the time during which this change occurred, we can determine the distance to the earthquake site. The dynamic spectrogram also shows a strong disturbance in the low-frequency range.

We must note that the results from processing the space monitoring data, which was carried out after this catastrophic tsunamigenic earthquake, showed that ionospheric anomalies were observed 4–5 days before it, which were registered by way of analyzing data from the GPS satellite navigation system. These anomalies manifested themselves as changes in the electron density profiles [11] and in changes in the total electron content (TEC) in the ionosphere in the Sumatra area [10]. At the same time, the use of a geomechanical model [18] for this region showed that the recorded increase in atmospheric pressure led to an increase in the stress–strain state of the Earth's crust and brought it closer to the strength limit before the Sumatra earthquake [10]. As an example of a non-tsunamigenic earthquake, let us consider the 52.5 m laser strainmeter recording from August 2000 (Figure 3, top). At that time there was a recording of the earthquake, which occurred on 4 August 2000 at 21:13:05 (hereinafter—universal time) at N48.85° and E142.42°, at a depth of 33 km, with a magnitude of 7.1. The recording showed oscillations of about 16 s, which were specific to an earthquake, although there was no deformation jump. The dynamic spectrogram in Figure 3 (bottom) shows that the amplitudes of oscillations in the period range of about 16 s are much greater than the amplitudes of oscillations in the low-frequency range.

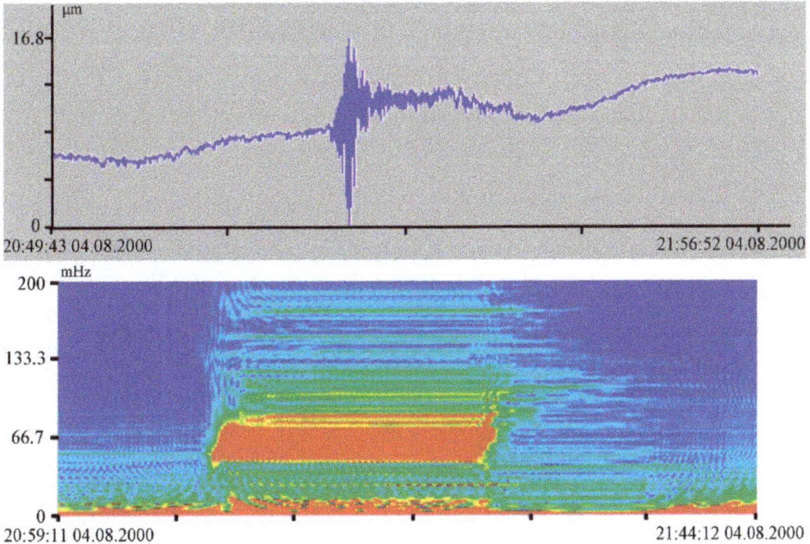

Figure 3. Fragment of the 52.5 m laser strainmeter recording from August 2000 (**top**) and a dynamic spectrogram of the laser strainmeter recording (**bottom**).

Analysis of the laser strainmeter records of tsunamigenic and non-tsunamigenic earthquakes showed that the deformation jump recording was specific to a tsunamigenic earthquake only.

Further, we will highlight some of the peculiarities of the appearance and development of deformation anomalies at the times of tsunami generation in the three tsunami-endangered regions, namely Indonesia, Chile, and the west coast of North America.

3.1. Earthquakes in Indonesia

The first powerful earthquake occurred on 11 April 2012 at 08:38:36 on the west coast of northern Sumatra, Indonesia, at 2.327° N, 93.063° E, at a depth of 20 km and a magnitude of 8.6. The maximum recorded tsunami wave height was 1.08 m. The distance from the earthquake epicenter to the location of the laser strainmeter was more than 5800 km. The laser strainmeter recorded this earthquake signal almost 18 min later at 08:55:39. The average propagation speed of the elastic wave was 5.66 km/s. In the dynamic spectrogram (Figure 4a), we can identify oscillations with periods ranging from 30 to 14 s that are peculiar to an earthquake. There is also a strong disturbance in the lower frequency range.

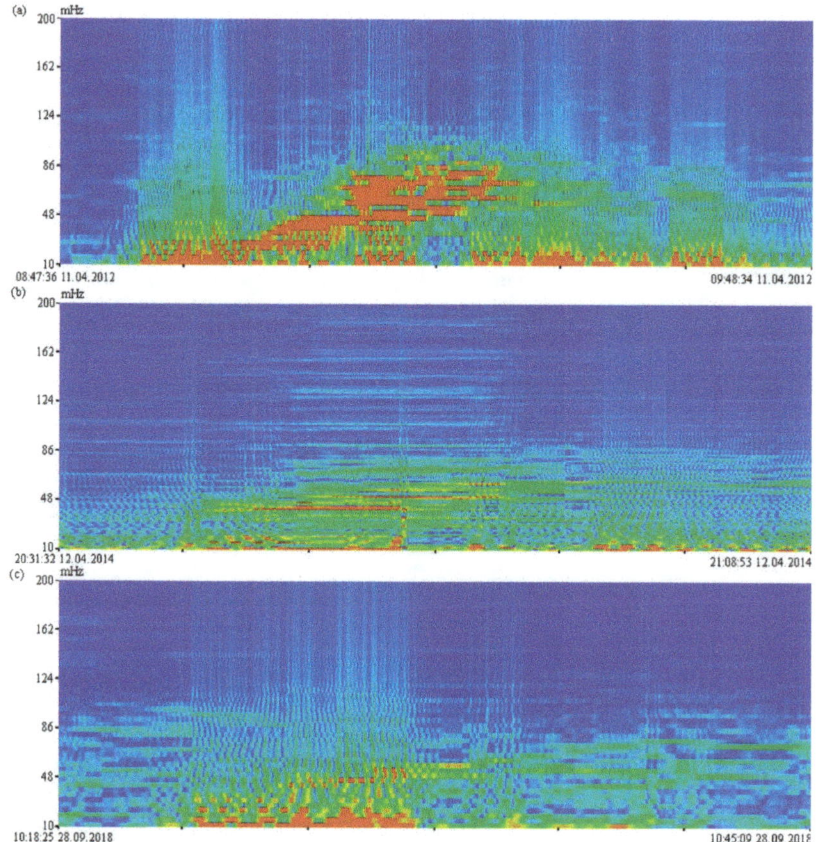

Figure 4. Dynamic spectrograms of 52.5 m laser strainmeter recordings from April 2012 (**a**), April 2014 (**b**), and September 2018 (**c**).

In the dynamic spectrograms of the laser strainmeter recording (Figure 4b), we can identify the signal of the earthquake, which occurred on 12 April 2014 at 20:14:39 at 11.270° S, 162.148° E, near the Solomon Islands, at a depth of 22.6 km and a magnitude of 7.6. In the coastal zone, the tsunami height reached 0.5 m. The laser strainmeter, located at a distance of over 6700 km, recorded the earthquake signal almost 20 min later at 20:33:58. For this earthquake, the average speed was 5.58 km/s. In the dynamic spectrogram of the laser strainmeter recording shown in Figure 4b, the earthquake signal amplitude is lower than in the previous case, but it also contains oscillations with periods ranging from 30 to 14 s.

The next earthquake under study occurred on 28 September 2018 at 10:02:45, with a magnitude of 7.5 and wave height of about 11 m. The earthquake epicenter was located at 0.256° S, 119.846° E, at a depth of 20 km and a distance of more than 4800 km from the laser strainmeter. The calculated average speed of the surface elastic wave was 5.49 km/s. On the dynamic spectrogram of the laser strainmeter record (Figure 4c), the earthquake signal was recorded 15 min later at 10:17:19. In the spectrogram, there were oscillations at periods of about 20 s peculiar to earthquakes of such magnitude. From the analysis of the dynamic spectrograms of the three earthquakes that occurred in Indonesia, it follows that along with the oscillations of the earthquake itself, which simply "shake" the Earth, there are disturbances in the lower frequency range.

Figure 5 shows the fragments of the laser strainmeter recordings at the times of registration of the three earthquakes in Indonesia. All figures show a deformation jump peculiar to tsunamigenic earthquakes. For example, in Figure 5a–c, the middle line of the direction of the laser strainmeter recording in the absence of a jump is marked in red, but at the time of the earthquake the recording deviated from its natural behavior (a deformation jump was observed), indicating the tsunamigenic nature of the earthquake.

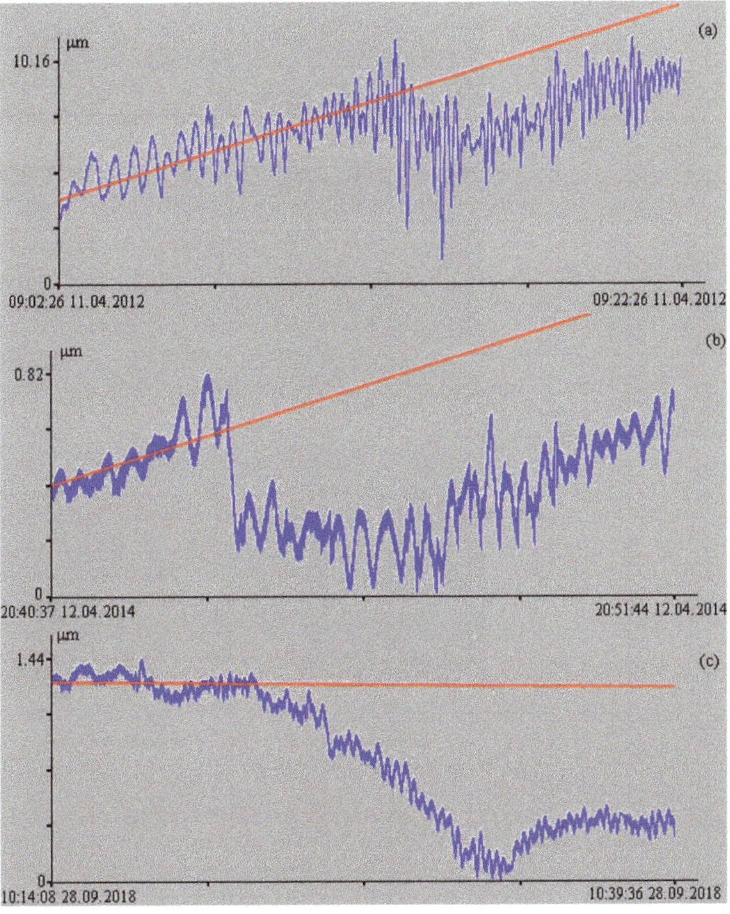

Figure 5. Fragments of 52.5 m laser strainmeter recordings from April 2012 (**a**), April 2014 (**b**), and September 2018 (**c**).

3.2. Earthquakes in Chile

The recordings from the 52.5 m laser strainmeter showed that three strong earthquakes occurred off the coast of Chile from 2010 to 2018. The first earthquake occurred on 27 February 2010 at 06:34:11 on the northwest coast of Chile at 36.122° S, 72.898° W, at a depth of 22.9 km; the maximum height of the catastrophic tsunami was 29 m. The distance from the earthquake epicenter to the location of the laser strainmeter was more than 17,800 km. The 52.5 m laser strainmeter recorded the signal of this earthquake at 07:19:00. Let us calculate the average speed of propagation of the elastic wave, which is equal to 6.77 km/s. When analyzing the dynamic spectrogram (Figure 6a) of the recording during this earthquake, we identified not only oscillations typical for an earthquake, but also a disturbance in the lower frequency range.

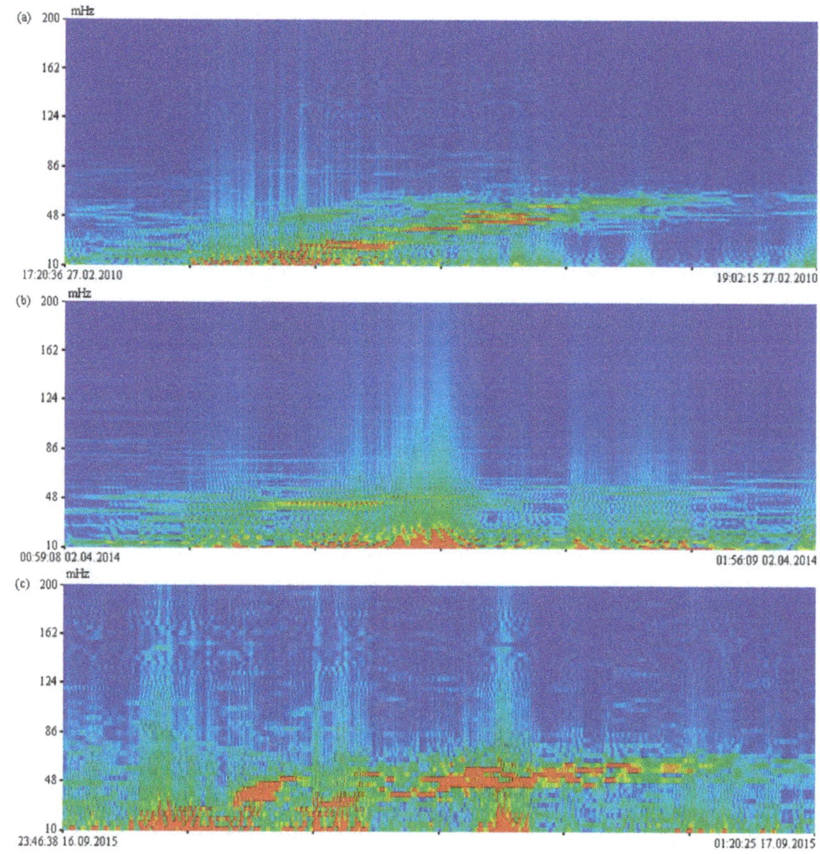

Figure 6. Dynamic spectrograms of 52.5 m laser strainmeter recordings from February 2010 (**a**), April 2014 (**b**), and September 2015 (**c**).

Let us analyze the dynamic spectrograms of the fragments of the 52.5 m laser strainmeter records from April 2014 and September 2015. During this period, there were two strong tsunamigenic earthquakes off the northwest coast of Chile. On 1 April 2014, at 23:46:47, a strong earthquake occurred at 19.610° S, 70.769° W, at a depth of 25 km, with a wave height of 4.6 m near the coast. The signal for this earthquake was recorded by the laser strainmeter at a distance of more than 16,700 km on 2 April 2014 at 00:24:10. On 16 September 2015 at 22:54:32, a strong earthquake occurred, with its epicenter at 31.573° S, 71.674° W, at a depth of 22.4 km. As a result of the earthquake, a tsunami with a height of 13.6 m was generated. The laser strainmeter located at a distance of about 17,650 km recorded the signal for this earthquake at 23:45:01. For these earthquakes, the average speeds of elastic wave propagation were 7.44 km/s and 6.47 km/s, respectively. In the dynamic spectrograms of the laser strainmeter recordings of these earthquakes (Figure 6b,c), oscillations occurred at periods of about 20 s, which are peculiar to earthquakes of this magnitude. Moreover, we noted disturbances in the lower frequency range.

When analyzing the recordings from the laser strainmeter at the times these earthquakes occurred, we identified deformation jumps. Figure 7a–c shows fragments of these earthquakes recordings, whereby the red line indicates the medium direction of the laser strainmeter recording and the deviation from this line at the time the seismic waves were registered indicates the tsunamigenic nature of the earthquakes (the presence of a deformation anomaly—a deformation jump).

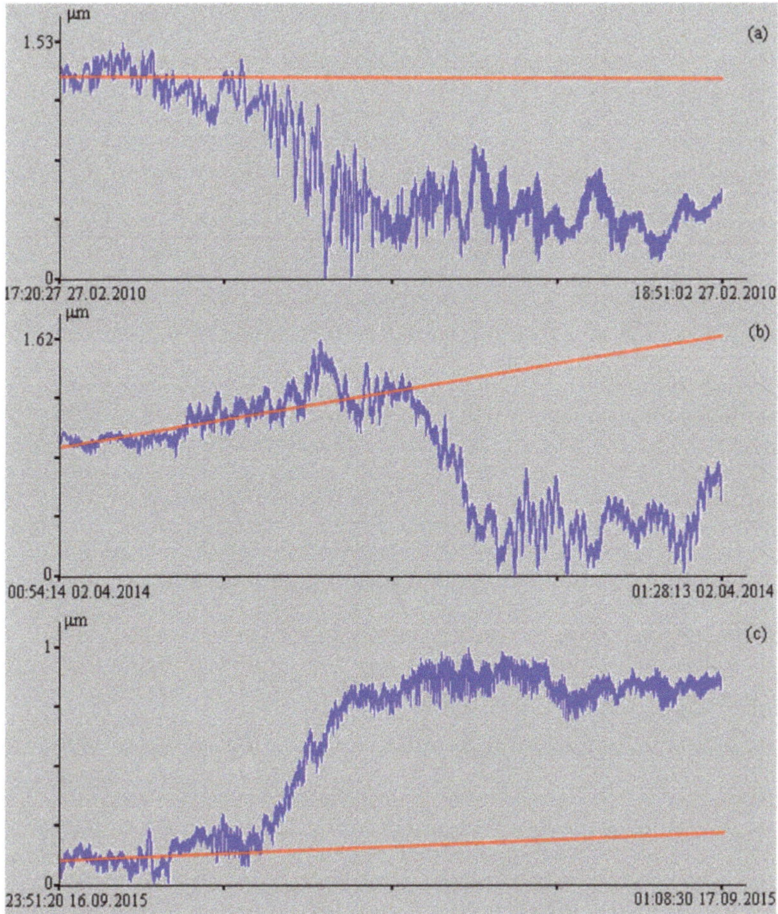

Figure 7. Fragments of 52.5 m laser strainmeter recordings from February 2010 (**a**), April 2014 (**b**), and September 2015 (**c**).

3.3. Earthquakes on the West Coast of North America

The first powerful earthquake occurred on 28 October 2012 at 3:04:08 on the southwest coast of Canada at 52.788° N, 132.101° W, at a depth of 14 km, with a magnitude of 7.8 and a tsunami height of 12.98 m on the shelf. The distance from the earthquake epicenter to the laser strainmeter was almost 6800 km. The laser strainmeter recorded this earthquake signal almost 19 min later at 03:23:13. For this earthquake, the average propagation speed of the elastic wave was 5.94 km/s. In the dynamic spectrogram (Figure 8a), we can see the oscillations ranging from 30 to 14 s, which are peculiar to earthquakes of this magnitude, as well as a strong disturbance in the lower frequency range.

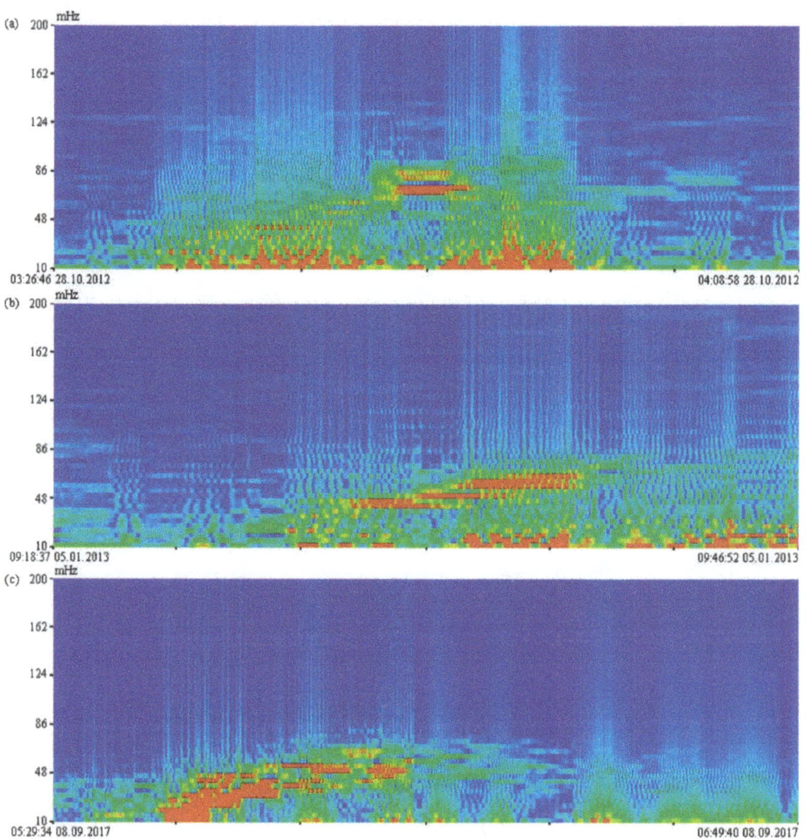

Figure 8. Dynamic spectrograms of 52.5 m laser strainmeter recordings from October 2012 (**a**), January 2013 (**b**), and September 2017 (**c**).

Figure 8b shows the dynamic spectrogram of the fragment of the laser strainmeter recording for 5 January 2013, in which we can identify the tsunamigenic earthquake that occurred at 8:58:14 off the coast of Alaska, USA. The earthquake, with a magnitude of 7.5, occurred at 55.228° N, 134.859° W, at a depth of 8.7 km, resulting in a tsunami with a maximum height of 1.5 m. The signal for this earthquake was detected in the laser strainmeter recordings at 09:16:31. The laser strainmeter was located 6500 km away from the epicenter. In the dynamic spectrogram of the laser strainmeter recording (Figure 8b), oscillations occurred from 30 to 14 s and disturbances occurred in the lower frequency region. Another earthquake occurred off the coast of Mexico on 9 August 2017 at 4:49:19, with a magnitude of 8.2. After this, a tsunami with a height of 2.7 m appeared. The earthquake occurred at 15.022° N, 93.899° W, at a depth of 47.4 km, at a distance of 12,150 km from the laser strainmeter. Let us calculate the average propagation speed of the elastic waves. For the 2013 earthquake the speed was equal to 5.92 km/s, while for the 2017 earthquake the speed was 5.48 km/s. In the dynamic spectrogram of the laser strainmeter record (Figure 8c), the earthquake signal was detected at 05:34:28. In the spectrogram, along with oscillations from the earthquake at periods ranging from 30 to 14 s, disturbances in the lower frequency range also occurred.

Figure 9a–c shows fragments of the laser strainmeter recordings at the times the three earthquakes occurred off the west coast of North America. All figures show deformation jumps peculiar to tsunamigenic earthquakes. In the figure, the middle line of the laser strainmeter recording direction in the absence of a jump is indicated in red, although at the times the earthquakes occurred, the recordings deviated from their natural behavior, showing the tsunamigenic nature of the earthquakes.

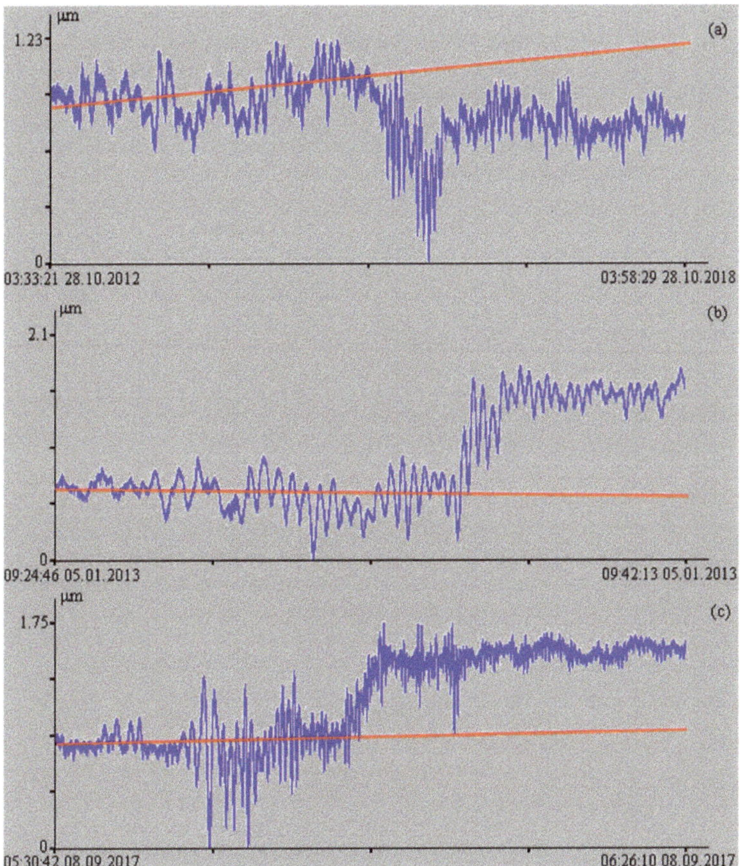

Figure 9. Fragments of 52.5 m laser strainmeter recordings from October 2012 (**a**), January 2013 (**b**), and September 2017 (**c**).

Analysis of the dynamic spectrograms of the laser strainmeter recordings of all earthquakes showed that along with the oscillations of the earthquakes themselves, with periods ranging from 30 to 14 s, there were also disturbances in the lower frequency range.

4. Analysis of Certain Characteristics of the Registered Deformation Anomalies

An earthquake that occurs near a geological fault provokes the displacement of geoblocks relative to each other. It may also break the connections between geoblocks and cause destruction of separate geoblocks, with powerful movement of the released individual geoblocks and their parts. It is this displacement, along with landslides, that is the cause of a tsunami. It is impossible to register this shift directly in the center; it can only be determined remotely. At large distances, the amplitudes of these slow displacements are very small. No instruments used in tsunami warning services are capable of registering such displacement; therefore, we can estimate this instead with various models, using continuous geodesic survey data, data from GPS receivers with low sampling rates, and tsunami data produced by the DART network. Models of finite faults from the USGS NEIC have an advantage. They use a kinematic approach based on the method proposed by Ji [19]. For calculations, both the body waves P and S and the Rayleigh and Love surface waves are used. To estimate the dissipative characteristics of deformation anomalies recorded by the 52.5 m unequal-arm laser strainmeter, we will use the calculated displacements of geoblocks (plates, parts of geoblocks, etc.) in the earthquake center according to this model. Table 1 lists the calculated displacements in the earthquake center and deformation anomaly values that the laser strainmeter registered at the moments the earthquakes were recorded.

Table 1. Calculated displacements of geoblocks and deformation anomaly values recorded by the laser strainmeter.

Date	Location	Calculated Displacement, m	Displacement on the Strainmeter, μm
27 February 2010	Chile	10.5	1.11
11 April 2012	Indonesia	5.4	2
28 October 2012	Canada	1.5	0.4
5 January 2013	USA	3	0.8
1 April 2014	Chile	8	1
12 April 2014	Solomon Islands	0.8	0.4
16 September 2015	Chile	3.2	0.6
8 September 2017	Mexico	4	0.5
28 September 2018	Indonesia	1.8	1

Since the intensity is proportional to the square of the amplitude, damping of the oscillation amplitude will be expressed by the law of intensity damping, whereby only the damping ratio will be two times smaller. To calculate the damping ratios, we will use the formula used to calculate the oscillation amplitude at the distance under consideration [20]:

$$A = A_0 e^{-\frac{1}{2}\mu x} \tag{1}$$

where A is the amplitude at the registration site, A_0 is the initial amplitude, μ is the damping ratio, and x is the distance.

Let us calculate the damping ratios for each earthquake using the displacement data from Table 1 and the distance from the earthquake epicenter to the laser strainmeter location.

From the obtained results (Table 2), it follows that the damping ratios for all considered tsunamigenic earthquakes are approximately identical. The average damping ratio for all earthquakes was 0.03.

Table 2. Earthquake damping ratios.

Date	Location	Damping Ratio, km^{-1}
27 February 2010	Chile	0.033
11 April 2012	Indonesia	0.026
28 October 2012	Canada	0.032
5 January 2013	USA	0.032
1 April 2014	Chile	0.032
12 April 2014	Solomon Islands	0.027
16 September 2015	Chile	0.026
8 September 2017	Mexico	0.034
28 September 2018	Indonesia	0.027

5. Discussion

Let us note some of the peculiarities mentioned above: (1) tsunamigenic earthquakes are characterized by the presence of a deformation anomaly—a deformation jump—and disturbances in the dynamic spectrogram in the lower frequency range, in comparison with the range of oscillations, which is peculiar to the earthquake source zone; (2) the damping ratios of deformation anomalies for all regions of the Earth are the same, within the measurement and calculation error ranges; (3) the durations of deformation anomalies for different tsunamigenic earthquakes vary from 15 s to 17 min.

Let us analyze some of the peculiarities of the above-mentioned tsunamigenic earthquakes. Let us focus on the presence in the dynamic spectrograms of disturbances in the low-frequency and ultra-low-frequency ranges that are not associated with oscillations, which are excited in the source zones of earthquakes. These disturbances are associated not with natural processes, but with the processing of records containing deformation jumps. The appearance of disturbances indicates only one thing, namely the presence of a deformation anomaly—a deformation jump—in the record. There are no oscillations in these areas. Let us demonstrate this using the example of processing an instrument recording containing oscillations caused by sea wind waves and a jump. Figure 10a shows an instrument recording containing a jump. Figure 10b shows its spectrum. Figure 10c shows the same instrument recording but without a jump, while Figure 10d shows its spectrum. All scientists involved in signal processing understand this effect. The increase in intensity in the low-frequency range is associated not only with the Gibbs phenomenon, but mainly with the presence of a jump in the recording; thus, the presence of disturbances in the spectrograms of tsunamigenic earthquake recordings containing deformation anomalies indicates only one thing—the influence of this deformation jump on the increases in intensity in the low-frequency and ultra-low-frequency ranges due to the processing effect, which is also remarkable. After all, looking only at the spectrogram, which was obtained in real time, one can notice the tsunamigenic nature of the earthquakes.

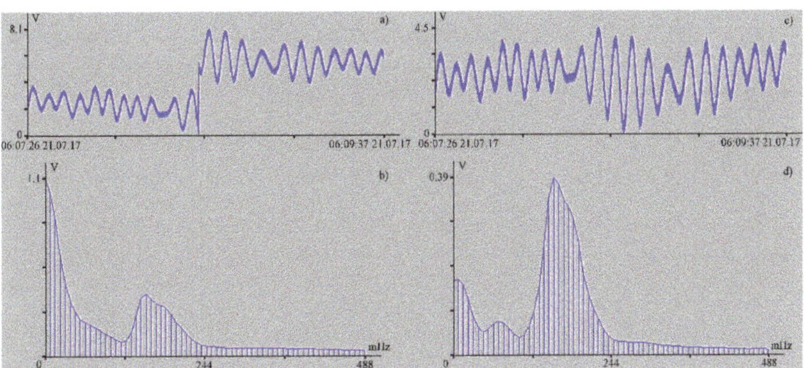

Figure 10. Fragment of the laser strainmeter recording showing hydrosphere pressure variations [21] when registering surface wind waves (**a**), along with its spectrum (**b**). The same recording is shown with the jump removed (**c**), along with its spectrum (**d**).

Let us pay some attention to the fact that for different regions of the Earth, the damping ratios of deformation anomalies are practically the same. This coefficient should consist of summands associated with divergence and absorption due to dissipative energy losses. If we consider only the cylindrical divergence, then the signal amplitude should decrease with distance in proportion to the square root of the distance. Regarding spherical divergence, the signal amplitude, with distance, should decrease in proportion to the distance. Solving the inverse problem, using Table 1, on the basis of the data from the laser strainmeter, for the spherical divergence we can obtain the following initial amplitudes of deformation anomalies (deformation jumps), which arose at the source of tsunami generation. In the sequence of column 3 in Table 1 (calculated displacement), these values are 19.6 (10.5 m), 16.7 (8 m), 10.6 (3.2 m), 11.6 (5.4 m), 2.7 (0.8 m), 4.8 (1.8 m), 2.7 (1.5 m), 5.2 (3 m), and 6.0 m (4 m). Taking into account the fact that the calculated data almost coincide with the model data (given in parentheses), we can state that the deformation anomaly arising at the source of tsunami generation, in the case of spherical divergence, moves similarly to the motion of a soliton; however, all calculations are correct when the signal moves along the surface of the Earth, i.e., in an arc rather than a chord. We do not know the path of the signal, so we take the limiting case, whereby the distances from the place of generation to the place of registration are equal to the lengths of the arcs of the circles, as determined by the coordinates of the points. Let us discuss the probable physical mechanism of formation of the deformation anomalies (deformation jumps) arising during the movement of geoblocks (joints) of the Earth's crust at the source of a tsunami generation. Let us assume that when an earthquake occurs, geoblocks are displaced relative to each other or one of the geoblocks (geological plate, joint of the Earth's crust) becomes out of balance and moves relative to the other geoblocks. This movement leads to the movement of huge masses of water, which subsequently degenerate into a tsunami. We are not interested in the origin point of the tsunami. We are only interested in the movement of the geoblock or geoblocks relative to each other. We can describe these geoblocks movements using the following equation:

$$\frac{\partial^2 u}{\partial t^2} + \sin u - \frac{\partial^2 u}{\partial x^2} = 0 \tag{2}$$

where u is the displacement of a geoblock. This is the classic sine-Gordon equation. One of the solutions for this equation in the factorized form, which is peculiar to solitons, allows us to obtain the following geoblock displacement:

$$u = 4 arctg\left[\exp\left(\pm\frac{x - Vt}{\sqrt{1 - V^2}}\right)\right] \tag{3}$$

where $V = 0.5$. A solution with "+" gives a kink, while a solution with "−" gives an anti-kink. Figure 11 shows the displacement of the geoblock in the form of an anti-kink.

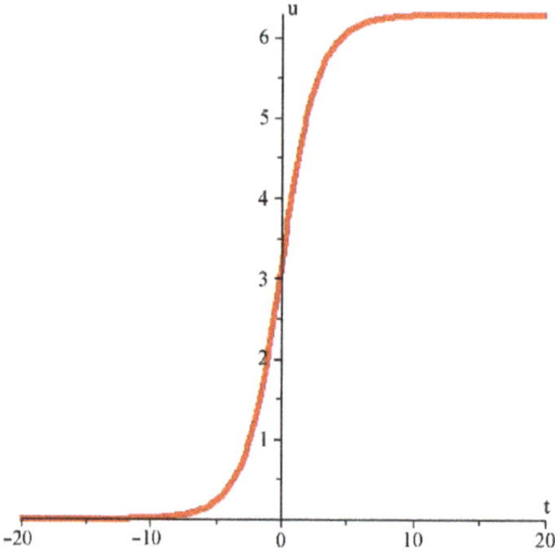

Figure 11. Displacement u at $V = 0.5$, shown as a function of time (anti-kink).

The displacement as a function of the coordinate and time at $V = 0.5$, demonstrating plastic deformation, is shown in Figure 12.

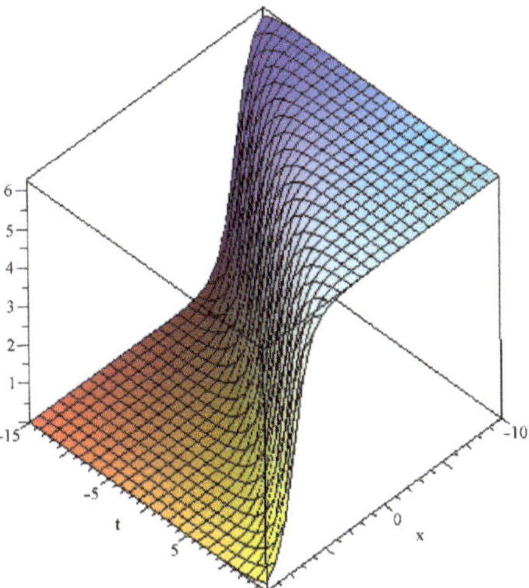

Figure 12. Geoblock displacement in the case of plastic deformation.

These geoblock displacements, when interacting with the environment, are transmitted to the surrounding space and propagate in the Earth in the form of a soliton—a deformation step, corresponding to a kink or anti-kink.

In addition to the exponential function, hyperbolic functions also satisfy Equation (2). In this case, the solution of Equation (2) gives a two-soliton solution. For this case, the displacement of the geoblock (homogeneity of the Earth's crust, a plate, etc.) depending on the coordinate and time for $V = 0.5$ is shown in Figure 13. This corresponds to elastic deformation under stretching. Plastic deformation can also occur under stretching only in the vicinity of $x = 0$.

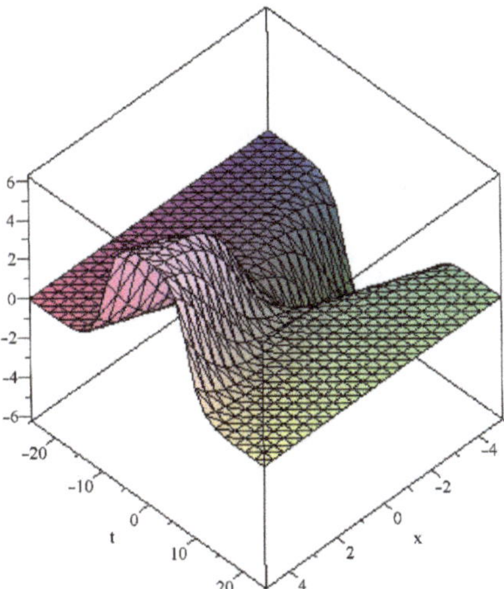

Figure 13. Geoblock displacement in the case of elastic deformation.

Under special conditions, such as for imaginary solutions, the solution to Equation (2) will give breathers. Figure 14 shows the geoblock displacement in the breather at $V = 0.5$.

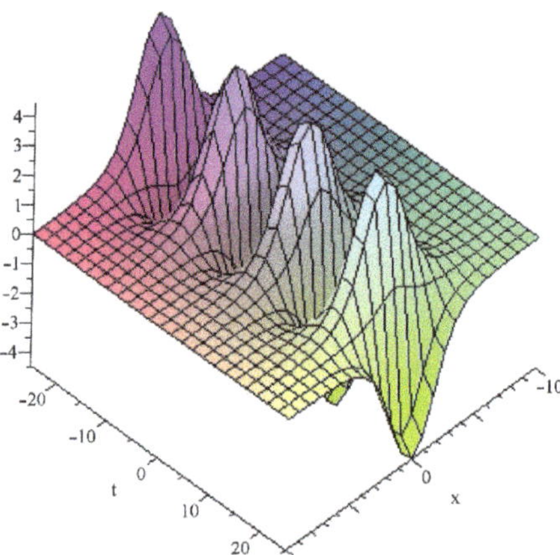

Figure 14. Geoblock displacement in the breather at $V = 0.5$.

These displacements (several deformation jumps) were observed during the 2011 Great East Earthquake in Japan (Tohoku). It is clear that any movements of a geoblock (or geoblocks) are transmitted to the neighboring environment, in which deformation anomalies propagate in the form of kinks, anti-kinks, and other similar disturbances.

Next, let us consider the behavior of an underwater landslide. At the advanced stage of the process, an underwater landslide can be represented as consisting of two parts: the head part (front) is a soliton (kink), while the tail part is a periodic wave. The landslide front is a same kink (bore) and can be described by a single-kink solution of the sine-Gordon equation, with "+" on the right-hand side of Equation (3). The formed soliton moves without experiencing any resistance from the medium. In expanding areas, the landslide front begins to blur. In such cases, the height of the soliton begins to decrease due to the energy conservation law. This process can be schematically represented in the following sequence during formation of an underwater landslide: (1) in the initial stage, a soliton (kink) and a periodic part are formed; (2) leaving the source of origin, the soliton begins to move in its environment with conservation of energy; (3) the motion of the soliton in the environment obeys the law of motion, ranging from cylindrical to spherical divergences. Taking into account the above, we must register a soliton with an ever-decreasing height and with increases in distance from the place of its registration. Figure 5 shows a laser strainmeter recording containing a deformation disturbance (a deformation jump) kink caused by an underwater landslide during the earthquake in Indonesia. Both earthquakes and underwater landslides can form the periodic oscillations observed in the figure.

6. Conclusions

During the processing of the experimental data from the laser strainmeter, we found that all tsunamigenic earthquakes are characterized by the presence of deformation anomalies—deformation jumps—in the instrument records. These deformation anomalies, leading to the formation of tsunamis in the vicinity of the earthquake source areas, occur during the relative movement of geoblocks (plates, joints) and underwater landslides. These geoblock movements can be described by the sine-Gordon equation, the one-kink and two-kink solutions of which explain the appearance of the observed deformation anomalies in laser strainmeter recordings. The behavior of deformation anomalies is the same as the behavior of solitons in non-linear medium. Considering that the signals from tsunamigenic earthquakes containing deformation anomalies propagate at speeds much higher than those of surface waves (from 5.48 to 7.44 km/s), we can assume that the signals do not propagate along the Earth's surface, with the divergences ranging from cylindrical to spherical. Further research should investigate the spatial behavior of the deformation anomalies from tsunamigenic earthquakes. For this, it will be necessary to place several laser strainmeters far from each other, along the assumed direction of movement of deformation anomalies (deformation jumps), kinks, anti-kinks, and breathers. These experimental studies will allow us to study the main parameters of the observed disturbances. Of particular interest is the conservation of the soliton shape with decreasing value due to divergence in space during movement. The development of this area of research, along with the application of the classical "magnitude geographical principle" used for determining the tsunami hazard of underwater earthquakes will bring us closer to short-term tsunami forecasting. Taking into account the above, with spherical divergence in accordance with Table 1, we can calculate the applicability of GPS receivers capable of registering displacement with an accuracy level of 1 mm for recording displacements (column 3 of Table 1). In this way, with an average displacement of 4.2 m (according to Table 1), a GPS receiver will be able to register a displacement of 2 mm at a distance of 2100 m under conditions of spherical divergence. It is clear that there is no point in discussing the prospects of using GPS receivers for registering displacements of geoblocks (plates, joints) leading to the occurrence of tsunamis. The cylindrical divergence of the signal does not help in situations.

Author Contributions: G.D.—problem statement, discussion, and writing of the article. S.D.—data processing, discussion, and writing of the article. All authors have read and agreed to the published version of the manuscript.

Funding: The work was carried out with financial support from the Russian Federation represented by the Ministry of Science and Higher Education of the Russian Federation, Agreement No. 075-15-2020-776.

Data Availability Statement: 3rd Party Data. Restrictions apply to the availability of these data.

Acknowledgments: We would like to express our deep gratitude to all employees of the Physics of Geospheres laboratory.

Conflicts of Interest: The authors declare no conflict of interest.

References

1. Stein, S.; Okal, E.A. Speed and size of the Sumatra earthquake. *Nature* **2005**, *434*, 581–582. [CrossRef] [PubMed]
2. Wei, Y.; Newman, A.V.; Hayes, G.P.; Titov, V.V.; Tang, L. Tsunami forecast by joint inversion of real-time tsunami waveforms and seismic or GPS Data: Application to the Tohoku 2011 tsunami. *Pure Appl. Geophys.* **2014**, *171*, 3281–3305. [CrossRef]
3. Pacific Tsunami Warning Center/International Tsunami Information Center (PTWC/ITIC). *User's Guide for the Pacific Tsunami Warning Center Enhanced Products for the Pacific Tsunami Warning System*; Revised Edition; IOC Technical Series; UNESCO/IOC: Paris, France, 2014; Volume 105.
4. Tatehata, H. The New Tsunami Warning System of the Japan Meteorological Agency. In *Perspectives on Tsunami Hazard Reduction. Advances in Natural and Technological Hazards Research*; Hebenstreit, G., Ed.; Springer: Dordrecht, The Netherland, 1997; Volume 9. [CrossRef]
5. Kamigaichi, O. Tsunami forecasting and warning. In *Encyclopedia of Complexity and Systems Science*; Meyers, R.A., Ed.; Springer: Berlin/Heidelberg, Germany, 2015. [CrossRef]
6. Kanamori, H.; Rivera, L. Source inversion of W phase: Speeding up seismic tsunami warning. *Geophys. J. Int.* **2008**, *175*, 222–238. [CrossRef]
7. Duputel, Z.; Rivera, L.; Kanamori, H.; Hayes, G.P.; Hirshorn, B.; Weinstein, S. Real-time W phase inversion during the 2011 off the Pacific coast of Tohoku Earthquake. *Earth Planets Space* **2011**, *63*, 535–539. [CrossRef]
8. Titov, V.V.; Song, Y.T.; Tang, L.; Bernard, E.N.; Bar-Sever, Y.; Wei, Y. Consistent estimates of tsunami energy show promise for improved early warning. *Pure Appl. Geophys.* **2016**, *173*, 3863–3880. [CrossRef]
9. Bondur, V.G.; Smirnov, V.M. Method for monitoring seismically hazardous territories by ionospheric variations recorded by satellite navigation systems. *Dokl. Earth Sci.* **2005**, *403*, 736–740.
10. Bondur, V.G.; Garagash, I.A.; Gokhberg, M.B.; Lapshin, V.M.; Nechaev, Y.V.; Steblov, G.M.; Shalimov, S.L. Geomechanical models and ionospheric variations related to strongest earthquakes and weak influence of atmospheric pressure gradients. *Dokl. Earth Sci.* **2007**, *414*, 666–669. [CrossRef]
11. Smirnov, V.M.; Bondur, V.G.; Smirnova, E.V. Ionospheric disturbances during of the thsunamigenic earthquake on navigation system data. In Proceedings of the Asian Association on Remote Sensing 26th Asian Conference on Remote Sensing and 2nd Asian Space Conference, ACRS, Vietnam, Ha Noi, 7–11 November 2005; pp. 1487–1494.
12. Perez del Postigo Prieto, N.; Raby, A.; Whittaker, C.; Boulton, S.J. Parametric Study of Tsunamis Generated by Earthquakes and Landslides. *J. Mar. Sci. Eng.* **2019**, *7*, 154. [CrossRef]
13. Dolgikh, G.I.; Dolgikh, S.G.; Kovalev, S.N.; Koren, I.A.; Ovcharenko, V.V.; Chupin, V.A.; Shvets, V.A.; Yakovenko, S.V. Recording of deformation anomaly of a tsunamigenous earthquake using a laser strainmeter. *Dokl. Earth Sci.* **2007**, *412*, 74–76. [CrossRef]

14. Dolgikh, G.I.; Dolgikh, S.G.; Kovalev, S.N.; Chupin, V.A.; Shvets, V.A.; Yakovenko, S.V. A deformation method for determining the tsunami potential of earthquakes. *Dokl. Earth Sci.* **2007**, *417*, 1261–1264. [CrossRef]
15. Dolgikh, G.I. Principles of designing single-coordinate laser strainmeters. *Tech. Phys. Lett.* **2011**, *37*, 204–206. [CrossRef]
16. Dolgikh, G.I.; Valentin, D.I.; Dolgikh, S.G.; Kovalev, S.N.; Koren, I.A.; Ovcharenko, V.V.; Fishchenko, V.K. Application of horizontally and vertically oriented strainmeters in geophysical studies of transitional zones. Izvestiya. *Phys. Solid Earth* **2002**, *38*, 686–689.
17. Dolgikh, G.I.; Batyushin, G.N.; Valentin, D.I.; Dolgikh, S.G.; Kovalev, S.N.; Koren', I.A.; Ovcharenko, V.V.; Yakovenko, S.V. Seismoacoustic hydrophysical complex for monitoring the atmosphere-hydrosphere-lithosphere system. *Instrum. Exp. Tech.* **2002**, *45*, 401–403. [CrossRef]
18. Bondur, V.G.; Gokhberg, M.B.; Garagash, I.A.; Alekseev, D.A. Revealing Short-Term Precursors of the Strong M > 7 Earthquakes in Southern California From the Simulated Stress–Strain State Patterns Exploiting Geomechanical Model and Seismic Catalog Data. *Front. Earth Sci.* **2020**, *8*, 571700. [CrossRef]
19. Ji, C.; Wald, D.J.; Helmberger, D.V. Source description of the 1999 Hector Mine, California earthquake; Part I: Wavelet domain inversion theory and resolution analysis. *Bull. Seism. Soc. Am.* **2002**, *92*, 1192–1207. [CrossRef]
20. Kitaygorodsky, A.I. *Introduction to Physics*; Science: Moscow, Russia, 1973; 685p.
21. Dolgikh, G.I.; Dolgikh, S.G.; Kovalev, S.N.; Chupin, V.A.; Shvets, V.A.; Yakovenko, S.V. Super-low-frequency laser instrument for measuring hydrosphere pressure variations. *J. Mar. Sci. Technol.* **2009**, *14*, 436–442. [CrossRef]

Journal of
*Marine Science
and Engineering*

Article

The Effect of Mesoscale Eddy on the Characteristic of Sound Propagation

Jiaqi Liu [1,2,3], **Shengchun Piao** [1,2,3], **Lijia Gong** [1,2,3,*], **Minghui Zhang** [1,2,3] **and Yongchao Guo** [1,2,3] **and Shizhao Zhang** [1,2,3]

[1] Acoustic Science and Technology Laboratory, Harbin Engineering University, Harbin 150001, China; liujiaqi@hrbeu.edu.cn (J.L.); piaoshengchun@hrbeu.edu.cn (S.P.); zhangminghui@hrbeu.edu.cn (M.Z.); guoyongchao@hrbeu.edu.cn (Y.G.); zhangshizhao@hrbeu.edu.cn (S.Z.)

[2] Key Laboratory of Marine Information Acquisition and Security, Ministry of Industry and Information Technology, Harbin Engineering University, Harbin 150001, China

[3] College of Underwater Acoustic Engineering, Harbin Engineering University, Harbin 150001, China

* Correspondence: lijia.gong@hrbeu.edu.cn; Tel.: +86-1510-456-2982

Abstract: A mesoscale eddy is detected and tracked in the western North Pacific region. Within the life cycle of the cyclonic eddies, the intensities of eddies make a difference. Satellite images indicate the oceanic eddy keeps westward-moving until it disappears. Oceanographic and acoustic characteristics of the eddy are studied. The acoustic energy distribution results from the different intensity of both modelled eddy and measured eddy are calculated. With sound propagation through the cyclonic eddy and anticyclonic eddy, the position of convergence zone moves away from and towards the acoustic source compared with the sound propagation through background hydrography. The coupling coefficient of different orders of normal modes changes significantly. The closer to the centre of the eddy, the stronger the coupling coefficient.

Keywords: mesoscale eddy; parabolic equation; normal mode

Citation: Liu, J.; Piao, S.; Gong, L.; Zhang, M.; Guo, Y.; Zhang, S. The Effect of Mesoscale Eddy on the Characteristic of Sound Propagation. *J. Mar. Sci. Eng.* **2021**, *9*, 787. https://doi.org/10.3390/jmse9080787

Academic Editor: Grigory Ivanovich Dolgikh

Received: 15 June 2021
Accepted: 17 July 2021
Published: 22 July 2021

1. Introduction

Ocean dynamic phenomena have been distributed in the global ocean. Among them, eddies are characterised by unstable, time-dependent water masses, which separate from their respective currents and enter water bodies with different physical, chemical and biological characteristics [1]. More than half of the kinetic energy of the ocean circulation is stored in the mesoscale eddies, with the remainder contained in the large-scale circulation, swirling motions of eddies mixed between layers and consequent mixing of nutrients, heat and salinity. Sound propagation through eddies has a crucial impact on humans, marine animals and invertebrates [2]. Under the influence of oceanic eddies, human underwater acoustic communication networks are impacted. The existence of eddies affects the detection and communication of marine creatures [3]. Some whales, seals and fishes use low-frequency sound to communicate, perceive the environment and respond to those sounds [4]. At present, there are some experiments to prove that the sound of deep-sea creatures is detected at a location of 15 km away. Fishes feed along with the density structure of the eddies, which indicates that eddies accelerate the transfer of energy and nutrients in the ocean. Anticyclonic eddies carry higher surface chlorophyll than cyclonic eddies. Because the mixed layer of cyclonic eddies is often shallower than that of anticyclonic eddies, the pycnocline depth of the anticyclonic eddies is deeper. More plankton in the deeper mixed layer therefore provides more nutrients for zooplankton [5]. The study of sound propagation through the eddies clearly plays a significant role in the life of human and marine organisms. Regarding the deep-sea sound propagation issue, we always focus on the convergent position of sound energy and the acoustic intensity distribution. Aiming at the phenomenon of how the oceanic eddy affects the position of the

71

convergence zone, this paper uses the parabolic equation method to solve sound field and combines it with the normal mode theory to analyse the difference of energy distribution of the acoustic wave at different positions.

It is of great significance for weather and marine scientific research to grasp the real-time movement of eddies. The nonlinear interaction of barotropic and baroclinic Rossby waves could lead to strong instability, which is a major source of the kinetic energy of eddies [6]. These eddies are affected by Coriolis, which lifts high-temperature cyclones from subsurface water to the surface of the sea [7]. Some smaller-scale eddies (tens of kilometres in diameter) are formed due to the instability of baroclinic at the horizontal density, while larger-scale eddies located near the warm current of Gulf Stream and the Kuroshio Current are generally due to strong horizontal shear motion, which intensely affects the temperature and salinity distribution in the depth direction. The dynamic model of the eddies was proposed as early as the 19th century. Taking the change of sea surface height and the temperature in the depth direction caused by the eddies into account, the sound speed shapes in space can be divided into the following types: bowl shape, lens-shaped, cone-shaped, moreover, the generation mechanism of eddies are divided into three types, sea surface wind, the interaction between ocean currents and bottom topography, and Kuroshio intrusion [8]. In the northern hemisphere, these cyclonic eddies rotate clockwise,whereas, in the southern hemisphere, they rotate counterclockwise. The low-temperature anticyclonic eddies cause the downwelling of the surface layer and dent formed in opposite directions of rotation. Eddies could be distinguished through the combination of sea surface height and sea surface temperature anomalies. Some scholars used buoys to capture the formation, the movement and the extinction of eddies. Mesoscale eddies in the Pacific generally last several weeks. Their motion trajectories are approximately circular with a diameter of 100–200 km. The effect of the mesoscale eddies gradually concerns many scholars and has great investigation result. The oceanic eddies not only affect the circulation structure of the ocean but also ocean temperature, salinity, sound speed profile. Mesoscale eddies keep rotating and moving every moment, and momentum and energy are transported, interacting strongly with oceanic circulations, affecting the vertical profile structure of the marine environment. In 1977, based on practical experience, Henrick proposed a quasi-elliptical eddy model. To detect and track mesoscale eddies [9], Dong used sea surface height fluctuations to study the distribution laws and characteristics of eddies from the perspective of oceanography [10]. We study from the perspective of ocean acoustics the influence of spatial and temporal distribution characteristics of eddies.

Since the 1970s, hydro-acoustics scientists have been concerned about the influence of ocean eddies on sound propagation. The ray method is used for analysing the receiver's time of arrival which is affected by the size, intensity and position of the eddy. By using the ray-tracing model, people could understand how rays refract and bend. Additionally, what people tracing is the ray stimulated from a particular source then propagating in free space [11]. In the classical ray theory, the sound energy in the waveguide is shown in the form of acoustic rays. With regards to sound rays from the point source traveling along certain paths to the receiving point, the acoustic field at the receiver is the consequence of the superposition of all types of sound rays. Furthermore, the receivers' signal phase shift is also calculated by Jian [12] combining the current model with the acoustic field model when an eddy is present. In addition to the frequency domain, scholars have also done some research on sound propagation through the eddies. Nysen [13,14] studied how the acoustic energy leaked into the deep sound channel from the subsurface sound channel with the dependence of frequency off the east coast of Australia. The strong coupling between the two ducts leads to the near-surface acoustic energy being trapped in the duct area affected by water mass. Lawrence [15] modelled a warm-core eddy propagation problem and expounded the widening of the acoustic convergence zone. Baer [16] bonded split-step parabolic-equation and used Henrick eddy model to calculate a non-typical three-dimensional structure of eddy and examined the signal amplitude of vertical line array. Specifically, the gain of the hydrophone array increased, and the energy flows horizontal angle changed by over 0.5 degrees. As underwater acoustic horizontal refraction has attracted attention because of the eddy, Weinberg [17] simulated acoustic propagation through ocean fronts and single mesoscale eddy. The received voltage amplitude was significantly dependent upon the position of the Gulf Stream ring by a fixed system experiment. Afterwards, the three-dimensional fully coupled parabolic equation method was applied to investigate the effect of the horizontal refraction of the sound field caused by mesoscale ocean dynamics on the impact of acoustic source localisation [18].

Even though the parabolic equation and the ray method could solve the sound field problem of horizontal ocean environment fluctuations, the normal mode theory still plays an irreplaceable role of analysing the effect of deep-sea eddies on the sound field. Dozier and Tapert [19] deduced the normal mode amplitude distribution in a random ocean, Colosi [20] used the coupling coefficient of acoustic normal mode to analyse the average energy and sound pressure correlation characteristics of sound propagation under the influence of internal waves. Similarly, we use the coupling coefficient derived by Tapert to apply it to the ocean environment with an eddy to analyse the coupling relationship of the normal mode of various orders in the deep ocean environment. As the sea depth increases, the total of normal mode order increases. The theory of deep-sea normal modes is more complicated than that of shallow-sea. High-order normal mode of the energy coupling is more intense, according to the simulation results. Consistent with the coupling and transmission characteristics of shallow sea normal mode, however, as the sound source is located far away from the channel axis and closer to the sea surface, the low-order normal mode no longer maintains a stable propagation path over a long distance. As a consequence, we focus on the energy distribution of higher-order normal modes in the deep sea. Incidentally, due to the surface and seabed boundary limitation of the impedance character, the confined depth direction is expressed as a specific form of a standing wave, whereas the unconfined direction is known as a form of a travelling wave, which is called normal mode in a given waveguide. The modal expansion is a sum of resonances or eigenfunctions for the waveguide. The eigenfunction mentioned here is limited to the formal solution of the Helmholtz equation under the conditions of a given waveguide section and a fixed edge condition, and eigenfunction, namely, mode shape function expansion is often done in the numerical models based on the normal mode approach [11]. Local eigenfunctions of different positions are used to analyse the redistribution of energy of the normal mode of each order in the eddy affecting the sound field and explain the root cause of the change of the convergence zone position. As follows in Section 2, an eddy environment was tracked and used to resolve the sound propagation and the forecasting process of the sound transmission based on the parabolic equation method, and then, the sound propagation experiments in the Kuroshio are introduced, and the comparison between the measured results and numerical results is carried out in Section 3. In Section 4, the numerical results of the modelled cyclonic and anticyclonic eddies acoustic propagation and coupling characteristics are described, the character of sound propagation of different normal modes is analysed. Summary of the discussions is drawn in Section 5.

2. Materials and Methods

2.1. Gaussian Eddy Hydrology Model

In the deep ocean, the speed of sound varies between 1450 and 1570 m per second. Notably, sound speed profiles have a remarkable influence on the variation of the temperature and salinity which is associated with eddies. Specifically, the intensity of eddy represents the difference between the acoustic properties at the eddy edge and eddy centre. Its intensity does also reflect how surface cool water rises or how deeply warm water descends. The upwelling and downwelling of the eddy (as shown in the Figure 1) could be simplified as an asymmetric model with Gaussian distribution. The temperature/salinity profile for a cold-core or warm-core eddy could be simply modelled as (referring to the theory section of [21,22]):

$$T(r,z) = T_k(z)(1 - e(-\frac{r}{R})) + T_c(z)e(-\frac{r}{R}),$$ (1)

where T_k is the background value of temperature/salinity without eddy, and T_c is the value of temperature/salinity of an eddy core. Typically, $R = 3R_0$, where R_0 is the radius of Rossby. Particular temperature and salinity formulation for an elliptical eddy feature model is simplified given by

$$T(r,z) = T_c(z)\alpha\beta(r),$$ (2)

where α is generally a delta function, $\beta(r)$ is an exponential function depending on the current characteristic. The empirical formula of sound speed profile could be determined by temperature and salinity. Gaussian eddy sound speed model has an advantage in considering underwater sound propagation related issues, and the overall sound speed is distributed as

$$c(r,z) = c_0(z) + \delta c(r,z) = c_0(z) + DC * exp(-(\frac{r - r_0}{dr})^2 - (\frac{z - z_0}{dz})^2)$$ (3)

where DC is the amplitude of variation of sound speed because of the existence of the eddy, r_0 and z_0 are the location of the core of eddy and dr and dz are the radius of semi-major axis and radius of semi-axis of eddy, respectively. Otherwise, background sound speed satisfies the Munk deep ocean sound speed profile, where $c_0(z) = 1500(1 + 0.0057(e^{-\zeta} - (1 - \zeta)))$, z_{c_0} is the depth of the sound channel axis in the equation $\zeta = \frac{2(z - z_{c0})}{z_{c0}}$.

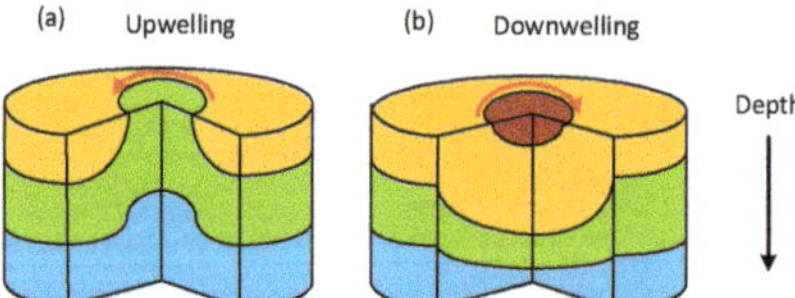

Figure 1. (**a**) Cyclonic eddy and (**b**) anticyclonic eddy three-dimensional schematic diagram. Shades of colour represent the water column temperature level, the red part has the highest temperature; by contrast the blue part has the lowest temperature, and the red arrow represents the direction of the eddy current in the northern hemisphere.

2.2. Parabolic Equation Modelling

To solve range dependent propagation problem for low-frequency sound propagation, a more appropriate model is the parabolic equation. The acoustical wave equation can be derived as serial spatial iterative equations [23–25]. Previously, we start with the Helmholtz equation with a point source for constant density sediment:

$$\frac{\partial^2 P}{\partial r^2} + \frac{1}{r}\frac{\partial P}{\partial r} + \rho\frac{\partial}{\partial z}\frac{1}{\rho}\frac{\partial P}{\partial z} + k^2 P = 0$$ (4)

where $k = \frac{\omega}{c} = \frac{\omega}{c_0}n = k_0 n$, k_0 is defined as reference wave number, ω is the angular frequency, c is the sound speed, c_0 is the reference sound speed of 1500 m per second, and n is the index of refraction. Simulating the acoustic field by frequency-domain finite difference method, selecting and keeping only the outgoing sound pressure component we consequently obtain

$$\frac{\partial[\phi H_0^{(1)}(k_0 r)]}{\partial r} = ik_0\sqrt{1 + X}[\phi H_0^{(1)}(k_0 r)]$$ (5)

$P = \phi H_0^{(1)}(k_0 r)$, where $H_0^{(1)}$ is zeroth-order Hankel function of the first kind, which satisfies the Bessel differential equation. Then it is displayed in the iterative form, $\phi(r + \triangle r) = e^{ik_0 \triangle r \sqrt{1+X}} \phi(r)$, where operator $X = \frac{1}{k_0^2}[\frac{1}{\alpha(z)}\frac{\partial}{\partial z}(\frac{\partial \alpha(z)}{\partial z}) + k^2 - k_0^2]$

This approximation equation is a one-way wave equation that could be estimated step by step in the recursion. Eventually, sound transmission loss is decibel quantities

$$TL = -20 lg \left| \frac{P}{p_0(r=1)} \right| \tag{6}$$

where p_0 is the pressure of the source in the free space. For a smoothed transmission loss result $(\sum_{j=1}^{J} |P(r,z)|^2)^{1/2}$, where J is the total number of wide-band covered frequency points.

2.3. Normal Mode Method

For the purposes of acquiring the primary causes of the redistribution of the sound energy, the mode decomposition method is applied. According to normal mode theory [11,26], the sound field is represented as the superposition of the normal modes.

$$P(r,z) = \sum_{n=1}^{\infty} a_n(r)\phi_n \tag{7}$$

where ϕ_n are eigenfunctions of depth separated equations, and $a_n(r)$ is amplitude of range dependence. The orthogonal property of ϕ_n is represented as

$$\int_0^D \frac{\phi_n \phi_m}{\rho} dz = \delta_{nm} \tag{8}$$

By multiplying local eigenfunctions to both sides of Equation (8), the amplitude distributions of different orders of normal modes are solved after integrating with the total depth direction. We then obtain the following form:

$$\int_0^D p(r,z)\phi_m(z)dz = \int_0^D \sum_{n=1}^{N} a_n(r)\phi_n(z)\phi_m(z)dz \tag{9}$$

To illustrate the effect of coupling between normal modes of each order, putting 7 into formation wave Equation (4) we obtained

$$\sum_{n=1}^{N}(\frac{\omega^2}{c^2}a_n(r)\phi_n(z) + \frac{\partial^2 a_n(r)}{\partial r^2}\phi_n + a_n(r)\frac{\partial^2 \phi_n(z)}{\partial z^2}) = 0 \tag{10}$$

The form expressed as a potential function is $\frac{\partial^2 \phi_n(z)}{\partial z^2} = (k_n^2 - k_0^2)\phi_n(z)$, where k_n is horizontal wave number of nth normal mode. The above formula is reorganised and obtained as follows,

$$\sum_{n=1}^{N}(\frac{\omega^2}{c^2}a_n(r)\phi_n(z) + \frac{\partial^2 a_n(r)}{\partial r^2}\phi_n + a_n(r)(k_n^2 - k_0^2)\phi_n(z)) = 0 \tag{11}$$

The orthogonal property of eigenfunction is used, and as a result we get

$$\frac{\partial^2 a_n(r)}{\partial r^2} + a_n(r)k_n^2 + \sum_{m-1}^{M} B_{nm}(r,t)a_m(r) = 0 \tag{12}$$

The coupling coefficient is described as

$$B_{mn}(r,t) = \frac{2\omega^2}{c_0^2}\int_0^D \frac{\delta c(r,z)}{c_0}\phi_n(z)\phi_m(z)dz \tag{13}$$

The range dependent amplitude component is obtained by using small angle single scattering approximation

$$\frac{\partial a_n}{\partial z} - ik_n a_n = -i \sum_{n=1}^{N} B_{nm}(r) a_m \tag{14}$$

The expression of coupling coefficient of the stochastic ocean is obtained

$$B_{mn} = \frac{k_0^2}{\sqrt{k_n k_m}} \int_0^D \frac{\delta c(r,z)}{c_0} \phi_n(z) \phi_m(z) dz \tag{15}$$

3. Eddy Tracking Experiments and Underwater Acoustic Propagation Experiment

From May to July 2019, a joint acoustic propagation/mesoscale eddy physics experiment was carried out near the Kuroshio. The purposes of the experiment were to explore eddy hydrology structure and the influence of eddy on sound propagation. The expendable conductivity temperature depth (XCTD), conductivity temperature depth (CTD) and moving vessel profiler (MVP) were used to measure temperature and salinity in the ocean. Observational networks were extended to cover practically the entire mesoscale eddy area, a longitude from 148.5° E to 150.5° E and latitude from 33.65° N to 34.15° N. Receiver hydrophone line array coordinate is located at 150°30′ E, 33°42′ N. Moreover, the survey depth fluctuates obviously from 5900 to 6040 m. The surface sound speed approximately is 1520 m/s; the sound speed at the lowest channel axis position is 1484 m/s and increases to around 1547 m/s near the seabed. The equipment used for the experiment is shown in Figure 2 to track and measure the distribution of sound speed which lasted four days since 16 June 2019. Two types of explosives were selected whose explosion depth is 100 m and 200 m, respectively. The deep-sea sound propagation experiment was conducted by 1 kg explosions, dropped evenly spaced at the constant latitude and along the direction of longitude. The broadband explosive was occurring every 3 km, and the ship sailing within 3 h at a speed of 5 knots on the survey line. Inasmuch as the impact of ocean currents, two depth meters are hung on the vertical hydrophone line array to determine the position of each hydrophone in actual time.

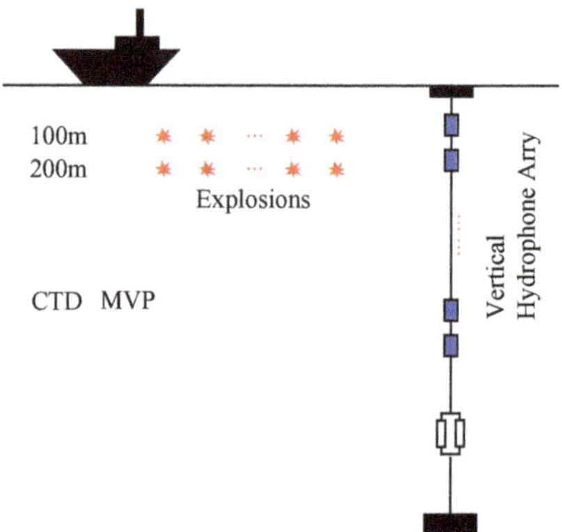

Figure 2. Sound propagation experiment configuration.

With the influence of sea conditions, the depth of the receiving hydrophones fluctuates up and down. After the depth of hydrophones is determined through depthometers, the signal of sound pressure at the same depths is corrected. The sampling frequency is 10 kHz. To guarantee the integrity of the received signal, the time window for intercepting the signal is selected according to the actual signal amplitude. What is more, the signal spectrum is distributed at 5–500 Hz. In a practical process of data processing, 320 Hz with the strongest received signal spectrum is truncated for one-third octave band filtering to process the voltage signal. The parabolic equation method is used for calculating the sound intensity at the receiving position, and the broadband average sound intensity is achieved by averaging the frequency points within the bandwidth to ensure the consistency of the processing result of the experiment and the propagation loss curve.

3.1. Hydrographic Data

The observation data and satellite remote sensing data are used for studying the evolution process of the vertical structure, such as the growth and extinction of this cyclonic eddy. The Figure 3 shows the temperature and salinity and the speed of sound profiles collected by MVP 33.7° N, 33.78° N, 33.95° N and 34.10° N, respectively. The sound propagation experiment is located at the latitude of 33.7° N section; the thermocline, halocline, sonic cline and sonic cline at each site are shown in the Figure 4.

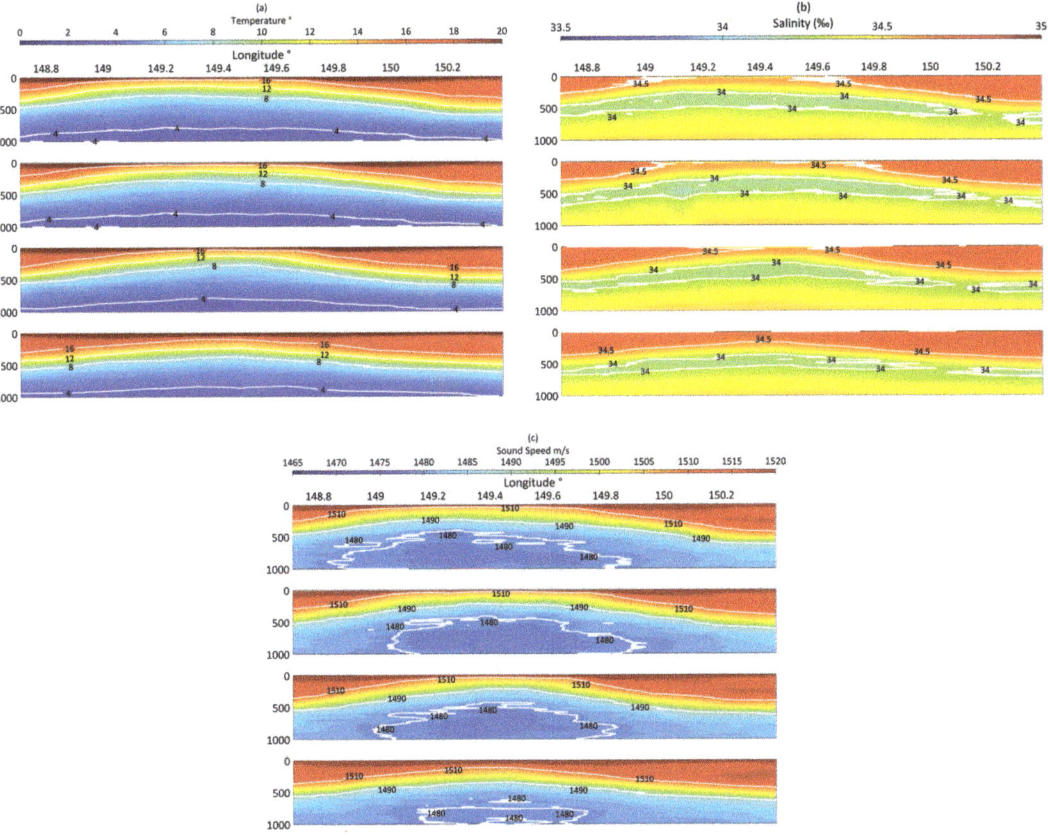

Figure 3. Measured data of (**a**) temperature, (**b**) salinity and (**c**) sound speed and each sub-figure from top to bottom was located at the latitude of 33.7°, 33.78°, 33.95° and 34.10° E.

It is known from Figure 4 that the depths of the thermocline, halocline and pycnocline came close. The position of saltation is important for human submersible vehicle and marine animal in underwater navigation and action. Above those depths, oceans mix whereby winds, turbulence and currents, the temperature, salinity, density and sound speed differ greatly. Mesoscale eddy redistributes heat and brings carbon and other elements from one part of a body of water to another. Below those depths, the temperature and sound speed gradually decreased. Taking an eddy in the North Pacific as an example, in the section of 33.7° in latitude, the depth of the maximum salinity layer (34.68 psu) is located at about 70 m, and the depth of the minimum salinity layer (33.9 psu) is about 500 m. Sound speed approximate minimum at 1480 m/s, and the depth of the sound channel is nearly 800–1000 m. Judging from Figure 3, from the surface to 300 m depth, along the latitude of the section, temperature, salinity, density and speed of sound have a peak and two valleys. The peak value position is near 149.3° E that also represents the centre of the cyclonic eddy, while the valley value locations are near 148.7° E and 150.5° E. Moreover, the difference in temperature and sound speed between the cyclonic centre and the surroundings is much weightier than that of the salinity and density. The temperature and sound speed at the eddy centre (0–1000 m vertical integral) is greater than the surrounding values by about 0.97 °C and 3.5 ms^{-1}. While the salinity and density differences are about 0.05 psu and 0.14 kgm^{-3}, respectively.

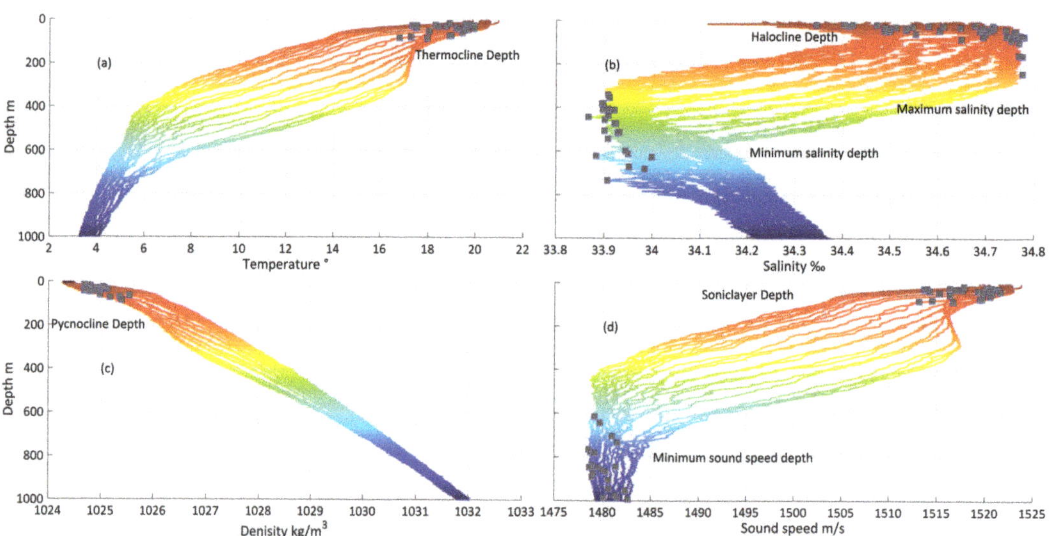

Figure 4. Measured data at the latitude of 33.7° N, (**a**) temperature, (**b**) salinity, (**c**) density and (**d**) sound speed profiles.

Satellite remote sensing data were obtained on the CMEMS website, and we tracked the position of the cyclonic eddy. The centre of the eddy moved westward by 18 min on the day of finishing the sound propagation experiment. The process of movement of the cyclonic was observed through sea surface height. The size, intensity and kinetic energy of the eddy increased at the early youth of the eddy's life cycle; these characteristics remain stable in the later 3/5 of its adult period, and then rapidly decrease in the final old age of eddy's life. As shown in Figure 5, the research area was from 148.5° to 151° east longitudes and from 33.5° to 34.5° north latitudes. The North Pacific cyclonic eddy moved westward over time and merged into Kuroshio after about three weeks. In Figure 6, the colour represents the satellite altimeters data and geostrophic velocity as the black arrow shows. Getting the information of marine and according to the sea surface temperature to determine the position of the eddy centre, we replan the acoustic propagation experiment, particularly the location of explosions. The scale of the tracked eddy is about 50 km. Combined with the distribution of hydrographic data, the eddy centre is predicted to be at 149.25° E and 34° N. Figure 7 shows the topography of the experimental area in the western North Pacific, the path of MVP300 (red dots) and the position of the vertical line array (green circle) on 18 June 2019.

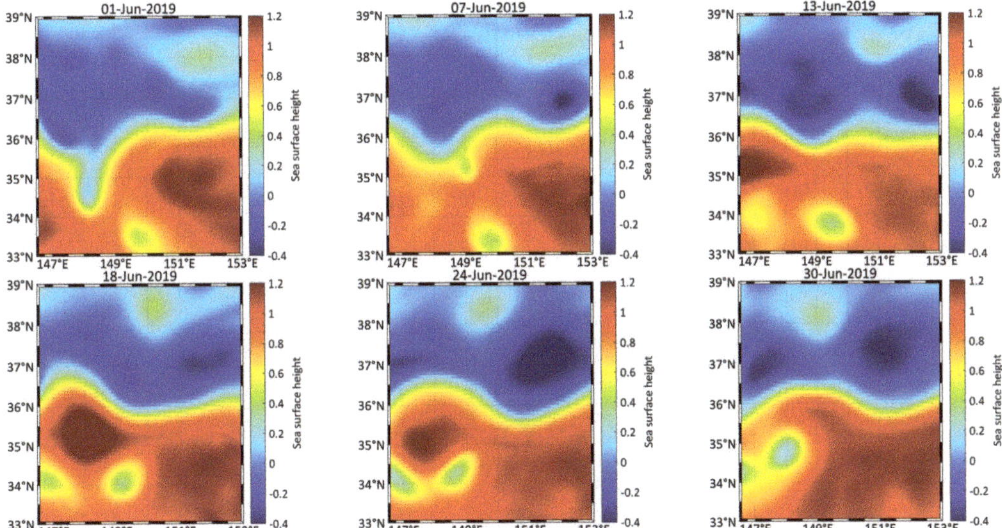

Figure 5. Sea surface height of the life cycle of the eddy.

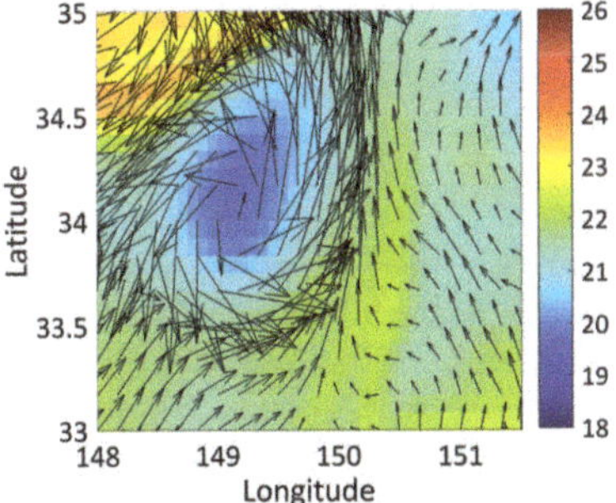

Figure 6. Sea surface temperature and geostrophic velocity of the eddy.

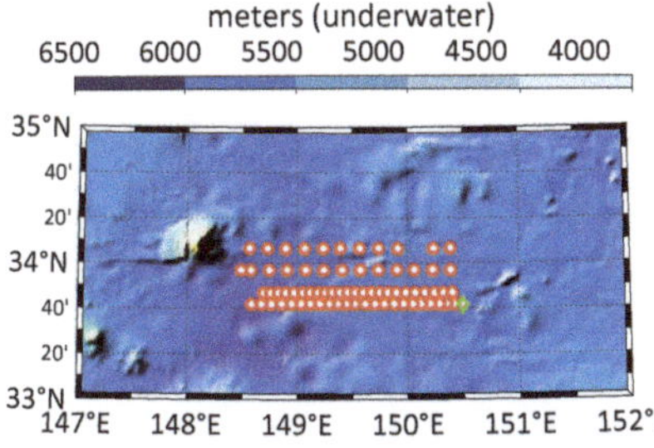

Figure 7. Topography of the western North Pacific. Colour shading indicates topographic relief from etopo1 data.

3.2. Acoustical Data

Considering multiple sources and a fixed receiver array, the acoustic reciprocity theorem was generalised. Moving explosions can be regarded as a receiver moving away from the sound source. Unfortunately, the measurement error was much higher because vertical spacing between the adjacent hydrophones is around 50 m. Only data whose standard deviation of receiver depth and hydrophones real depth are sufficiently small are selected when compared with the simulation of sound pressure transmission loss. Therefore, fewer receivers acoustic data are picked, calculated and analysed.

The whole depth of sound pressure transmission loss diagram is illustrated in Figure 8, and the simulated acoustic source was located in depths of 200 m. The sediment parameter is as follows: sound speed is 1650 m/s, density is 1.5 g cm^{-3}, and the bottom attenuation coefficient is 0.5 dB/λ. The average depth of the ocean is 6137 m, which is approximately horizontal. Owing to the change of thermocline arising from cold-core eddy downwelling, surface sound channel (ranging from 0 m to 400 m in depth) is formed. New sound path markedly appears, as can be seen in Figure 8, marked as "new path".

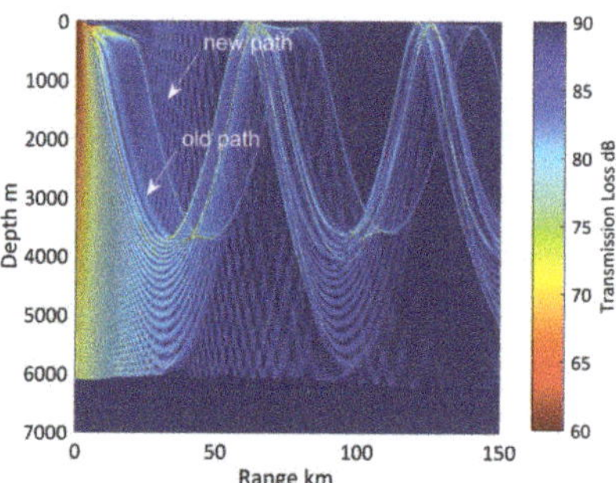

Figure 8. Sound transmission loss propagation through the eddy.

According to the eddy structure discussed in the previous section, we select the position of placing the receiver hydrophones array and the place of dropping the explosions. Through the prior simulation results of sound propagation, we know that the convergence zone was expected to be the most affected on both the energy distribution and arrival time structure. Ultimately, we dropped explosions within the scope of the first and second convergence zones, removed the locations where the ocean current has a great impact on the depth of the hydrophones, and calculated the propagation loss distribution using the remaining locations. Compared with sound pressure transmission loss impact by oceanic eddy, topography could be ignored. Figure 9 shows the comparison of the the numerical results with experimental data when the sound source was located in the mixed layer (200 m). The agreement of experiment and modelling can get a satisfactory result.

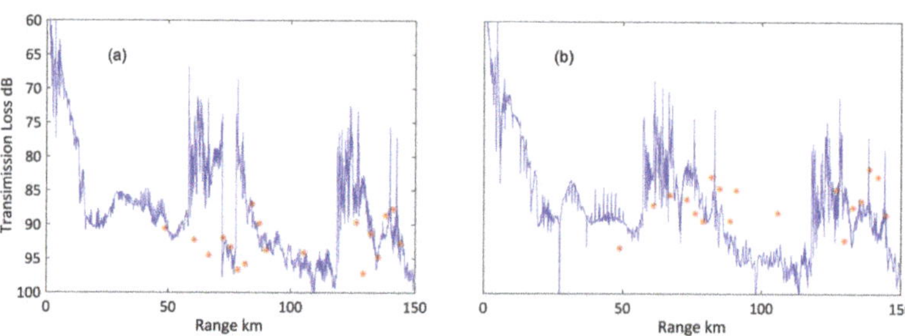

Figure 9. Transmission loss and experimental results in depths of 200 m (red asterisk), 100 m explosions (**a**) and 200 m explosions (**b**), and the blue line is simulated sound pressure transmission loss with a different range.

The curve of diagrams Figure 9 best illustrates the existence of a "new path", which can be seen in the splitting of convergent zone approximately at 70 and 140 km. It is a pity that we have not been able to measure the sound propagation at the same location in the absence of the eddy situation. Eventually, the comparison of the two simulations of travel times results is given in Figure 10, which respectively are travel through the eddy and without the eddy.

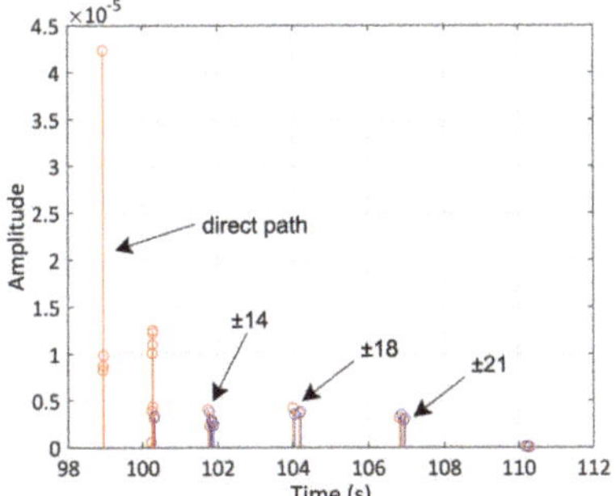

Figure 10. Arrivals in blue travel through the eddy and in red is travel without the eddy. The arrow points out sound exit angle from the source.

For receiver at the depth of 200 m and at the range of 150 km, arrivals structure contain the number of echoes or arrivals. The direct path is missing under the circumstance of sound traversing the cyclonic eddy. An acoustic signal arrives earlier in range independent horizontal sound speed stratification environment than that of in cold-core eddy environment.

4. Discussion

4.1. Analysing the Influence of Mesoscale Eddy on the Sound Field

The analysis of effect of eddies on acoustic properties implied that the movement of the sound field converging zone is caused by the vertical gradient of the sound speed. Setting up the location of minimum sound speed profile as 1000 m, coefficient DC in Equatoin (3) is −25 m/s and 25 m/s for cyclonic and anticyclonic eddy, respectively. Furthermore, the semi-major axis of the Gaussian eddy model is 40 km, while the semi-minor axis is 400 m. Figure 11 shows sound speed contours of the cold core eddy and the warm core eddy. The bottom of the western North Pacific's sound speed is 1650 m per second, and density is 1.5 g per cubic centimetre. The sediment parameters of deep water have little effect on the sound intensity in the distance. We select the parameters based on historical data in the Pacific. The parabolic equation method is applied to sound propagation problems to discuss how sound waves traverse eddies. The harmonic point source is located 200 m below the surface of the ocean. Subsequently, sound pressure redistribution is investigated.

A sound source produced waves of frequency 300 Hz and wavelength 5 m, and the first convergence zone typically moved 2.5 km closer to the sound source compared to that of the sound field excited in the background sound speed environment. The convergence area of the sound field influenced by the anticyclonic eddy has moved 3.3 km away from the sound source. In addition, the warm eddy has resulted in an obvious widening of the convergence zone. The sound rays go deeper into the influence region of eddies, and sound rays were no longer limited into the deep-ocean sound channel which could propagate for a great distance. As the sound rays will deflect towards a negative gradient direction of the SSP, due to the cold-core of the cyclonic eddy, more rays tend to deflect to the centre of the eddy like the convex lens effect. The rays refract gradually and deflect upwards exceeding the position where they should horizontally reverse. The width of the convergence zone became narrower, and the location of it moved toward the sea surface. Eddy intensity is characterised by the sound speed of eddy currents. Figure 12 shows the influence of intensity of eddy on transmission loss of the enlarged picture of the position of the first convergence zone. With a larger absolute intensity of the eddy, the convergence zone position moves farther away from the source. When the eddy intensity increases from 20 to 25 m/s, the sound speed changes by 1 m/s per 10 km in a horizontal direction, the location of the cyclonic eddy convergence zone moves towards the point source by 0.25 km, and the location of the anticyclonic eddy convergence zone moves away from the point source by 0.85 km.

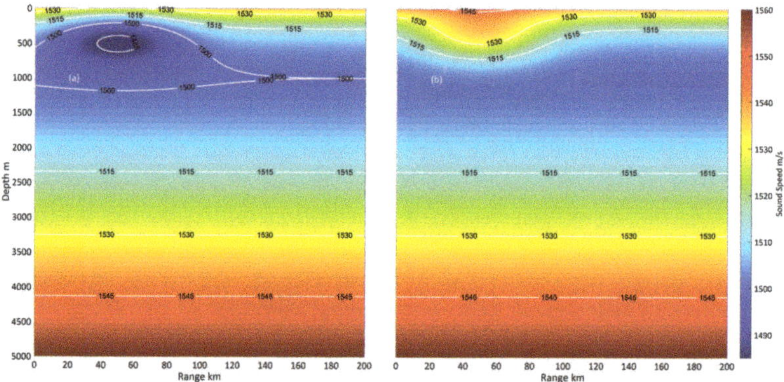

Figure 11. Sound speed distribution of cyclonic eddy (**a**) and anticyclonic eddy (**b**).

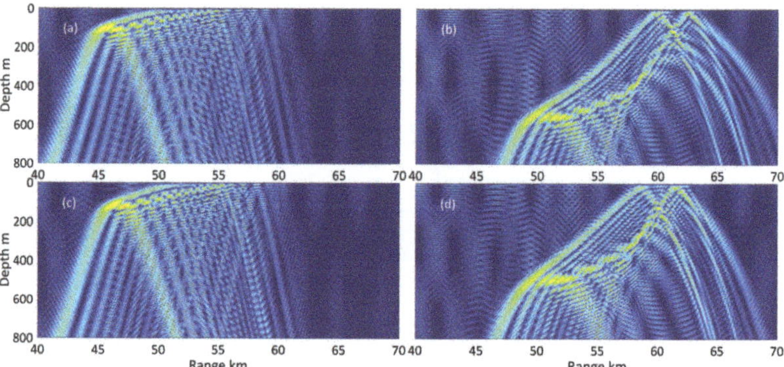

Figure 12. Transmission loss of convergence zone 25 m/s intensity of cyclonic eddy (**a**) and anticyclonic eddy (**b**) and 20 m/s intensity of cyclonic eddy (**c**) and anticyclonic eddy (**d**).

Figure 13 shows sound propagation through cyclonic eddy and anticyclonic eddy. To analyse the affection of ocean eddy on sound transmission, typical acoustic propagation paths are drawn by white curves. Among them, SRBR is surface reflected/bottom-reflected rays, RSR is refracted/surface-reflected rays, and RR is refracted/refracted rays. SRBR rays are closely linked to seafloor parameters and are valid for short ranges (<20 km). When the ocean is deeper than the critical depth, the remainder of the depth causes the formation of a convergence zone. Except for the SRBR rays, other types of rays are influenced by the critical depth. In addition, comparing these two pictures, the change of RSR rays caused by eddies is the greatest. At the location of the first converge zone, the depth of ray reversal changes dramatically. Near the sound source, those acoustic rays have a small grazing angle corresponding to the lower-order normal mode being excited. When it comes to the position of rays convergence in Figure 13a, the sound should be reversed at a small angle; however, reversal occurs at larger angles near the sea surface and vice versa. Instead of using the ray method, analysing from the normal mode perspective can get a similar conclusion. Figure 14 illustrates, on the one hand, the whole depth function (eigenfunction) moving up at the centre of the cold eddy and, on the other hand, the whole depth function (eigenfunction) moving down at the centre of warm eddy. Indeed, the closer the receiver position gets to the centre of the eddy, the more violently the eigenfunction moves. Above all, with the process of sound propagation, the position of the seabed reversal points moves toward the source. Then, the rays reverse near the surface more quickly, which causes the position of the convergent area to move toward the source. As a consequence, the eigenfunction at the location of the convergence zone moves both upwards and forwards. Equally, with each order eigenfunction influenced by warm-core eddy, firstly, the position of seabed reverses downwards; secondly, surface reversals occur in a position further away; thirdly, the convergence zone keeps away from the source and to the surface.

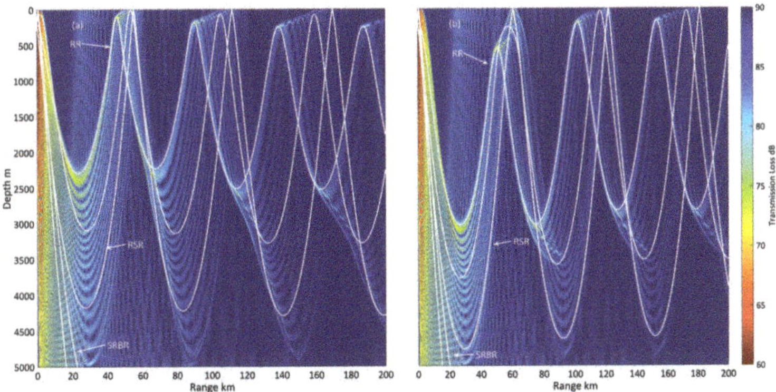

Figure 13. Transmission loss distribution of cyclonic eddy (**a**) and anticyclonic eddy (**b**).

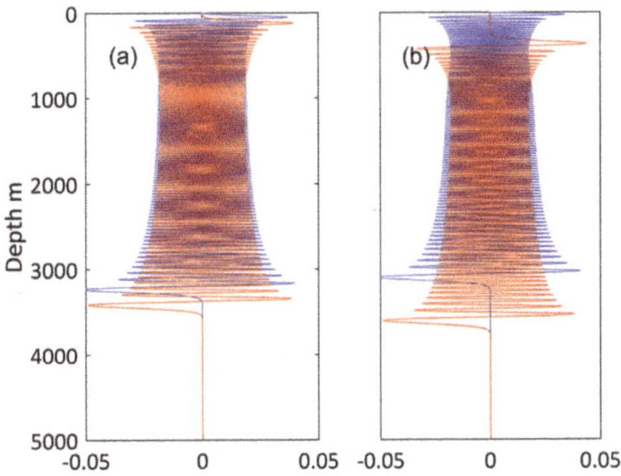

Figure 14. The eigenfunctions vary with depth influenced by eddy; red line represents warm-core eddy, and blue line represents cold-core eddy. (**a**) 40 km away from the center of eddy; (**b**) 0 km away from the center of eddy.

4.2. Analysing the Acoustic Characteristics of Mesoscale Eddy Using Normal Mode Method

Based on the Equation (9) in the second section, the amplitude of each order normal mode is calculated. The Figure 15 displays the normalised amplitude distribution of the normal mode of various orders at different positions away from the sound source.

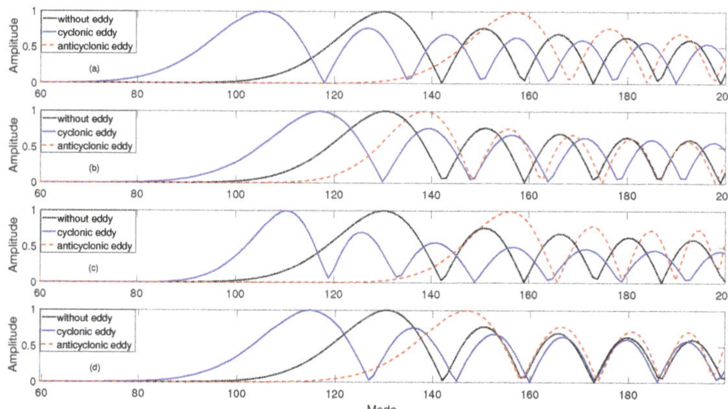

Figure 15. Normalised amplitude distribution of normal mode at different distances from the sound source, which are (**a**) 10 km, (**b**) 50 km, (**c**) 90 km and (**d**) 150 km away from the source.

The blue line illustrates the normal mode amplitude distribution under the influence of cyclonic eddy, and the red dotted line shows the normal mode amplitude distribution under the influence of anticyclonic eddy. Amplitude distribution of normal mode at various distances is calculated. At different horizontal distances, after passing through the cold eddy, that is, 100 to 200 km, the amplitude of the lower-order is generally increased. On the contrary, after passing through the warm eddy, the amplitude distribution curve of the normal mode will move towards higher-order as a whole. Affected by the eddy, the amplitude distribution of the normal mode of each order varies greatly from 10 to 90 km away from the source. Under the influence of the cold eddy, when the receiver position moves from the eddy edge to the position close to the eddy centre, more energy is distributed in the higher-order, corresponding to more sound rays emitted from the source at a larger grazing angle. Those rays have a shorter sound path, and a nearby convergence zone occurs at a close distance. However, under the influence of the warm eddy, when the receiver position moves from the edge of the eddy to the position close to the eddy centre, more energy is distributed in the lower order, corresponding to more sound rays emitted from a comparatively smaller grazing angle. These rays have longer sound paths, and energy will converge at a greater distance.

Figure 16 shows the coupling matrix at 10, 50 and 90 km of the main orders of normal mode. The left panel is the distribution of the coupling coefficient affected by the cyclonic eddy, and the right panel is the distribution of the coupling coefficient affected by the anticyclonic eddy. Because the eddy centre is located above the background sound channel axis, sound speed continues to decrease at 500 m below sea level, causing the change of the depth of the axis of the sound channel. For the cold-core eddy, the order of the normal mode coupling is more than that of the warm-core eddy. In addition, the reasons for the variation of the coupling strength of the normal mode at different distances from the eddy center can be influenced by the variation rate of sound speed.

Measurement of sound speed of the eddy is used for taking everything into account. From Figure 17 we can learn that, with the movement of the distance to the centre of the eddy, normal mode theory is used for getting the amplitude of local modes at different distances to the centre of the eddy. The same as the simulation of the cold eddy, the measured eddy shows the same conclusion that the amplitude of higher-order normal modes increases when the receiver position is to the centre of the eddy.

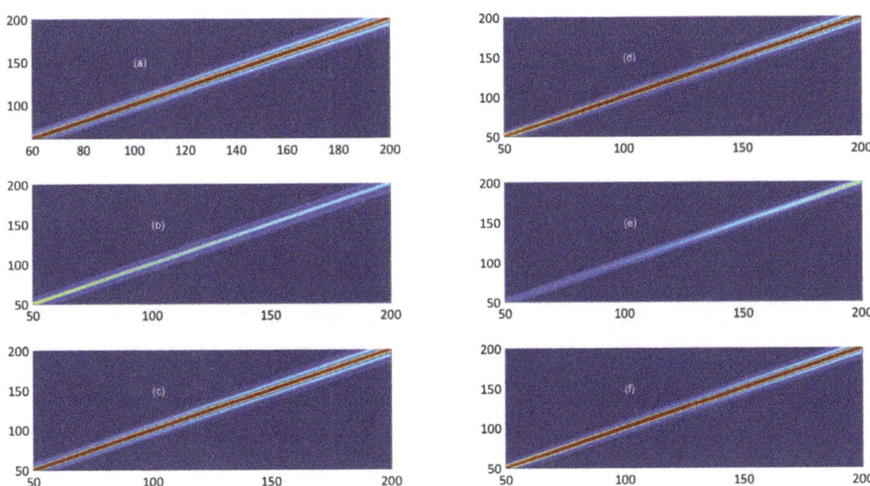

Figure 16. Coupling coefficient affected by cold eddy: (**a**) 10 km, (**b**) 50 km and (**c**) 90 km away from the source and coupling coefficient effected by warm eddy (**d**) 10 km, (**e**) 50 km and (**f**) 90 km away from the source.

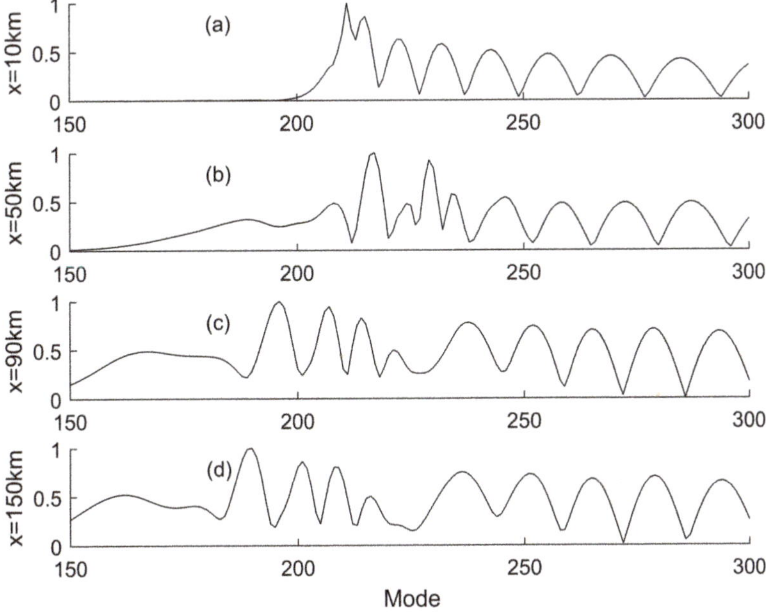

Figure 17. Normalised amplitude distribution of normal mode affected by real eddy at different positions from the sound source, which are (**a**) 10 km, (**b**) 50 km, (**c**) 90 km and (**d**) 150 km away from the source.

5. Conclusions

Mesoscale eddy will lead to changes in the location of the convergence zone relative to the acoustic source position. The convergence zone shifts toward the acoustic source as the cyclonic eddy occurs on the acoustic propagation path, and the anticyclonic eddy causes the convergence zone to move away from the acoustic source and increases the width of the convergence zone. Each order eigenfunction moves up influenced by cold-core eddy (moves down influenced by warm-core eddy); the seabed reversal points move toward (away from) the source, and then, the rays reverse near the surface moves toward (away from) the source. The eddy causes the amplitude of normal modes to be different at different positions away from the source. For the cyclonic eddy, the amplitude of higher-order normal modes increases when the receiver is close to the centre of the eddy; on the contrary, the amplitude of lower-order normal modes increases when the receiver is close to the centre of the anticyclonic eddy. We researched both source and receiver near the sea surface sound propagation problem with the effect of eddies. It brought with it a clarity of purpose, an easy understanding of human submersible vehicles and large marine creatures. They are known to produce low-frequency sounds that can propagate over long distances and receive echoes, detect underwater objects and locate food. Eddies could interfere with them in their prediction of distance and range.

Author Contributions: Conceptualization, S.P.; methodology, J.L. and S.P.; software, J.L. and Y.G.; validation, J.L.; formal analysis, J.L., S.P. and Y.G.; investigation, J.L., L.G. and M.Z.; resources, S.P.; writing—original draft preparation, J.L. and L.G.; writing—review and editing, J.L., S.P., L.G., M.Z., Y.G. and S.Z.; visualization, S.P. and L.G.; supervision, S.P.; project administration, S.P.; funding acquisition, S.P. All authors have read and agreed to the published version of the manuscript.

Funding: This research was funded by National Natural Science Foundation of China, grant number 11904065.

Acknowledgments: The measuring of oceanic front data of the north-western ocean and the experiments of acoustic propagation were accomplished with the help of College of Meteorology and Oceanography in National University of Defense Technology.

Conflicts of Interest: The authors declare no conflict of interest.

References

1. Lochte, K.; Pfannkuche, O. Cyclonic cold-core eddy in the eastern North Atlantic. II. Nutrients, phytoplankton and bacterio-plankton. *Mar. Ecol. Prog. Ser.* **1987**, *39*, 153–164. [CrossRef]
2. National Research Council. *Marine Mammals and Low-Frequency Sound: Progress Since 1994*; National Academies Press: Washington, DC, USA, 2000.
3. Godø, O.R.; Samuelsen, A.; Macaulay, G.J.; Patel, R.; Hjøllo, S.S.; Horne, J.; Kaartvedt, S.; Johannessen, J.A. Mesoscale eddies are oases for higher trophic marine life. *PLoS ONE* **2012**, *7*, e30161. [CrossRef] [PubMed]
4. Tyack, P. Acoustic communication under the sea. In *Animal Acoustic Communication*; Springer: Berlin/Heidelberg, Germany, 1998; pp. 163–220.
5. Waite, A.M.; Raes, E.; Beckley, L.E.; Thompson, P.A.; Griffin, D.; Saunders, M.; Säwström, C.; O'Rorke, R.; Wang, M.; Landrum, J.P.; et al. Production and ecosystem structure in cold-core vs. warm-core eddies: Implications for the zooplankton isoscape and rock lobster larvae. *Limnol. Oceanogr.* **2019**, *64*, 2405–2423. [CrossRef]
6. Kobashi, F.; Kawamura, H. Seasonal variation and instability nature of the North Pacific Subtropical Countercurrent and the Hawaiian Lee Countercurrent. *J. Geophys. Res. Ocean.* **2002**, *107*, 6-1–6-18. [CrossRef]
7. Goh, G.; Noh, Y. Influence of the Coriolis force on the formation of a seasonal thermocline. *Ocean Dyn.* **2013**, *63*, 1083–1092. [CrossRef]
8. Lin, X.; Dong, C.; Chen, D.; Liu, Y.; Yang, J.; Zou, B.; Guan, Y. Three-dimensional properties of mesoscale eddies in the South China Sea based on eddy-resolving model output. *Deep Sea Res. Part I Oceanogr. Res. Pap.* **2015**, *99*, 46–64. [CrossRef]
9. Henrick, R.; Jacobson, M.; Siegmann, W.; Clark, J. Use of analytical modeling and limited data for prediction of mesoscale eddy properties. *J. Phys. Oceanogr.* **1979**, *9*, 65–78. [CrossRef]
10. Dong, C.; Mavor, T.; Nencioli, F.; Jiang, S.; Uchiyama, Y.; McWilliams, J.C.; Dickey, T.; Ondrusek, M.; Zhang, H.; Clark, D.K. An oceanic cyclonic eddy on the lee side of Lanai Island, Hawai'i. *J. Geophys. Res. Ocean.* **2009**, *114*. [CrossRef]
11. Jensen, F.B.; Kuperman, W.A.; Porter, M.B.; Schmidt, H. *Computational Ocean Acoustics*; Springer Science & Business Media: Berlin/Heidelberg, Germany, 2011.
12. Jian, Y.; Zhang, J.; Liu, Q.; Wang, Y. Effect of mesoscale eddies on underwater sound propagation. *Appl. Acoust.* **2009**, *70*, 432–440. [CrossRef]
13. Nysen, P.A.; Scully-Power, P.; Browning, D.G.; Bannister, R.W. Project ANZUS Eddy: Acoustic measurement of a warm core ocean eddy. *J. Acoust. Soc. Am.* **1975**, *58*, S31. [CrossRef]
14. Nysen, P.A.; Scully-Power, P.; Browning, D.G. Sound propagation through an East Australian Current eddy. *J. Acoust. Soc. Am.* **1978**, *63*, 1381–1388. [CrossRef]

15. Lawrence, M.W. Modeling of acoustic propagation across warm-core eddies. *J. Acoust. Soc. Am.* **1983**, *73*, 474–485. [CrossRef]
16. Baer, R.N. Calculations of sound propagation through an eddy. *J. Acoust. Soc. Am.* **1980**, *67*, 1180–1185. [CrossRef]
17. Weinberg, N.; Clark, J. Horizontal acoustic refraction through ocean mesoscale eddies and fronts. *J. Acoust. Soc. Am.* **1980**, *68*, 703–705. [CrossRef]
18. Heaney, K.D.; Campbell, R.L. Three-dimensional parabolic equation modeling of mesoscale eddy deflection. *J. Acoust. Soc. Am.* **2016**, *139*, 918–926. [CrossRef] [PubMed]
19. Dozier, L.; Tappert, F. Statistics of normal mode amplitudes in a random ocean. I. Theory. *J. Acoust. Soc. Am.* **1978**, *63*, 353–365. [CrossRef]
20. Colosi, J.A.; Morozov, A.K. Statistics of normal mode amplitudes in an ocean with random sound-speed perturbations: Cross-mode coherence and mean intensity. *J. Acoust. Soc. Am.* **2009**, *126*, 1026–1035. [CrossRef]
21. Spall, M.A. Dynamics of downwelling in an eddy-resolving convective basin. *J. Phys. Oceanogr.* **2010**, *40*, 2341–2347. [CrossRef]
22. Calado, L.; Gangopadhyay, A.; Da Silveira, I. A parametric model for the Brazil Current meanders and eddies off southeastern Brazil. *Geophys. Res. Lett.* **2006**, *33*. [CrossRef]
23. Collins, M.D. Applications and time-domain solution of higher-order parabolic equations in underwater acoustics. *J. Acoust. Soc. Am.* **1989**, *86*, 1097–1102. [CrossRef]
24. Tappert, F. Parabolic equation method in underwater acoustics. *J. Acoust. Soc. Am.* **1974**, *55*, S34. [CrossRef]
25. Tappert, F.D. The parabolic approximation method. In *Wave Propagation and Underwater Acoustics*; Springer: Berlin/Heidelberg, Germany, 1977; pp. 224–287.
26. Pierce, A.D. Extension of the method of normal modes to sound propagation in an almost-stratified medium. *J. Acoust. Soc. Am.* **1965**, *37*, 19–27. [CrossRef]

Journal of
Marine Science and Engineering

Article

Analysis of Underwater Acoustic Propagation under the Influence of Mesoscale Ocean Vortices

Sartaj Khan [1,2,†], Yang Song [1,2,*,†], Jian Huang [1,2] and Shengchun Piao [1,2]

[1] College of Underwater Acoustic Engineering, Harbin Engineering University, Harbin 150001, China; sartaj@hrbeu.edu.cn (S.K.); a370699479@gmail.com (J.H.); shengchunpiao@hrbeu.edu.cn (S.P.)

[2] Acoustic Science Technology Laboratory, Harbin Engineering University, Harbin 150001, China

* Correspondence: song_yang@hrbeu.edu.cn; Tel.: +86-158-4650-8908

† Sartaj Khan and Yang Song contributed equally to this work.

Abstract: Mesoscale ocean vortices are common phenomenon and fairly distributed over the global oceans. In this study, mesoscale vortex in the South China Sea is identified by processing of AIPOcean data. The characteristic parameters of the identified vortex are extracted by using Okubo-Weiss (OW) method. The empirical sound velocity formula and interpolation method are used to obtain the spatial characteristics of temperature and sound velocity of the mesoscale vortex. After this, a theoretical model based on the Gaussian method is established to fit and simulate the vortex parameters. Using this model, the influence of mesoscale vortex strength, cold and warm vortex, vortex center position and sound source frequency on sound propagation are analyzed in COMSOL software. Finally, the actual parameters of the identified vortex are compared with the ideal Gaussian vortex model. It is found that different types of mesoscale vortices have different effects on the underwater sound propagation characteristics. Cold vortices, for example, cause the sound energy convergence zone to move toward the sound source, reducing the convergence zone's span, whereas warm vortices cause the sound energy convergence zone to move away from the sound source, increasing the convergence zone's span. Furthermore, the stronger the mesoscale vortices, the greater the impact on the sound field. Our COMSOL-based results are consistent with previous research, indicating that this model could be useful for studying underwater acoustic propagation in vortices.

Keywords: mesoscale vortex; acoustic propagation; AIPOcean; OW method; COMSOL software

Citation: Khan, S.; Song, Y.; Huang, J.; Piao, S. Analysis of Underwater Acoustic Propagation under the Influence of Mesoscale Ocean Vortices. *J. Mar. Sci. Eng.* **2021**, *9*, 799. https://doi.org/10.3390/jmse9080799

Academic Editor: Grigory Ivanovich Dolgikh

Received: 28 June 2021
Accepted: 21 July 2021
Published: 24 July 2021

Publisher's Note: MDPI stays neutral with regard to jurisdictional claims in published maps and institutional affiliations.

1. Introduction

More than 90% of the surface kinetic energy in the ocean lies in the entire ocean circulation [1], which can be seen as cyclones, anticyclones and typhoons in the ocean. The physical properties of mesoscale vortices are similar to those of cyclones and anticyclones in the atmosphere [2,3]. Mesoscale ocean vortices (named eddies) are common and fairly distributed over the global oceans. They usually existing for 10 to 100 days and have medium spatial scales in the horizontal direction ranging from 50 to 500 km [4–6]. These kinds of vortices are widely distributed in the South China Sea. Earlier, Li et al. [7] reported an anticyclonic ring detached from Kuroshio in the South China Sea. Using observation data, Wu et al. [8] analyzed sound speed field in the same region of Ref. [7] by using 2D parabolic equation (PE) algorithm. So, it is of great importance to study the ocean vortices as they largely influence the vertical structure of water column in the sea. In this way, acoustic propagation in the vortex region will be changed significantly [9]. Previous studies have determined that the loss of acoustic signal strength of sonar can be up to about 40 db in the presence of vortices [10]. The study of mesoscale vortex, therefore, has important application value for underwater acoustic equipment, underwater weapons, submarine warfare and anti-submarine warfare. Due to the importance of underwater acoustic propagation to the sonar detection and submarine navigation, it is significant for the oceanographers and acousticians to understand the mechanisms of vortex and its influence on underwater acoustic propagation.

Vortex recognition is the first step in the study of vortices in seawater. Since the 1970s, oceanographers and acousticians have studied and developed many methods for the identification of vortices. In the field of vortex recognition algorithms based on physical parameters, one of the most widely used is based on the properties of Okubo-Weiss (OW) parameter q (Okubo 1970 [11]; Weiss 1991 [12]). This parameter is used to determine the spatial structure of the vortex and to identify regions with different mixing properties [13]. The OW method is the earliest proposed and most widely used method for vortex identification [14,15], because it has a simple principle and a small amount of calculation. Chaigneau [16] used the altitude data obtained from high-resolution satellite measurements in the past 15 years to identify and compare the vortices in the waters of Peru using both the OW method and the WA method [17]. Their results showed that the average radius of the sea vortex is 80 km, and it gradually increases to the north, and the average vortex life is about 1 month. In this area, the warm vortex tends to move to the northwest, and the cold vortex tends to move to the southwest. Doglioli et al. [18] proposed a wavelet analysis method, based on the high-resolution numerical model of the seas surrounding South Africa, focusing on the study of anticyclones and cyclones in the Kepler Basin, which are believed to be actively involved in the Indian Ocean Atlantic Ocean exchange, including vortices. The calculation of the vortex trajectory and the time evolution of the vortex characteristics provide objective tools for the identification and tracking of the 3D vortex. Francesco et al. [19] developed a new method of identifying vortices based on the geometric shape of the velocity vector. The vortex is automatically identified through the spatial distribution of the velocity vector near the center of the vortex, which is a good algorithm for high-frequency radar surface velocity data for the Gulf of Southern California.

In the past, much work has been done to investigate the physical characteristics of mesoscale ocean vortices and their influence on acoustic propagation. Vastano and Owens [20] used numerical simulation technique based on the geometric acoustic model to study the effect of cold vortices on the sound propagation law of the deep-sea sound channel when the sound source was located at the center of the vortex. They used ray tracking method to describe the effect of vortex on transmission loss. A coherent ray technique was used by Weinberg and Zabalgogeazcoa [21] to investigate the effect of cold vortices on sound propagation when the sound source and receiver positions are both outside the vortex, and the results are expressed in a time series. These previous studies of acoustic transmission were based on the direct oceanographic measurements of vortex. Baer [22] used the PE method to study the propagation characteristics of sound waves in the ocean vortex field. It was found that in a vortex field, using a 100 Hz non-directional sound source, the received acoustic signal energy within the grazing angle increases by 5 to 50%. When the vortex moves along the line connecting the sound source and the receiver, the difference in the position of the vortex relative to the sound source can cause a 20 db decrease in the propagation loss. Li et al. [5] used an underwater acoustic model-MMPE to examine the acoustic propagation under the influence of different types, intensities and positions of eddies, and different frequencies and depths of sources. In recent years, Chen et al. [6] used UMPE model to analyze the acoustic propagation characteristics under the influence of cold-core vortex. For this, they selected the cold vortices based on the summer hydrological environment in the Kuroshio extension sea region. Chen et al. [23] examined the influence mechanism of eddies on acoustic propagation from the perspective of surface waveguide by Argo floats data and typical 2D ray algorithm. A fully 3D PE model was used by Heaney [24] to discover the horizontal refraction due to mesoscale eddy at low frequency acoustics in the South Indian Ocean.

This study provides an objective tool for the identification and tracking of 3D vortices. The principle and process of mesoscale vortex recognition by OW method is introduced, and a cold vortex is identified at position 15–18° N, 115–117° E in the South China Sea (Section 2). In a disparate departure from previous research, this study applies the OW method to extract and identify vortices using real time data from the AIPOcean 1.0 numerical product. A theoretical model of sound field calculation under the influence of vortex is then developed in COMSOL software [25] based on the finite element method (Section 2). Utilizing this model, the influence of mesoscale ocean vortex on sound propagation using low frequency is analyzed and discussed (Section 3). Finally, main findings and future work are summarized in Section 4.

2. Data and Methods

2.1. AIPOcean Region and Argo Data

China joined the international Argo program in 2002 and deployed 100–150 Argo buoys between 2002 and 2005 in the South China Sea, the East China Sea and the Yellow China Sea to build a regional observation network in the Pacific Northwest along the coast of China. After that, 20–30 buoys have been deployed every year to maintain the normal operation of the local observation network. At present, China has become an important member of the Argo project, providing important assistance for ocean observation in the Northwest Pacific [26]. The ocean reanalysis data set of the joined area Asia-Indian-Pacific Ocean (hereafter AIPOcean; [27]) used in this paper is the numerical product of China's Argo project, and its range is 30°–180° E, 28° S–44° N as shown in the Figure 1, the spatial resolution is about $1/4° \times 1/4°$. This data set contains the average 3D temperature, salinity and flow field from 1 January 1993 to 31 December 2006, as well as the 2D sea surface height. The original data is interpolated on the standard vertical plane on the isodensity coordinates.

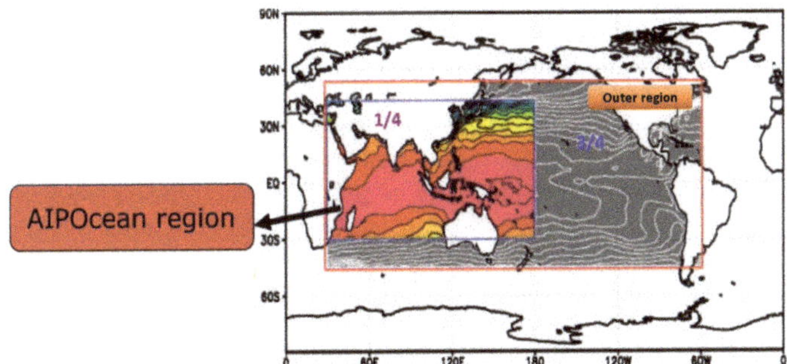

Figure 1. The geographical location of the AIPOcean region and the peripheral area used in the establishment of the grid model.

2.2. Vortex Identification and Extraction Based on OW Method

2.2.1. Vortex Identification Based on OW Method

The zonal velocity and meridional velocity data were provided by Argo buoys. These two datasets can be combined with the OW method to identify mesoscale vortices. According to the definition of OW function [11,12,28]:

$$w = \frac{\partial v}{\partial x} - \frac{\partial u}{\partial y} \tag{1}$$

$$S_n = \frac{\partial u}{\partial x} - \frac{\partial v}{\partial y} \tag{2}$$

$$S_s = \frac{\partial v}{\partial x} + \frac{\partial u}{\partial y} \tag{3}$$

where S_n, S_s, and w are the normal and the shear components and vorticity of the relative vortex. (u, v) indicates the horizontal velocity, with positive values representing the east (x) and northward (y) directions. We can define the parameter q (OW parameter) as:

$$q = S_s^2 + S_n^2 - w^2 \tag{4}$$

Then a suitable q_0 is given as a threshold to determine the position of the vortex center of the mesoscale vortex. Here, $q_0 \leq -0.2\sigma_q$, which is commonly used in the research, is selected as the threshold to judge the position of the vortex center of the mesoscale vortex, where σ_q is the standard deviation of the q value.

After determining the position of the vortex center of the mesoscale vortex, it is necessary to combine the velocity field and the sea surface height field to calculate the change trend of the velocity and sea surface height in different directions. Select the location of the maximum value of the velocity in different directions within a reasonable radius as the mesoscale. The edge estimation position of the vortex, the average value of the distance between the edge position in each direction and the identified vortex center position as the radius, and finally the average gradient of the region is calculated as the intensity of the mesoscale vortex. The data related to zonal flow velocity, meridional flow velocity, temperature, salinity, etc. are obtained from AIPOcean 1.0 product. The flow chart of the OW algorithm is shown in Figure 2. As shown in the figure, the entire identification process can be divided into the following steps:

(1) Extraction of Argo buoy data: As the data of the Argo buoy are available in Network Common Data Form (NetCDF) format, it is necessary to use the ncread function in the Matlab [29] to extract the zonal flow velocity, meridional flow velocity, temperature and salinity required for identification.

(2) Delineate the scope of calculation: The extracted zonal flow velocity and meridional flow velocity exist in the form of 3D variables of latitude, longitude and height. The surface velocity at a depth of 100 m is selected and a matrix is formed in the range of $10° \times 10°$ for calculation.

(3) Calculate the value and threshold of the q parameter: After obtaining the matrix of the zonal flow velocity u and the meridional flow velocity v, the gradient function can be used to calculate the derivatives of u and v in the x and y directions, and then Equations (1)–(4) can be used to calculate the value of q. The standard deviation function is used to calculate the threshold q_0.

(4) Determine the position of the center of the mesoscale vortex: After obtaining the value of q and q_0, the position of the vortex center can be obtained by comparing the two values. After obtaining the position of the vortex center, the maximum value of the flow velocity in different directions can be selected as the edge estimation value using the flow velocity field in the Argo numerical product, and the average value of the distance from the vortex center can be used as the radius estimation value of the mesoscale vortex.

To illustrate the steps of this procedure, we will apply it to processing a selected ocean mesoscale vortex (Table 1). According to the temperature and salinity and flow velocity profiles of the Argo buoy, appropriate processing has been performed to obtain the spatial distribution characteristics of the mesoscale vortex such as temperature and sound velocity.

Table 1. Mesoscale vortex information extracted by the OW method.

Vortex Center Position	Type	Volex Radius	Maximum Flow Rate
16.74° N, 115.72° E	Cold vortex	27.055 km	15.6 cm/s

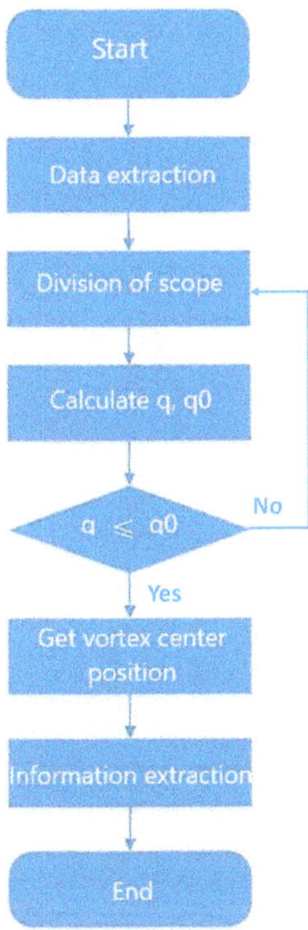

Figure 2. Flow chart of the OW algorithm.

2.2.2. Extraction of Mesoscale Vortex Spatial Characteristics

After completing the identification process of the mesoscale vortex, we start to extract the spatial characteristic information such as the temperature and sound velocity of the mesoscale vortex. The speed of sound in seawater varies from about 1500 to 1580 m/s. By extracting the temperature and salinity data in the mesoscale vortex region, the temperature and salinity distribution at a depth of 0–1000 m is obtained. Usually, the speed of sound is obtained by a special sound speed measuring tool. However, since there is no sound speed data, an empirical formula for sound speed is used here (Mackenzie, 1981 [30]):

$$
\begin{aligned}
C = {} & 1448.96 + 4.591T - 5.304 \times 10^{-2}T^2 + 2.374 \times 10^{-4}T^3 + 1.340(S - 35) + 1.630 \times 10^{-2}D \\
& + 1.675 \times 10^{-7}D^2 - 1.025 \times 10^{-2}T(S - 35) - 7.139 \times 10^{-13}TD^3
\end{aligned}
\tag{5}
$$

where C = sound speed (m/s), T = temperature (°C), S = salinity (parts per thousand) and D = depth (m). Ranges of validity of this formula cover:

$$
0 \leq T \leq 30, 30 \leq S \leq 40, 0 \leq D \leq 8000
\tag{6}
$$

The temperature, salinity and depth data, extracted from the Argo profiles, are entered into the sound velocity empirical Formula (5) for calculation, and the distribution of the vortex sound velocity can be obtained. Figure 3a shows the distribution of isovelocity lines map at 500 m depth obtained from the Argo profiles. As can be seen from the figure, outward spreading trend in the sound velocity is observed on the right side of the vortex. The sound velocity changes are more gradual at the outer side and more intense near the center of the vortex. The isovelocity line distribution of the entire vortex can be approximated as an irregular elliptical shape, and we can further process the vortex on this basis to facilitate subsequent simulation processes. As shown in the figure, at the same depth, the sound velocity distribution of the mesoscale vortex is approximately elliptical, with a major axis radius of about 70 km.

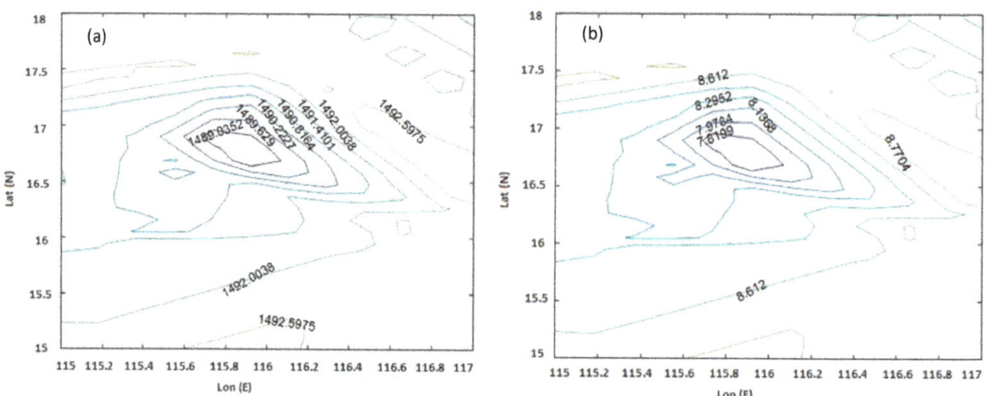

Figure 3. (**a**) Isovelocity (m/s) and (**b**) Isotemperature (°C), contour lines map at 500 m depth.

The temperature distribution of the mesoscale vortex at a depth of 500 m is shown in Figure 3b. It can be seen from the figure that the vortex center is the region with a minimum temperature. By comparing the temperature distribution and the sound velocity distribution, it is observed that the temperature distribution and the sound velocity distribution trends tend to be consistent. This is because of the three factors T, S and D that affect the speed of sound, the temperature T has the greatest effect on the speed of sound. The temperature distribution can be approximated to describe the sound velocity distribution characteristics of mesoscale vortices.

The temperature and sound velocity distribution of the mesoscale vortex obtained from the vertical profiles at depth range from 250 to 600 m from Argo data is given in Figure 4. The temperature and speed of sound are consistent with the temperature and speed of sound at the edge of the vortex. The vortex parameters obtained from Argo buoy data are shown in Table 2.

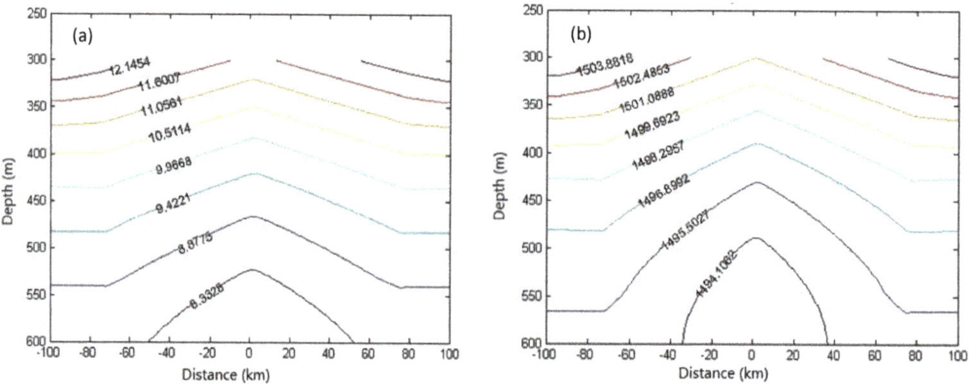

Figure 4. (**a**) Isotemperature (°C) and (**b**) Isovelocity (m/s), contour lines map at a depth range of 250 to 600 m.

Table 2. Vortex parameters based on Argo Buoy data.

Depth z/m	Vortex Center Position		Vortex Size at Different Depths		Vortex Center Temp (T$_0$/°C)	Vortex Edge Temp (T$_1$/°C)	Temperature Difference between Edge and Center (°C)
	x_0/m	y_0/m	x_m/km	y_m/km			
300	270	−135	96.9	71.7	11.45	12.69	1.24
400	170	−185	76.0	59.6	9.68	10.48	0.80
500	160	−147	73.6	52.4	8.51	9.25	0.74
600	453	−204	63.0	49.5	7.77	8.46	0.69

A linear interpolation method is applied to obtain the temperature of the mesoscale vortex region. The required mesoscale vortex parameters are: vortex center position (x_0, y_0) relative to the coordinates given in Table 1, vortex size (x_m, y_m), vortex center temperature $(T_0/°C)$ etc. and the environment temperature $T(x,y,z)$ is:

$$T(x,y,z) = T_0(z) + [T_1(z) - T_0(z)]r(x,y) \tag{7}$$

And,

$$\frac{[x - x_0(z)]^2}{x_m{}^2(z)} + \frac{[y - y_0(z)]^2}{y_m{}^2(z)} = r^2(x,y) \tag{8}$$

where $0 \leq r(x,y) \leq 1$, $z_{min} \leq z \leq z_{max}$, $r(x,y) < 1$ present the effective area of the vortex, and $r(x,y) > 1$ is the area in which the temperature and sound velocity are consistent with the temperature and sound velocity of the vortex edge.

T_1 and z represent depth of vortex and temperature at vortex edge, respectively. x_0 and y_0 represent the position of vortex center, they are determined by maximum or minimum temperature horizontal position of vortex. Using interpolation methods (Equations (6) and (7)) and empirical sound velocity formula (Equation (5)), the simulated sound velocity distribution and temperature distribution of the mesoscale vortex at different depths are shown in Figure 5a–f. By comparing Figures 4 and 5, it can be seen that even under different depth planes, the sound velocity distribution and temperature distribution still tend to be consistent, which validates Equation (6).

2.3. Modeling of Mesoscale Ocean Vortex

2.3.1. Fitting of Vortex Parameters Based on Gaussian Method

After obtaining the spatial characteristic parameters of the mesoscale vortex in the previous section, in this study we take the actual parameters at a depth of 400 m in Table 2, as an example, to fit the ideal Gaussian vortex model [31], and establish a mesoscale vortex model based on this. Table 3 shows the main parameters of mesoscale vortices selected from Table 2.

Table 3. Mesoscale vortex parameters.

Vortex Depth	Short Half Axis	Long Half Axis	Temperature Difference between Vortex Edge and Vortex Center	Sound Velocity Difference between Vortex Edge and Vortex Center
400 m	29.8 km	38 km	0.8 °C	3.6 m/s

Taking an ideal Gaussian vortex as an example, the effect of the existence of a vortex on a 2D plane in the vertical direction on the speed of sound is [5,20,31]:

$$\delta_c(x,y,z) = DC * \exp[-(\frac{r - r_e}{DR})^2 - (\frac{z - z_e}{DZ})^2] \tag{9}$$

In the formula, DC represents the intensity of the vortex, and the value of DC is the maximum difference in sound velocity from the center of the vortex to the edge of the vortex.

The value of *DC* is negative for the cold vortex, and *DC* is positive for the warm vortex. *DR* is the horizontal radius and *DZ* is the vertical radius of the vortex. The horizontal and vertical positions of the vortex center are represented by r_e and z_e, respectively. By putting the main parameters of the mesoscale vortex in Table 3 into the ideal Gaussian vortex model Equation (8), we will obtain the Gaussian vortex model parameters that conform to the actual sound velocity distribution. The sound velocity distribution obtained by this method is shown in Figure 6.

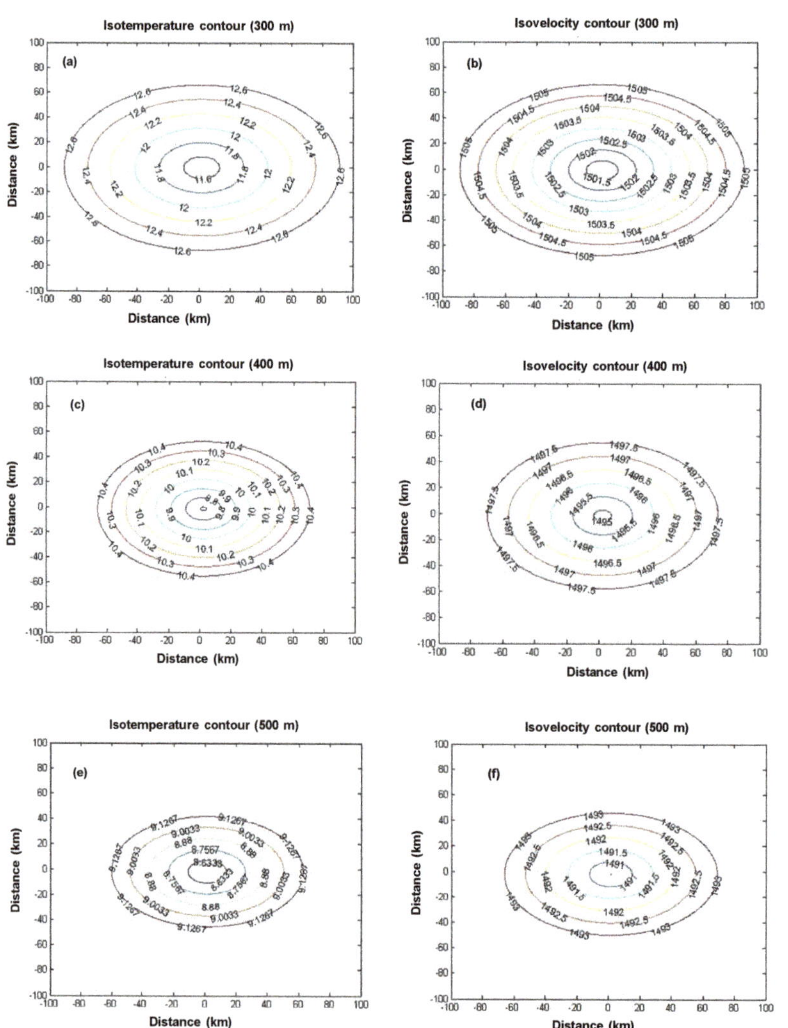

Figure 5. Interpolation plots of temperature (°C) distribution in the horizontal direction at different depths and the corresponding interpolation distribution plots of sound velocity (m/s): (**a,c,e**) Isotemperature contour is 300, 400 and 500 m, (**b,d,f**) Isovelocity contour is 300, 400 and 500 m.

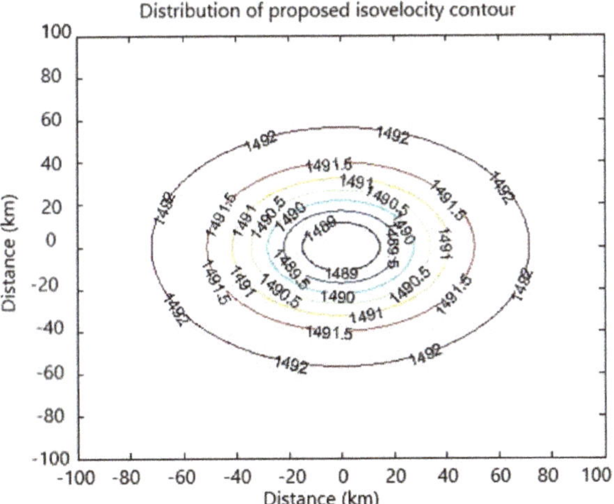

Figure 6. Sound velocity (m/s) distribution image obtained by Gaussian fitting method.

For comparison, we superimpose the sound velocity distribution field obtained from the in-situ data (Figure 3a) on the sound velocity distribution field obtained by the fitting data (Figure 6). By qualitatively comparing Figures 3a and 6, it is found that the distribution of the sound velocity gradient tends to be roughly the same. The distribution trends show the mesoscale vortex in the ideal state, but it also reflects the characteristics of the actual data, such as the rising trend of sound velocity from large to small, and the slower the change of sound velocity is closer to the edge. It can be considered that the fitting data better reflects the actual data, and the vortex model can be used for simulation.

2.3.2. Sound Field Modeling Based on COMSOL Finite Element Method

According to the law of sound propagation, at the deep sound speed channel, when the sound waves emitted by a sound source placed near the sea surface, they will reverse at a certain depth and return to the vicinity of the surface again and form a convergence zone. This reversal depth is generally between 4000 and 5000 m, and the position of the first convergence zone is about 40 km from the source. Based on these factors such as the inversion depth, the distance of the convergence zone, and the size of the vortex, Gaussian vortex model [30] is adopted to describe the ocean mesoscale vortex (Figure 7a,b).

The upper layer of the model is a seawater medium, and the lower seafloor is set as a liquid seafloor, which directly simulates the presence of mesoscale vortices in the seawater medium in the form of sound velocity distribution. COMSOL will select the physical field before building the model. Since this model uses a liquid seabed and only considers pressure acoustics, the physical field is set to pressure acoustics, frequency domain and steady state is selected for the research direction.

Due to the existence of mesoscale vortices, the sound velocity of seawater in this model is:

$$c(x,y,z) = c_0(z) + \delta_c(x,y,z) \tag{10}$$

In the formula, $c_0(z)$ is the standard model of SOFAR sound velocity profile proposed by Munk [32]:

$$c_0(z) = 1500\{1 + \varepsilon[e^{-\eta} - (1 - \eta)]\} \tag{11}$$

where $\eta = 2 * (z - z_0)/B$, z is depth, z_0 is the minimum depth of the sound velocity model and taken as 1000 m, B is the waveguide width and its value is 1000 m, ε is the magnitude of deviation from the minimum value and its value is 0.0057, proposed by Munk [32]. According to the Formula (10), the vertical distribution of the deep-sea sound channel, when there is no vortex, is shown in Figure 7b. In Equation (8), $\delta_c(x,y,z)$ indicates the influence of the existence of mesoscale vortices on the distribution of sound velocity in seawater.

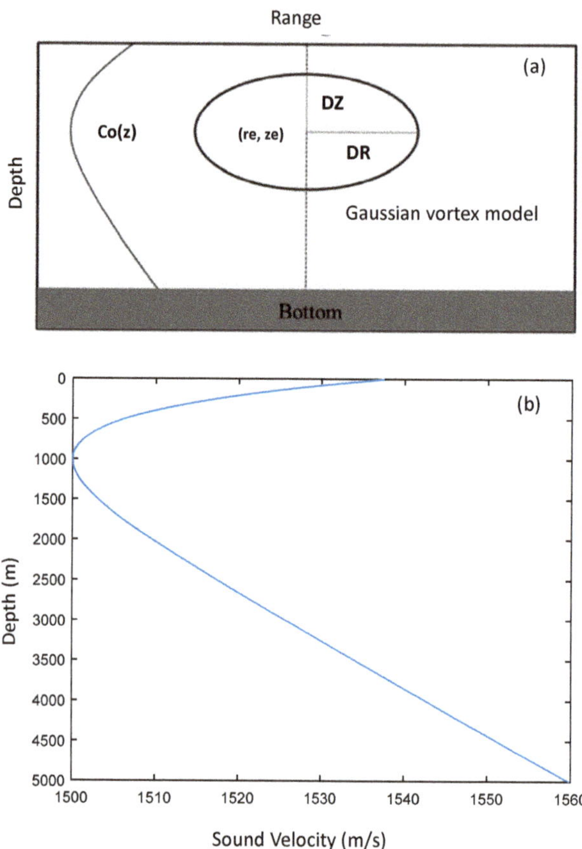

Figure 7. (**a**) Schematic diagram of the 2D model of Gaussian Ocean Vortex, (**b**) Vertical distribution of deep-sea sound channel.

3. Results and Discussion

3.1. The Influence of Mesoscale Vortices on Sound Propagation

3.1.1. The Influence of Vortex Intensity on Sound Propagation

Consider the impact of mesoscale vortex intensity on sound propagation first. Based on the mesoscale vortex model established in Section 2, we choose 5 vortices with varying vortex strengths for simulation experiments. We take a non-directional point sound source with a frequency of 20 Hz, a depth of 200 m, and DC values of −20 m/s, −10 m/s, 0, 10 m/s, and 20 m/s, respectively. A negative DC value indicates a cold vortex, a positive DC value indicates a warm vortex, and a DC value of 0 indicates there is no vortex. For this, the vortex's center is set to be the same distance as the sound source, i.e., Rx = 20,000 m, while all other conditions remain constant. Figures 8–10 show the simulation results of the sound pressure level distribution for cold vortices, without vortices and warm vortices, respectively.

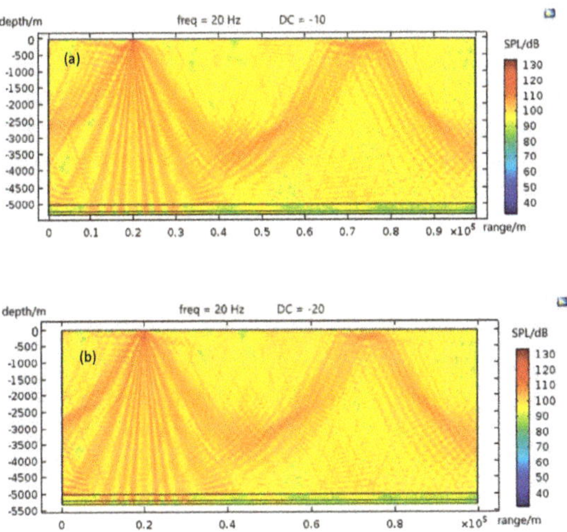

Figure 8. The distribution of sound pressure level under different intensities of cold vortex, (**a**) DC = −10 m/s and (**b**) DC = −20 m/s.

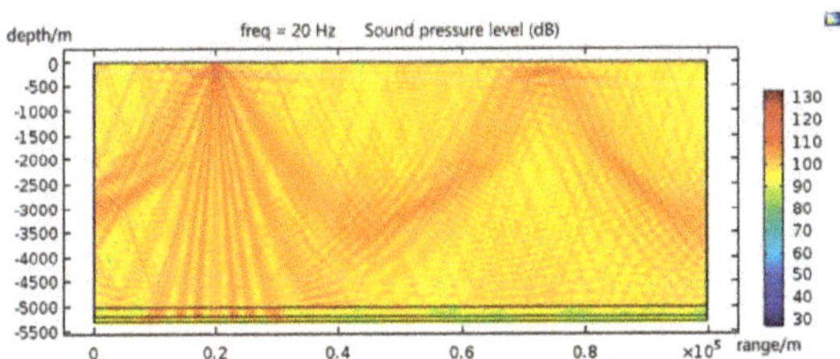

Figure 9. Sound pressure level (dB) distribution when there is no vortex (DC = 0).

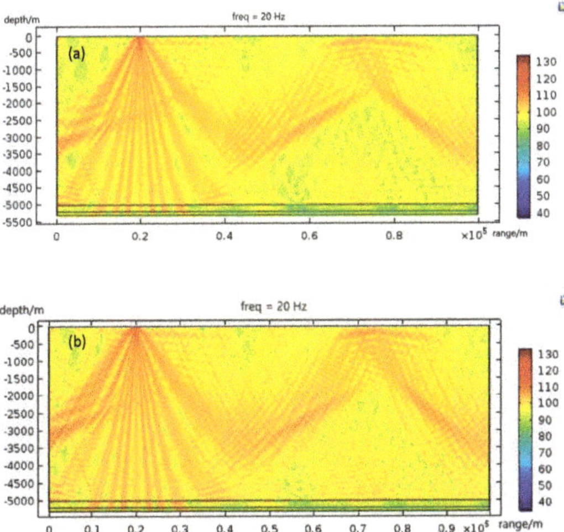

Figure 10. Distribution of sound pressure levels (dB) under different intensities of warm vortices, (**a**) DC = 20 m/s and (**b**) DC = 10 m/s.

The transmitting sound wave bends in the direction of the lower sound velocity region due to the uneven environment of sound velocity distribution in seawater. As a result, in the deep-sea channel, the sound rays are inverted and propagated upward, forming a 'sound energy convergence zone' near the sea surface in the positive sound velocity gradient region below the channel axis. The deep-sea channel's first convergence zone is frequently found at 45–50 km. Figure 10 shows that a significant amount of sound energy is concentrated about 50 km away from the sound source, which is significantly higher than in other regions.

In Figures 8–10, the convergence zone first appears at 45 km from the sound source when DC = −10 m/s (Figure 8a); when DC = 0 (Figure 9), the convergence zone appears at 50 km from the sound source; when DC = 10 m/s (Figure 10b), the convergence zone appears at about 55 km away from the sound source; and when DC = 20 m/s (Figure 10a), the convergence zone appears at about 60 km away from the sound source. We can conclude from this that when there is a cold vortex, the converging zone moves in the direction of the sound source; when there is a warm vortex, the converging zone moves away from the sound source. Furthermore, the convergence effect of sound rays is weakened when there is a warm vortex. The width of the entire convergence area is enlarged, and in an environment where there is a cold vortex, the sound rays converge more densely and appear darker in the sound pressure level distribution image. Figure 11 depicts the vertical distribution of sound velocity in seawater with different intensities of mesoscale vortices. Figure 12 depicts the deep-sea channel distribution of the vertical section drawn through the vortex's center. It can be seen from figures that the effect of the vortex on the speed of sound mainly exists in the region close to the sea surface, which is about 0–1000 m deep.

The sound velocity distribution of the vertical section of the model without and with mesoscale vortex is located (see Figure 11), are obtained by applying our COMSOL-based model. From these figures, we can observe that the sound velocity is significantly changed after the mesoscale vortex appears: the isovelocity line bends upwards with the vortex core in the center, and forms an elliptical low-sound velocity zone which is more dominant in the central region. The distribution of iso-sound velocity lines is consistent with the actual sound velocity distribution obtained by interpolation method in Section 2.3.1, which can prove that this method can better simulate the sound velocity distribution of the sound field in the presence of mesoscale vortices. It is demonstrated that when DC = 20 m/s (Figures 11a and 12), the warm vortex will cause a depth of 0–1000 m, i.e., the sound velocity gradient of the upper part of the channel axis will become smaller near the sea surface, and the rapid change in the part close to the channel axis will be positive. When DC = −20 m/s (Figure 11c and Figure 12), the cold vortex even causes a sound velocity gradient to appear at a depth of about 500 m that is lower than the sound velocity of the channel axis, reducing the positive sound velocity gradient from 1000 m to 500 m. The mesoscale vortices are thought to change the properties of the sound field and influence the distribution of the sound velocity gradient, thereby influencing sound propagation. Some of the stronger vortices even altered the nature of the sound velocity gradient, such as a cold vortex with DC = −20 m/s (Figure 11c and Figure 12), which moves the positive gradient up and expands the negative gradient.

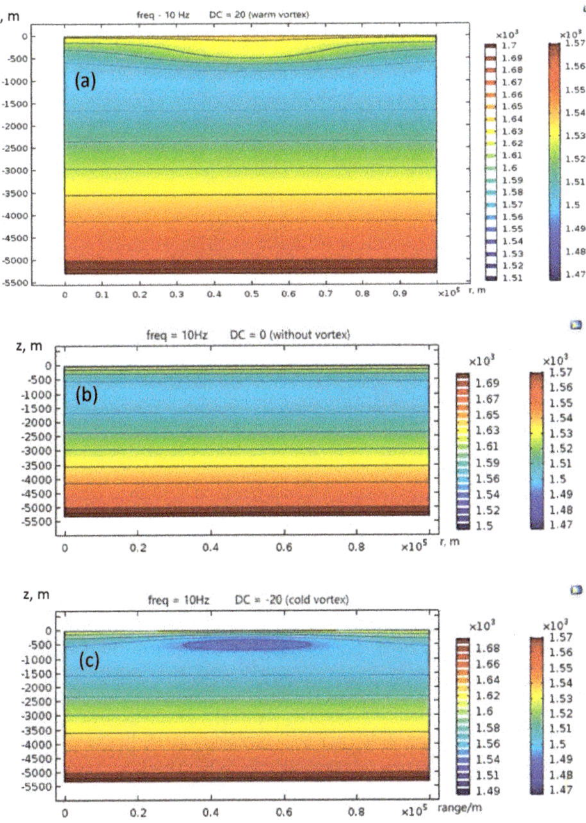

Figure 11. The vertical plane of sound velocity (m/s) distribution when vortices with different DC values exist (**a**) DC = −20 m/s, (**b**) DC = 0, and (**c**) DC = 20 m/s.

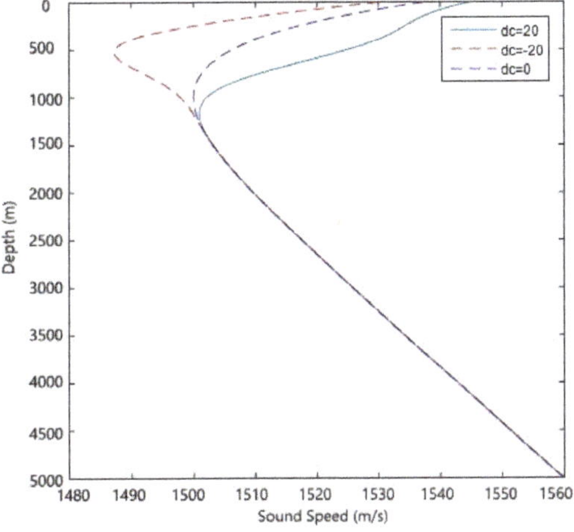

Figure 12. Distribution of the deep-sea sound channel in the vertical section of the vortex center.

3.1.2. The Influence of Vortex Position on Sound Propagation

Given that the vortex affects sound propagation by changing the sound velocity distribution in the seawater, the vortex's influence on the sound velocity distribution at different positions in the sound field must be different as well. This section discusses the effect of the vortex on sound propagation at various positions. The sound pressure level distribution when the sound source is located at the center of the vortex is shown in Figure 13. The sound velocity distributions of the cold vortex at different positions in the sound field are presented in Figure 14a,b. We can see from Figure 14 that when the sound source is set at the center of the vortex, the convergence zone moves away from the sound source. Furthermore, the depth of the convergence zone has decreased by about 100 m from its original depth.

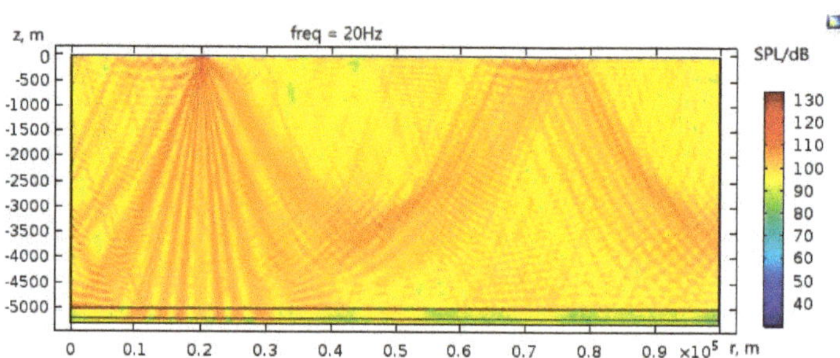

Figure 13. Sound pressure level distribution when the sound source is located at the center of the vortex (DC = −20 m/s).

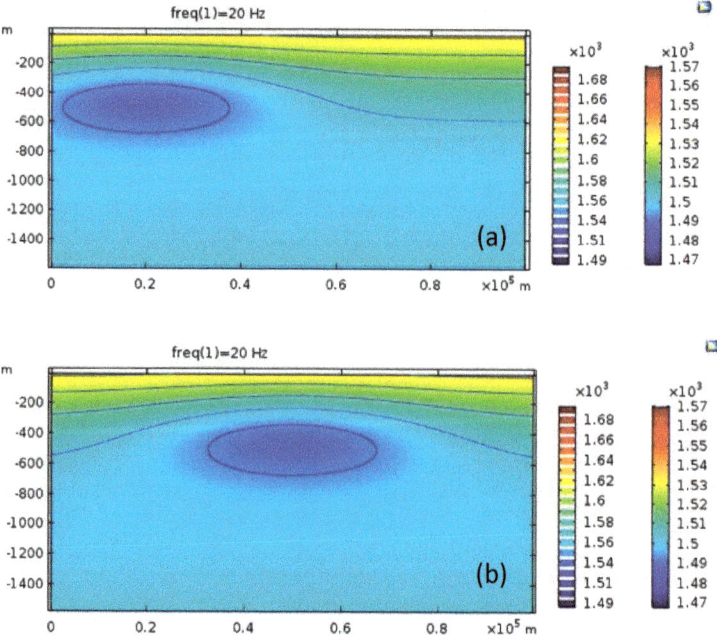

Figure 14. The sound velocity distribution of the cold vortex at different positions in the sound field, (**a**) Rx = 20,000 m, DC = −20 m/s, (**b**) Rx = 50,000 m, DC = −20 m/s.

Now, using the vortex center as the origin, we draw the change in sound velocity gradient at various distances from the vortex center. Variation in sound velocity gradient at different distances are shown in Figure 15a–e. The sound velocity gradient changes from a negative gradient in the upper half to a positive gradient in the lower half, and then gradually changes to a uniform negative gradient, which is the distribution of the sound velocity gradient in the absence of a vortex. As the sound waves are affected by the sound velocity gradient as they propagate, the sound waves bend to the lower sound velocity location, and the height of the convergence zone decreases.

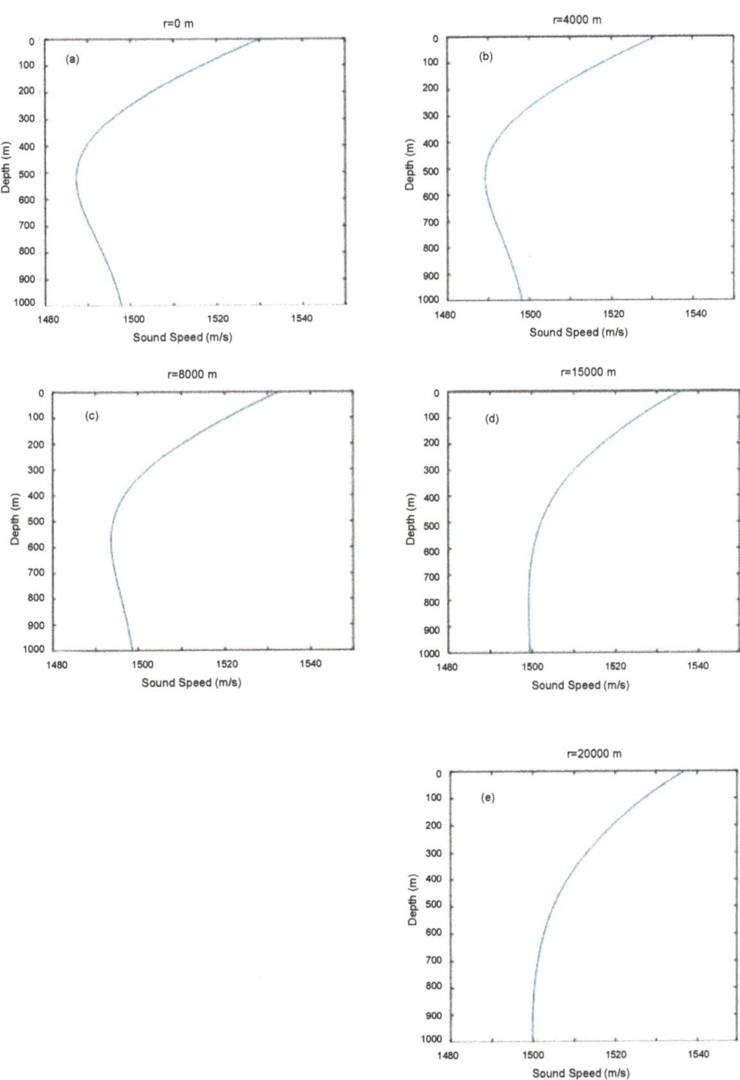

Figure 15. Variation in sound velocity gradient at different distances, (**a**) r = 0 m, (**b**) r = 4000 m, (**c**) r = 8000 m, (**d**) r = 15,000 m, (**e**) r = 20,000 m.

3.1.3. The Influence of Sound Source Frequency on Sound Propagation

In this study, we tested different frequencies to see how they affected sound propagation in seawater. Figure 16 compares the sound pressure level distribution at two frequencies, 10 Hz and 20 Hz, when DC = 20 m/s. The lower sound source frequency has no effect on the appearance of the convergence zone, but the influence of the vortex on the convergence zone becomes smaller, making it difficult to distinguish what changes have occurred. Another reason for this phenomenon is that when the frequency is reduced, the convergence effect of the sound waves appears to be smaller, and the distribution of the sound waves on the entire plane becomes more uniform, making the convergence zone less obvious.

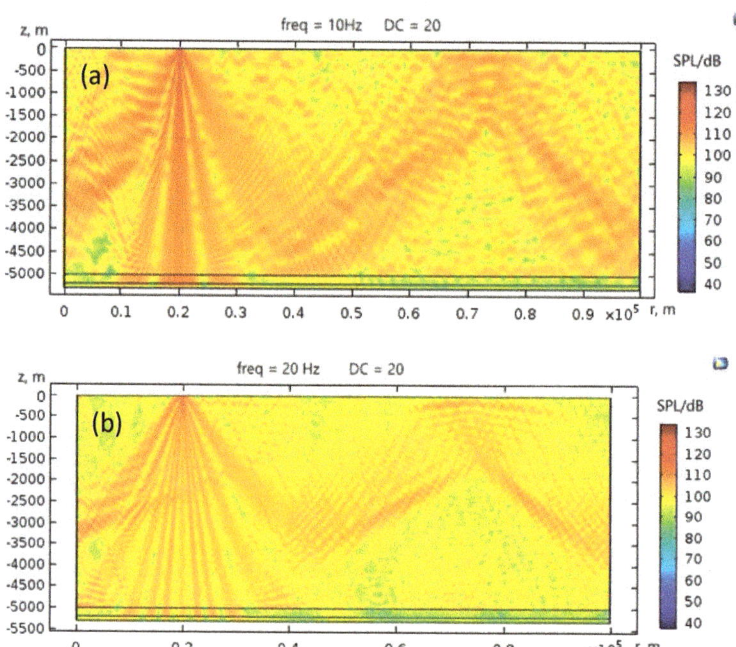

Figure 16. Sound pressure level distribution image at different frequencies when DC = 20 m/s, (**a**) f = 20 Hz, (**b**) f = 10 Hz.

Figure 17 shows related images of sound transmission loss and distance at a depth of 400 m obtained with COMSOL software at 10 Hz and 20 Hz to better explore the effect of sound source frequency on sound propagation when the vortex exists. The change trend of the sound transmission loss image tends to be the same at lower frequencies such as 10 Hz and 20 Hz, but the convergence zone effect is less obvious. Low frequencies, such as 100 Hz and 200 Hz (Figure 18a,b), have no significant effect on the convergence zone; however, as the sound source frequency rises to 1000 Hz, the convergence effect becomes more apparent than at 200 Hz (Figure 18c). The transmission loss is approximately 30 db lower than in other locations. It is possible to conclude that changing the frequency of the sound source will affect sound propagation in the vortex area, and the convergence effect of higher frequency sound waves will become stronger in this area.

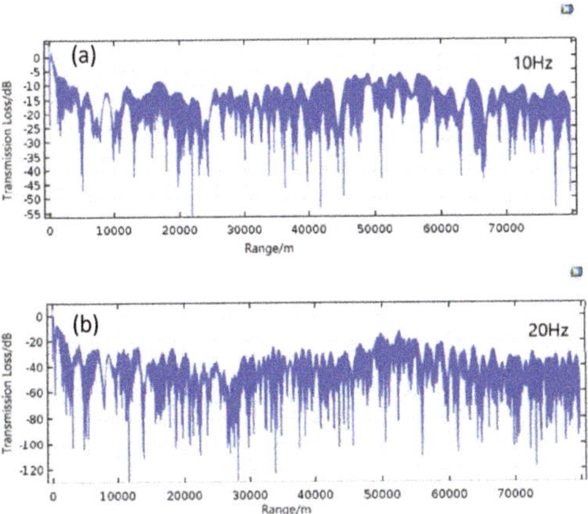

Figure 17. Sound transmission loss at different frequencies, (**a**) f = 10 Hz, (**b**) f = 20 Hz, depth 400 m (COMSOL model).

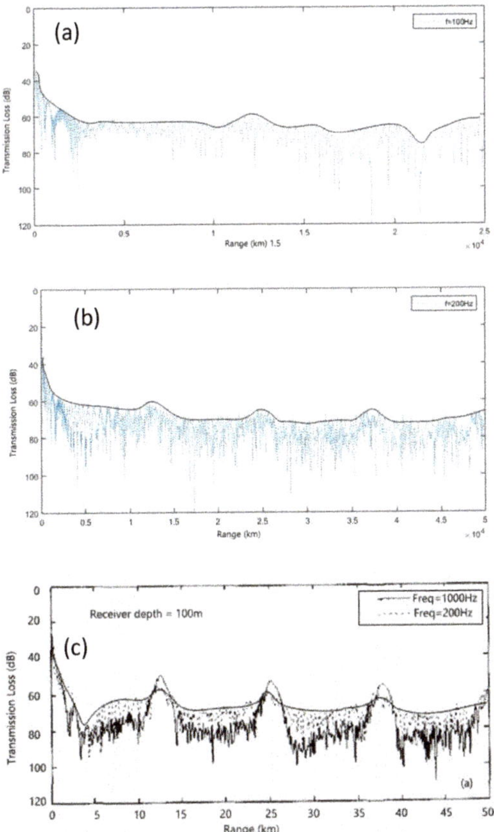

Figure 18. Sound transmission loss at different frequencies, (**a**) f = 100 Hz, 100 m depth (RAM PE Model), (**b**) f = 200 Hz, 100 m depth (RAM PE Model), (**c**) f = 200 Hz, 1000 Hz, depth 100 m (MMPE model).

4. Conclusions and Future Work

Long-term mesoscale vortices exist in the South and East China Seas, and their presence affects the sound propagation characteristics of seawater. The relevant sound propagation characteristics and sound propagation rules were studied and summarized. This has a high application value in future practical work areas such as underwater acoustic equipment research and development and tactical use. The main focus of this study was to use real time data from the AIPOcean 1.0 numerical product to identify the mesoscale vortices using the OW method, then observes the impact of mesoscale vortex on sound velocity distribution, and finally simulates the actual vortex parameters with the Gaussian vortex model in COMSOL. Finally, using this model, the impact of the presence of mesoscale vortices on the sound propagation law was investigated, yielding the following results:

(1) In a deep-sea environment where the sea surface is an absolute soft boundary and the lower seabed is a liquid seabed, the sound waves radiated by sound sources near the sea surface will reconverge at a certain distance and form a sound energy convergence zone. The cold vortex will cause the convergence zone region to move toward the sound source, reducing the width of the convergence zone; a warm vortex will cause the convergence zone region to move away from the sound source, increasing the width of the convergence zone.

(2) If the sound source is in the center of the vortex, the vortex's effect on sound propagation is more obvious, and the depth of the convergence zone is shifted downward. The reason for this is that the mesoscale vortex has the greatest impact on the sound velocity distribution at the vortex center, which has a greater impact on sound propagation. These findings also support the hypothesis that the mesoscale vortex influences sound propagation by influencing the sound velocity distribution.

The above-mentioned results are consistent with previous research using different models, confirming that our COMSOL-based model worked correctly and could be a useful tool for investigating underwater acoustic propagation in vortices. Due to the model's limitations, this study only completed the low-frequency, 2D sound field modelling and simulation. When using the COMSOL software for high-frequency or even 3D simulation, the amount of calculation increases drastically, making it difficult to process. In future research, we can try to study the effect of the vortex on the 3D sound field while solving the computational constraints in order to better fit the scale of the actual vortex.

Author Contributions: Conceptualization, Y.S., S.K. and J.H; methodology, J.H.; software, J.H.; validation, Y.S. and S.K.; formal analysis, J.H. and S.K.; investigation, S.P. and Y.S.; resources, S.P. and Y.S.; data curation, J.H. and S.K.; writing—original draft preparation, J.H. and S.K.; writing—review and editing, Y.S. and S.K.; visualization, J.H.; supervision, S.P.; project administration, S.P.; funding acquisition, YS.P. and Y.S. All authors have read and agreed to the published version of the manuscript.

Funding: This research was funded by National Key Research and Development Program of China, grant number 2016YFC1400100 and The APC was funded by Yang Song.

Acknowledgments: We acknowledge the National Key Laboratory, College of Underwater Acoustic Engineering, Harbin Engineering University, China for supporting in this research. The authors would like to thank two anonymous reviewers for their insightful and constructive comments and suggestions.

Conflicts of Interest: The authors declare no conflict of interest.

References

1. Wunsch, C.; Schmittner, A.; Chiang, J.C.H.; Hemming, S.R. The past and future ocean circulation from a contemporary perspective. *Extrem. Events* **2007**, *173*, 53–74. [CrossRef]
2. Zhang, Z.; Wang, W.; Qiu, B. Oceanic mass transport by mesoscale eddies. *Science* **2014**, *345*, 322–324. [CrossRef] [PubMed]
3. Chen, G.; Hou, Y.; Chu, X. Mesoscale eddies in the South China Sea: Mean properties, spatiotemporal variability, and impact on thermohaline structure. *J. Geophys. Res. Space Phys.* **2011**, *116*. [CrossRef]

4. Faghmous, J.H.; Frenger, I.; Yao, Y.; Warmka, R.; Lindell, A.; Kumar, V. A daily global mesoscale ocean eddy dataset from satellite altimetry. *Sci. Data* **2015**, *2*, 1–16. [CrossRef] [PubMed]

5. Li, J.; Zhang, R.; Liu, C.; Fan, H. Modeling of ocean mesoscale eddy and its application in the underwater acoustic propagation. *Mar. Sci. Bull.* **2012**, *14*, 1–15.

6. Chen, X.; Hong, M.; Zhu, W.; Mao, K.; Ge, J.J.; Bao, S. The analysis of acoustic propagation characteristic affected by mesoscale cold-core vortex based on the UMPE model. *Acoust. Aust.* **2019**, *47*, 33–49. [CrossRef]

7. Li, L.; Nowlin, W.D., Jr.; Jilan, S. Anticyclonic rings from the Kuroshio in the South China Sea. *Deep Sea Res. Part I Oceanogr. Res. Pap.* **1998**, *45*, 1469–1482. [CrossRef]

8. Wu, P.M.; Guo, X.G.; Wu, R.S. Analyses of sound velocity field in southwest sea area off Hengchun of Taiwan. *J. Oceanogr. Taiwan Strait* **2001**, *20*, 286–290. (In Chinese)

9. Available online: https://www.researchgate.net/publication/348706507_Influence_of_Mesoscale_Vortex_on_Underwater_Low-Frequency_Sound_Propagation (accessed on 22 June 2021).

10. Jian, Y.; Zhang, J.; Liu, Q.; Wang, Y. Effect of Mesoscale eddies on underwater sound propagation. *Appl. Acoust.* **2009**, *70*, 432–440. [CrossRef]

11. Okubo, A. Horizontal dispersion of floatable particles in the vicinity of velocity singularities such as convergences. *Deep Sea Res. Oceanogr. Abstr.* **1970**, *17*, 445–454. [CrossRef]

12. Weiss, J. The dynamics of enstrophy transfer in two-dimensional hydrodynamics. *Phys. D Nonlinear Phenom.* **1991**, *48*, 273–294. [CrossRef]

13. McWilliams, J.C. The emergence of isolated coherent vortices in turbulent flow. *J. Fluid Mech.* **1984**, *146*, 21–43. [CrossRef]

14. Morrow, R.; Birol, F.; Griffin, D.; Sudre, J. Divergent pathways of cyclonic and anti-cyclonic ocean eddies. *Geophys. Res. Lett.* **2004**, *31*. [CrossRef]

15. Isern-Fontanet, J.; Font, J.; García-Ladona, E.; Emelianov, M.; Millot, C.; Taupier-Letage, I. Spatial structure of anticyclonic eddies in the Algerian basin (Mediterranean Sea) analyzed using the Okubo-Weiss parameter. *Deep Sea Res. Part II Top. Stud. Oceanogr.* **2004**, *51*, 3009–3028. [CrossRef]

16. Chaigneau, A.; Gizolme, A.; Grados, C. Mesoscale eddies off Peru in altimeter records: Identification algorithms and eddy spatio-temporal patterns. *Prog. Oceanogr.* **2008**, *79*, 106–119. [CrossRef]

17. Sadarjoen, I.A.; Post, F.H. Detection, quantification, and tracking of vortices using streamline geometry. *Comput. Graph.* **2000**, *24*, 333–341. [CrossRef]

18. Doglioli, A.M.; Blanke, B.; Speich, S.; Lapeyre, G. Tracking coherent structures in a regional ocean model with wavelet analysis: Application to Cape Basin eddies. *J. Geophys. Res. Space Phys.* **2007**, *112*. [CrossRef]

19. Nencioli, F.; Dong, C.; Dickey, T.; Washburn, L.; McWilliams, J.C. A vector geometry-based eddy detection algorithm and its application to a high-resolution numerical model product and high-frequency radar surface velocities in the Southern California Bight. *J. Atmos. Ocean. Technol.* **2010**, *27*, 564–579. [CrossRef]

20. Vastano, A.C.; Owens, G.E. On the acoustic characteristics of a gulf stream cyclonic ring. *J. Phys. Oceanogr.* **1973**, *3*, 470–478. [CrossRef]

21. Weinberg, N.L.; Zabalgogeazcoa, X. Coherent ray propagation through a Gulf Stream ring. *J. Acoust. Soc. Am.* **1977**, *62*, 888–894. [CrossRef]

22. Baer, R.N. Calculations of sound propagation through an eddy. *J. Acoust. Soc. Am.* **1980**, *67*, 1180–1185. [CrossRef]

23. Chen, C.; Yang, K.-D.; Ma, Y.-L.; Duan, R. Comparison of surface duct energy leakage with bottom-bounce energy of close range propagation. *Chin. Phys. Lett.* **2016**, *33*, 104302. [CrossRef]

24. Heaney, K.D.; Campbell, R.L. Three-dimensional parabolic equation modeling of mesoscale eddy deflection. *J. Acoust. Soc. Am.* **2016**, *139*, 918–926. [CrossRef] [PubMed]

25. COMSOL Multiphysics® v.5.6. Available online: www.comsol.com (accessed on 20 August 2019).

26. Liu, Z.; Wu, X.; Xu, J.; Li, H.; Lu, S.; Sun, C.; Cao, M. China Argo project: Progress in China Argo ocean observations and data applications. *Acta Oceanol. Sin.* **2017**, *36*, 1–11. [CrossRef]

27. Yan, C.; Zhu, J.; Xie, J. An ocean data assimilation system in the Indian Ocean and west Pacific Ocean. *Adv. Atmos. Sci.* **2015**, *32*, 1460–1472. [CrossRef]

28. Cui, F.J. Mesoscale Eddies in the Sound China Sea Identification and Statistical Characteristics Analysis. Master's Thesis, Ocean University of China, Qingdao, China, 2015; pp. 20–21.

29. *Version R2017a, Copyright 1984–2015*; The MathWorks, Inc.: Natick, MA, USA, 2017; Available online: https://jp.mathworks.com/company/newsroom/mathworks-announces-release-2017a-of-the-matlab-and-simulink-pro.html (accessed on 20 August 2019).

30. MacKenzie, K.V. Nine-term equation for sound speed in the oceans. *J. Acoust. Soc. Am.* **1981**, *70*, 807–812. [CrossRef]

31. Henrick, R.F.; Siegmann, W.L.; Jacobson, M.J. General analysis of ocean eddy effects for sound transmission applications. *J. Acoust. Soc. Am.* **1977**, *62*, 860–870. [CrossRef]

32. Munk, W.H. Sound channel in an exponentially stratified ocean, with application to SOFAR. *J. Acoust. Soc. Am.* **1974**, *55*, 220. [CrossRef]

Journal of
Marine Science and Engineering

Article

The Approach for Studying Variability of Sea Wave Spectra in a Wide Range of Wavelengths from High-Resolution Satellite Optical Imagery

Valery Bondur [1] and Alexander Murynin [1,2,*]

[1] AEROCOSMOS Research Institute for Aerospace Monitoring, 105064 Moscow, Russia; vgbondur@aerocosmos.info

[2] Federal Research Center "Computer Science and Control", Russian Academy of Sciences, 119333 Moscow, Russia

* Correspondence: amurynin@bk.ru

Abstract: The development and validation of a method for remote measurement of the spectra of sea waves, which significantly expands the capabilities to study surface waves in large water areas in a wide range of wavelengths, is described. The applied approach is based on the use of retrieving operators, which are constructed by the method of numerical simulation, taking into account the nonlinear nature of brightness field modulation by the slopes of the sea surface. Retrieving operators have a set of parameters that are adapted to the real conditions of aerospace imaging. To assess the adequacy of the retrieving of wave spectra recorded from satellite images, they are compared with the spectra obtained by ground-based means under controlled conditions. The studies have shown the adequacy of remote measurement of the spectra of slopes and elevations of sea waves with wavelengths in the range 0.1–1 m. The possibility of using the developed method for studying the variability of sea waves in the coastal zone under conditions of limited fetch, including in the presence of anthropogenic disturbances, is shown.

Keywords: sea surface; wave spectra; satellite imagery processing; aerospace monitoring; sea waves; wave spectra; retrieving operator

Citation: Bondur, V.; Murynin, A. The Approach for Studying Variability of Sea Wave Spectra in a Wide Range of Wavelengths from High-Resolution Satellite Optical Imagery. *J. Mar. Sci. Eng.* **2021**, *9*, 823. https://doi.org/10.3390/jmse9080823

Academic Editor: Grigory Ivanovich Dolgikh

Received: 1 July 2021
Accepted: 26 July 2021
Published: 30 July 2021

Publisher's Note: MDPI stays neutral with regard to jurisdictional claims in published maps and institutional affiliations.

1. Introduction

The spatiotemporal structure of surface waves changes randomly under the influence of many natural factors (internal waves [1], current fields [2,3], and anthropogenic impacts [4,5]) on the water environment. For a compact presentation of information about the structure of surface waves and its changes under the influence of various processes, it is advisable to use spatial and frequency spectra of sea waves, characterizing the distribution of wave energy over wave components of different lengths and periods [6,7]. An effective way to obtain information on the spatial spectra of sea waves in vast areas of the seas and oceans is the spatial spectral analysis of various aerospace images [4,7,8].

Satellite optical images of high spatial resolution significantly expand the capabilities to study the spectra of surface waves in large water areas in a wide range of spatial frequencies.

For a reliable assessment of the spectra of surface waves from satellite optical images, special methods should be used to reconstruct the characteristics of sea waves from the characteristics of the brightness fields recorded by space sensors [9–14]. In this case, retrieving operators are used, which can be built on the basis of numerical modeling of various imaging conditions and characteristics of remote sensing equipment, including taking into account the nonlinear modulation of brightness fields by surface slopes [9,13,14].

To assess the adequacy of the retrieval of wave spectra recorded from space images, it is necessary to compare them with the spectra obtained by standard wave recorders under controlled conditions.

109

This paper describes a method for studying the spectra of sea waves in a wide range of wavelengths, retrieved from the spectra of optical satellite images of high spatial resolution (tens of cm) using specially developed retrieving operators. They are compared with the spectra obtained by the array of string wave recorders by processing stereo images taken from an offshore platform, as well as with the results of measurements made using drifting wave buoys.

2. Research Methods

Let us consider an approach to studying the spectra of elevations and slopes of the sea surface.

Rough sea surface is a random field of elevations (wave applicates):

$$z = \zeta(x, y, t) \tag{1}$$

where (x, y, z) is the rectangular Cartesian coordinate system, in which the plane (x, y) coincides with the level of a calm (no waves) water surface, and t is the time.

Fixing (1) at $t = t_0$, we obtain a two-dimensional random function:

$$z = \xi(x, y) = \zeta(x, y, t)|_{t=t_0} \tag{2}$$

when studying surface waves using optical images, two-dimensional fields of signal are recorded corresponding to a fixed moment in time. Therefore, in this work, we will consider the representation of waves in the form (2).

Optical images of the sea surface are formed as a result of reflection and refraction of light according to the laws of geometric optics, as will be discussed below. Therefore, when analyzing the structure of the sea surface, along with the field of elevations $\xi(x, y)$, it is convenient to characterize the fields of slopes (or gradients) along the axes [9]:

$$\xi_\alpha(x, y) = \frac{\partial \xi(x, y)}{\partial \alpha}, \ \alpha = x, y \tag{3}$$

The field of slopes (gradients) of the sea surface in an arbitrary direction φ, taking into account (3), can be expressed by the following equation:

$$\beta_\varphi(x, y) = \cos \varphi \xi_x(x, y) + \sin \varphi \xi_y(x, y) \tag{4}$$

Taking into account the properties of the Fourier transform, it is possible to relate the power spectrum $\Phi_\varphi(\mathbf{k})$ of the slope field (4) with the elevation field spectrum $\Psi(\mathbf{k})$

$$\Phi_\varphi(\mathbf{k}) = (\cos \varphi k_x + \sin \varphi k_y)^2 \Psi(\mathbf{k}) \tag{5}$$

This relationship (5) can be used to retrieve the elevation spectrum $\Psi(\mathbf{k})$ if the slope spectrum is measured experimentally, for example, from data extracted from an aerospace image:

$$\Psi(\mathbf{k}) = \frac{\Phi_\varphi(\mathbf{k})}{(\cos \varphi k_x + \sin \varphi k_y)^2} \tag{6}$$

However, Equation (6) restores the spectrum of elevations not for all values $\mathbf{k}$, since for:

$$k_x / k_y = -tg\varphi \tag{7}$$

the denominator goes to zero. Therefore, in the vicinity of the line (7) in the spectrum of slopes, there is a lack of information on the spectrum of elevations of the sea surface. For complete retrieval of two-dimensional wave spectra $\Psi(\mathbf{k})$ from optical images, a multi-position method is used [13,14].

The method is based on the use of M (M $\geq$ 2) fragments separated within a large image, or using several images that differ in shooting conditions. From M fragments, slope spectra corresponding to different directions of the gradient can be retrieved.

With the spatial stationarity (ergodicity) of waves, the selected fragments correspond to the same spectrum of sea surface elevations $\Psi(\mathbf{k})$, which makes it possible to create a system of M equations:

$$\Phi_m(\mathbf{k}) = (\cos\varphi_m k_x + \sin\varphi_m k_y)^2 \Psi(\mathbf{k}) \quad m = 1,\ldots,M \tag{8}$$

From this system, one can determine the desired two-dimensional wave spectrum [13,14]:

$$\Psi(\mathbf{k}) = \frac{\sum\limits_{m=1}^{M} \Phi_m(\mathbf{k})}{\sum\limits_{m=1}^{M} (\cos\varphi_m k_x + \sin\varphi_m k_y)^2} \tag{9}$$

Thus, to retrieve a two-dimensional spectrum of sea surface elevations, it is necessary to have slope spectra for two or more directions.

The required set of slope spectra can be obtained in various ways:

- Stereo imaging, when images from two (or more) cameras with different viewing directions of the surface are simultaneously recorded;
- Processing of several fragments of images obtained during aerial photography with a wide shooting angle, on the assumption of spatial stationarity (some kind of ergodicity) of waves within the surface area captured by the camera's field of view;
- Processing of multi-temporal surveys of the same area of the surface from different angles using the assumption of the stationarity of waves.

When processing a one-shot satellite image, none of the listed sets of conditions are realized. Therefore, the retrieval of the two-dimensional spectrum of sea surface elevations is possible only outside the angular zone in the vicinity of the straight line defined by the condition (7).

In order to obtain the spectra of the slopes of the sea surface, retrieving operators $R(\mathbf{k})$ should be used, which make it possible to switch from the spectrum $S(\mathbf{k})$ of a fragment of the optical image to the spectrum of slopes $\Phi(\mathbf{k})$:

$$\Phi(\mathbf{k}) = R(\mathbf{k})S(\mathbf{k}) \tag{10}$$

In the early works devoted to this problem, a simple representation of the retrieving operator was based on the concept of a linear relationship between the surface slopes and the brightness recorded in the pixels of the images [15–17]. The concept of linear modulation of the brightness field by the slopes of the sea surface makes it possible to obtain a retrieving operator that does not depend on the wave vector: $R(\mathbf{k}) = const.$

In reality, the modulation of brightness by slopes can be significantly nonlinear, especially in the presence of sun glare.

From a physical point of view, the most important point is the distortion of the power exponent (slope) of the spectrum. Numerical estimates of such distortions were obtained in [14,18]. The adequacy of the proposed method for retrieving wave spectra from the spectra of optical images, which takes into account the nonlinear modulation of the recorded brightness field by the slopes of the rough sea surface [13,14], was first experimentally confirmed by experiments on an oceanographic platform near the village of Katsiveli (Black Sea) [19] and confirmed in subsequent works as applied to satellite images of high spatial resolution [9–11]. The adaptation of the method to the conditions of analysis of vast ocean areas from space images was carried out using the results of sea truth buoy measurements [20].

When constructing operators that retrieve the spectra of the sea surface slopes, it is necessary to take into account all the known physical mechanisms of non-linear distortions of the sea surface slope spectra [13,18,21–24]. Since it is difficult to use the relation between slope fields, sea surface brightness fields and their spectra for analytic solution of the problem, it is reasonable to conduct direct numerical modeling of the sea surface images and the calculation of the resulting distortions of the wave spectra.

A diagram illustrating the interaction of various algorithms for numerical modeling and image processing is shown in Figure 1.

To obtain the retrieving operators by the numerical method, the following sequence of operations is performed.

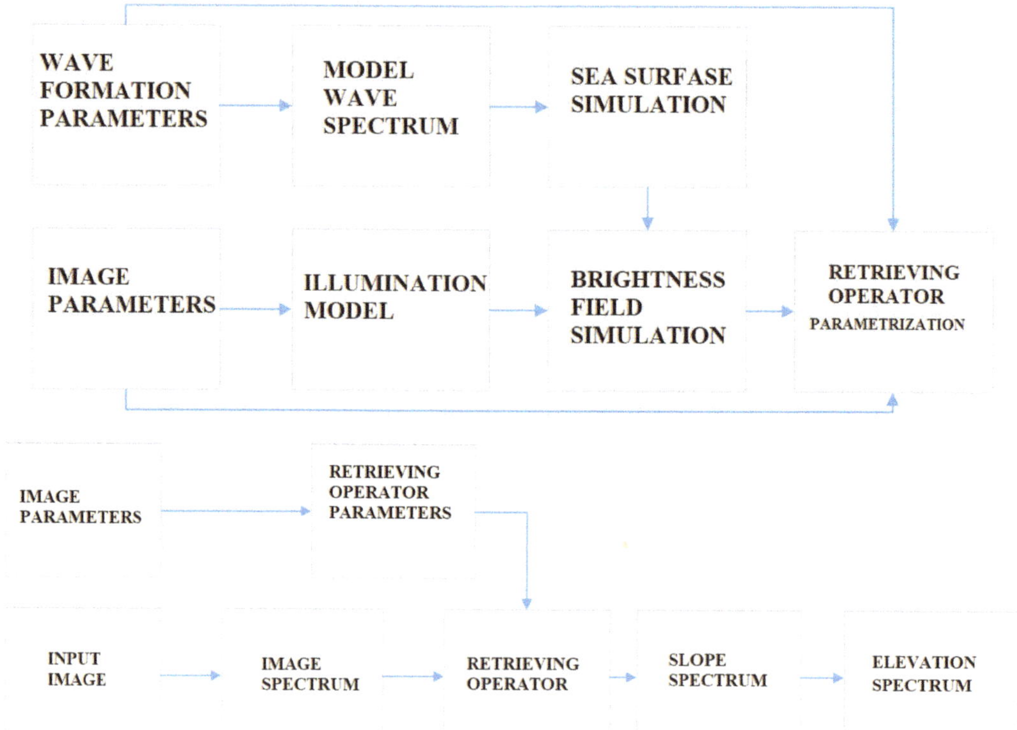

Figure 1. Diagram of interacting algorithms used to construct the retrieving operator by numerical simulation and processing satellite images using this operator.

Two-dimensional random fields of sea surface slopes $\zeta_\alpha(x,y)$, $\alpha = x,y$ are synthesized for a given spatial spectrum of elevations $G(\mathbf{k})$ and a phase spectrum given by a random function with a uniform distribution of phases [9,13,14].

For the given illumination and viewing conditions, the brightness field, $L(x,y)$ is calculated using the synthesized slope field and taking into account various physical processes involved in its formation. When modeling optical images, a model is used that describes the reflection and refraction of light on the water surface according to the laws of geometric optics, as shown in Figure 2a. Figure 2a shows vectors of unit length that determine the reflection and refraction of light on an element of the sea surface according to the laws of geometric optics:

- the vector $\mathbf{v}$ defines the normal to the sea surface element and depends on the gradient of the elevation field;
- vectors $\mathbf{r}_1$ and $\mathbf{r}_2$, indicate the directions in the upper and lower hemispheres, whence comes the light that enters the receiver after reflection and refraction, respectively;
- the vector $\mathbf{r}_\odot$ determines the direction to the Sun;
- vector $\mathbf{r}_s$ defines the direction to the sensor.

To take into account the nonlinearity of the modulation of the brightness field by the surface slopes due to the presence of sun glare, light scattering by small-scale surface roughness within the area, the size of which is determined by the spatial resolution of the sensor registering the image, is of significant importance. The higher the spatial resolution a satellite image has, the smaller the site size is. Accordingly, the dispersion of roughness is less and the effect of sun glare is greater.

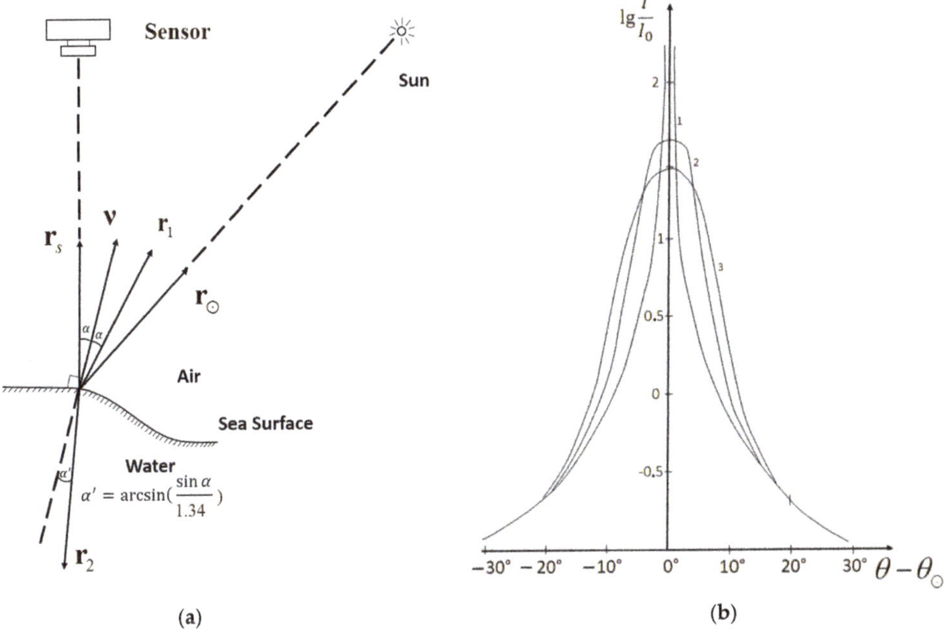

Figure 2. (a) Scheme of reflection and refraction of light at the air-water interface. The vectors of unit length indicate the directions in the upper and lower hemispheres from which light comes, falling into the receiver after reflection and refraction; (b) Angular distribution of the effective intensity of light in the vertical plane of the Sun, reflected by an element of the sea surface, for different sizes of this element (explained in the text).

Figure 2b shows examples of the angular dependence of the effective light intensity in the plane of the solar vertical, corresponding to different image pixel sizes Δ and rms roughness σ_Δ: 1—$\Delta \to 0$, $\sigma_\Delta = 0$; 2—$\Delta = 0.1m$, $\sigma_\Delta = 0.07rad$; 3—$\Delta = 1m$, $\sigma_\Delta = 0.09rad$ [18]. Light intensity is normalized using a multiplier used to approximate the sky light intensity distribution on a cloudless day [18,25].

The brightness field, $L(x,y)$ is converted into a model image signal, $L_m(x,y)$, taking into account the specified characteristics of the recording equipment.

The retrieving operator R is defined by comparing the spectrum of the model optical image, $S_M(\mathbf{k})$ to the spectrum of slopes $\Phi_G(\mathbf{k})$ corresponding to the direction φ_c determined by the specified illumination and viewing conditions. In particular, the retrieving operator can be represented in the form of a spatial-frequency filter [9,13,14].

$$R(\mathbf{k}) = \frac{(\cos \varphi_c k_x + \sin \varphi_c k_y)^2 G(\mathbf{k})}{S_M(\mathbf{k})} \tag{11}$$

With this definition of the retrieving operator, the retrieval of the sea surface slope spectrum $\Phi(\mathbf{k})$ is reduced to multiplying the spectrum of the image fragment, $S(\mathbf{k})$ by the spatial frequency filter, $R(\mathbf{k})$ according to Equation (10).

When numerically simulating images of the sea surface, parameters must be set that determine the conditions for the formation of these images, and therefore, to one degree or another, affect the type of the retrieving operator $R(\mathbf{k})$. When processing real images of the sea surface, these parameters are taken directly from the conditions for obtaining an optical image.

When analyzing large amounts of information to determine the retrieving operator, $R(\mathbf{k})$, it is advisable to perform preliminary numerical modeling of sea surface images for various combinations of parameters. For each specific combinations of parameters retrieving operator $R(\mathbf{k})$ is approximated by an analytical formula $R(\mathbf{k}) = R(\mathbf{k}, \mathbf{a})$ associated with a certain set of parameters $\mathbf{a}$.

To parameterize the retrieving operator constructed by the numerical simulation, the power function of the wave number with parameters depending on the wave azimuth can be used as follows [11]:

$$R(\mathbf{k}, \mathbf{a}) = R_{lo}(\mathbf{k}) \left(\cos{(\phi - \phi_c)}^{a_3} k^{-(a_1 + a_2 \cos(\phi - \phi_c))} \right) \tag{12}$$

where the parameters a_1 a_2 a_3 characterize the region of high spatial frequencies, where a power-law decay of the spatial wave spectrum is observed. For typical conditions of high-resolution satellite imagery at different positions of the Sun, the values of these parameters were obtained by numerical simulation of physical processes of formation of optical images of waves [9,18];

$R_{lo}(\mathbf{k})$ is the retrieving operator in the low-frequency region, the construction technique of which is described in [11]. In the same work [11], the results were obtained on preliminary parameterization of $R_{lo}(\mathbf{k})$.

- Thus, the practical use of the retrieving operator is as follows. From the conditions for obtaining images, a multi-dimensional vector of conditions is constructed, the components of which serve to determine the set of parameters $\mathbf{a}$ from the results of numerical simulation carried out in advance. The spatial-frequency filter constructed from the set of parameters $\mathbf{a}$ is then applied to the spatial spectra of the selected image fragments to calculate the corresponding spectra of the sea surface slopes. The obtained slope spectra are used to retrieve the elevation spectra:
- by Equation (9) in the presence of several spectra of slopes corresponding to different directions;
- or by Equation (6), taking into account the restrictions imposed by Equation (7), if there is no possibility of obtaining slope spectra for several different directions.

3. Results

To validate the developed method, an experimental check of the adequacy of its work should be carried out. The main focus of attention should be the validation of the method with the variability of waves associated with the presence of natural and anthropogenic anomalies.

To validate the retrieving operators constructed by the method of numerical simulation, we will use an approach based on the comparison of satellite and in situ measurements obtained in complex experiments [10]. Complex experiments are carried out under controlled conditions and include a space survey of test areas of the sea surface and synchronous sea truth measurements using wave recorders, as well as stereophotogrammetric survey of the sea surface at low altitude. Spatial spectra of sea surface elevations, retrieved from satellite images under fixed conditions, are compared with the spectra of elevations obtained from in situ measurements with string wave recorders using a special technique that takes into account the different nature of remote and in situ measurements [10].

To solve the set tasks, the results of complex experiments were used, which were carried out near the stationary oceanographic platform of the Black Sea hydrophysical sea truth test site (Katsiveli village, Crimea), installed at a distance of about 600 m from the coast.

Complex experiments of two types were carried out:

The main complex experiment (experiment #1) consisted of a space survey and synchronous in situ measurements of wave spectra from the oceanographic platform.

An additional complex experiment (experiment #2) consisted of conducting the research in the short-wave region of wave spectra ($\Lambda = 0.04$–1.0 m) using string wave recorders and stereo imaging from the deck of an oceanographic platform under wave formation conditions close to those observed in the main complex experiments.

To measure the spatio-temporal wave spectra, we used in situ data obtained using a wave measuring complex based on an array of string wave recorders, which are a set of six resistive wave recorders measuring sea surface elevations at points located in the center and vertices of a pentagon. The distance from the central string to each of the outer strings is 25 cm. The main technical characteristics of this complex: the maximum height of the measured waves—up to 4 m; resolution—not worse than 3 mm; number of measuring channels—6; and channel sampling frequency—10, 20, 50, and 100 Hz.

3.1. The Results of Studies of the Variability of Waves in the Complex Experiment #1

The purpose of the main complex experiments is to validate the method of remote measurement of sea wave spectra from satellite images using in situ measurements on a stationary platform.

The image was recorded by high spatial resolution equipment from the GEOEYE satellite on 24 September 2015 at 11:52 local time.

The details of processing the satellite image obtained in the main complex experiment are described in [10,11].

Satellite images were processed using the following sequence of computational operations:

- calculation of the spatial spectrum of the selected image fragment $S(\mathbf{k})$;
- retrieval of the spectrum of slopes $\Phi(\mathbf{k})$ using the retrieving operator $R(\mathbf{k})$, built for a set of conditions determined by the conditions for obtaining a space image;
- calculation of the spatial spectrum of elevations $\chi(k)$ from the retrieved spectrum of slopes in a given range of wave azimuths (± 60 angular degrees relative to the azimuth of the Sun) according to formula (6);
- recalculation into the frequency spectrum $\psi(\omega)$ taking into account the dispersion ratio of gravitational waves by the formula:

$$\psi(\omega) = \chi(k(\omega))\frac{dk(\omega)}{d\omega} \tag{13}$$

An example of the comparison of spectra is shown in Figure 3. Figure 3a shows a fragment of a high-resolution satellite image, registered from the GEOEYE satellite. The oceanographic platform from which in-situ measurements were carried out is marked with the yellow circle. The red rectangle marks a sea surface area used for the wave spectra retrieval.

Figure 3b shows a sequence of spectra based on the data of a string wave-recorder, which demonstrates the variability of waves in the experimental area for three hours before satellite imagery. Each given spectrum was obtained with a half-hour averaging. The time evolution of the frequency spectra for three hours preceding the satellite's pass shows the development of wind waves—an increase in their energy and a decrease in the frequency of the spectral peak. At the same time, the system of swell waves remained dominant. Swell waves came from an azimuth of 130°, wind waves propagated along the direction of the east wind.

Figure 3c shows the spectrum of slopes, retrieved by the method described above from high-resolution satellite imagery data, and also shows the angular sector in which the spectrum of sea surface elevations was reconstructed for subsequent comparison with in situ measurements.

Figure 3d combines the frequency spectra of elevations obtained during processing of the satellite image, taking into account the dispersion relation and the calculation by Equation (13), as well as the spectra obtained from the data of the array of string wave recorders. The measure of the discrepancy between the estimates of the spectra was $\Delta \approx 0.07$ [10].

The red dashed lines show the limits within which the spectrum of developing wind waves should lie according to the Toba approximation of the wave frequency spectrum [26]:

$$S(\omega) = \alpha g u_* \omega^{-4} \tag{14}$$

where u_* is the dynamic speed in air, and α is the coefficient for two lines, equal to 0.06 and 0.11, respectively.

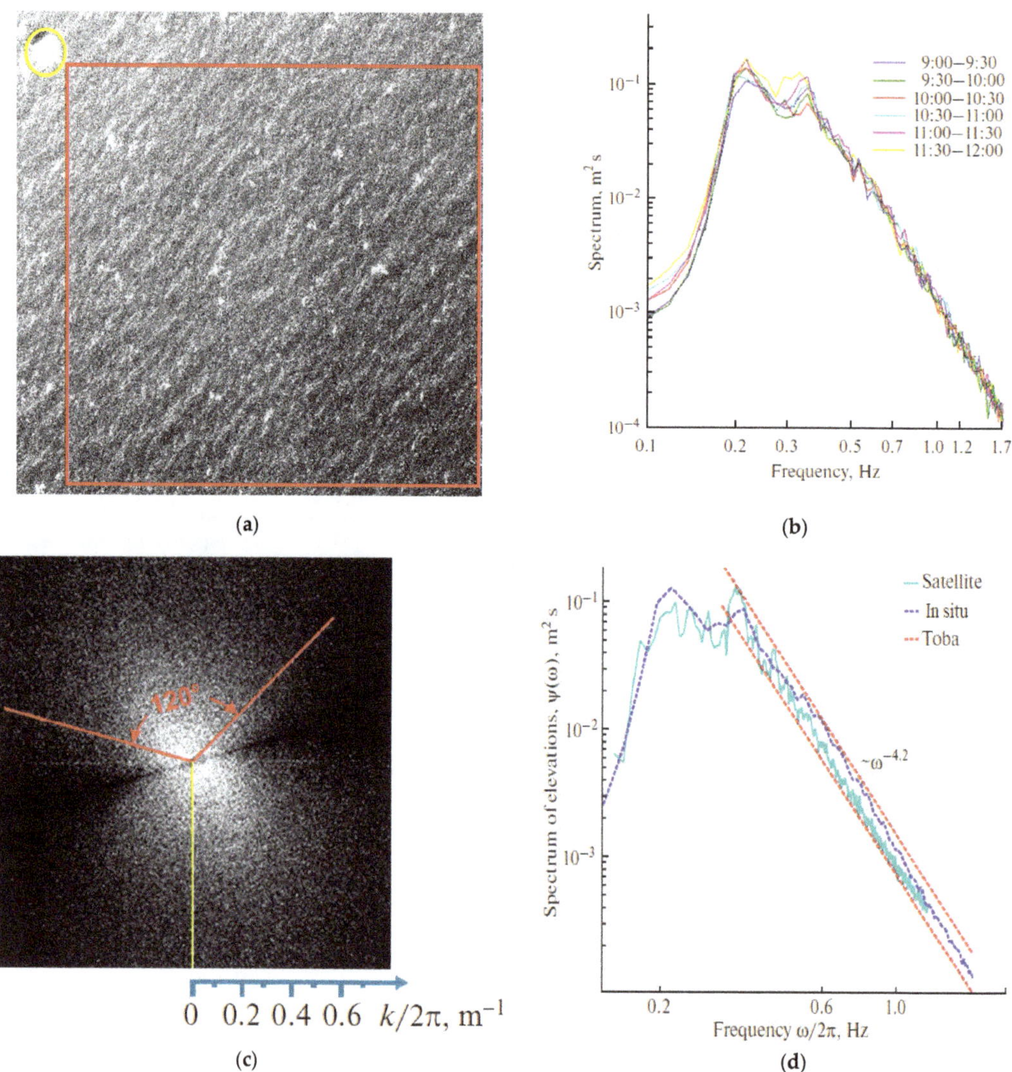

Figure 3. Comparison of sea wave spectra: (**a**) a tile of a satellite image, where a red rectangle marks a sea surface area with a linear size of 1 km used for the wave spectra retrieval, and the yellow circle marks the oceanographic platform from which in-situ measurements were carried out (**b**) a sequence of spectra based on wave-recorder data for three hours with a half-hour averaging; (**c**) a slope spectrum with an outlined corner sector to retrieve a two-dimensional elevation spectrum; (**d**) comparison of spectra converted to frequency coordinates.

Figure 4 shows an example of comparing the frequency spectrum of elevations obtained by a half-hour averaging of the data of the array of string wave recorders with the spectra of elevations reconstructed by processing satellite image fragments of various sizes. The given plots show that when retrieving wave spectra from an image of a sea surface area with a linear size of 0.5 km (Figure 4a), it is not possible to retrieve the spectrum of elevations in the low-frequency region due to the lack of data.

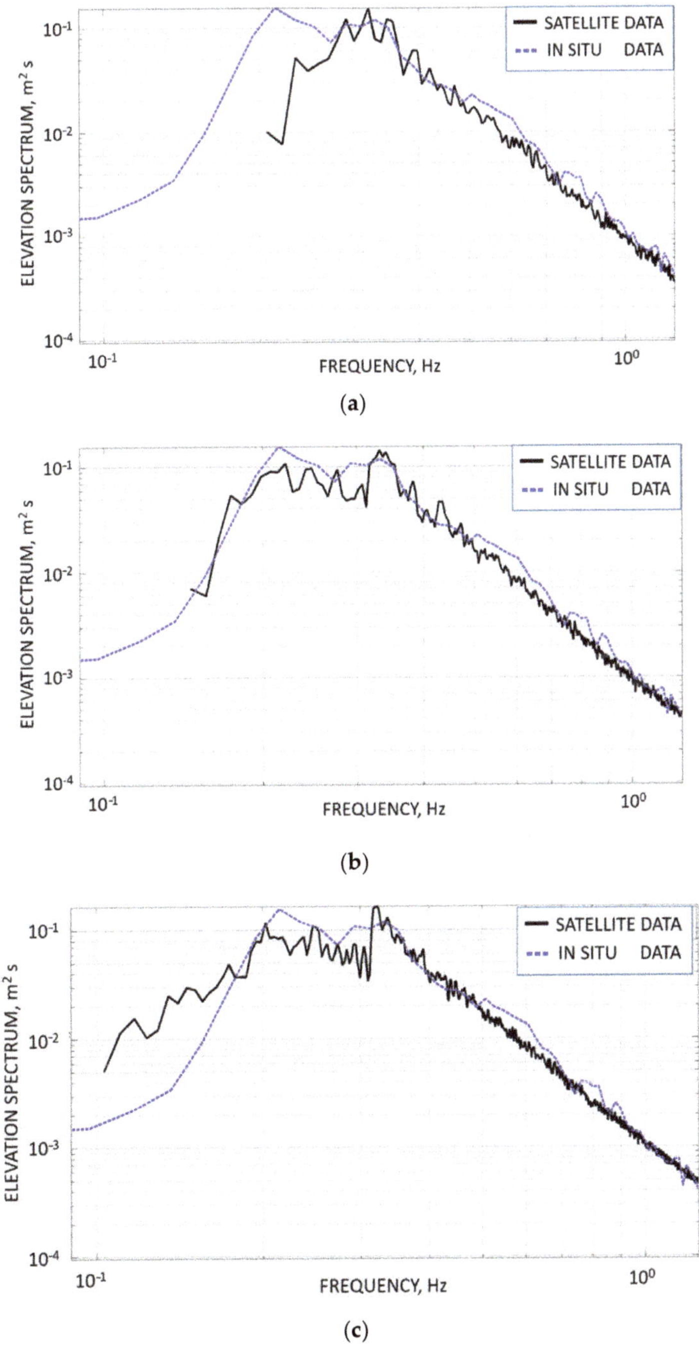

Figure 4. Comparison of the frequency spectra of sea wave elevations measured using the data of the string wave recorder with averaging over 30 min (immediately before satellite imagery) with the spectra retrieved from two-dimensional spectra reconstructed from satellite images at various linear dimensions of the sea surface area: (**a**) 0.5 km, (**b**) 1 km (**c**) 2 km.

When the spectrum of elevations is reconstructed from a region of the surface with a linear size of 2 km, the retrieved spectrum turns out to be broader in the low frequency region than the spectrum obtained from the data of the array of string wave recorders. In this case, the processed area of the sea surface covers a zone remote from the coast, where the waves are more developed than in the zone where the string wave recorders are located. The best coincidence of the shape of the spectra is observed at a fragment linear size of 1 km.

3.2. The Results of Studies of the Variability of Waves in Complex Experiment #2

The purpose of the additional complex experiments is to study the applicability of the developed method for remote measurement of the spectra of sea waves in the region of short waves.

In additional experiments, along with in situ measurements carried out by string wave recorders, a stereo survey was carried out on the platform from a low altitude with a high spatial resolution.

Stereophotogrammetric processing of synchronously recorded optical images of the sea surface from two different points of view (stereopairs) makes it possible to directly measure the realizations of two-dimensional fields of sea surface heights, and from them to calculate the spatial spectra of elevations. The stereophotogrammetric method makes it possible to measure the shape of a rough sea surface and represent it as a function of Equation (2).

Stereo photography of the sea surface was carried out from the working platform, which was located at an altitude of 16 m above sea level with a baseline of 10.2 m using stereo cameras with a focal length of 99 mm [10].

Measurements of coordinates x, y, z of sea surface elements by stereo-pairs. From the elevation field measurements, the spectra of the sea surface elevations were calculated $\Psi_\varphi(k_\varphi)$ using the interpolation procedures and the discrete Fourier transform.

Simultaneously with the stereophotography on the platform, measurements of the frequency spectra of the sea surface elevations were carried out with the help of a spectral wave recorder, which provided the registration of the spectrum in the frequency range, and the recording time of one spectrum in this range was about 20 min.

Experimental conditions: the zenith angle of the Sun was 30° (the images were recorded in a cloudless sky), and the wind speed was 5 m/s. During the experiments, the wind was blowing off the coast, so the fetch of the wind waves did not exceed the distance from the platform to the coast, which is ~600 m.

Images of the sea surface obtained during stereo photography were also used to retrieve the spectrum of the sea surface elevation using a developed retrieving operator. At the same time, fragments of two optical images of a stereopair were analyzed, in which the same area of the sea surface is presented, which was processed stereophotogrammetrically. This makes it possible, when comparing the estimates of the spectrum of sea surface elevations obtained by different techniques, to study the accuracy of these techniques.

Synchronous wave-recorder measurements were carried out to analyze the selected stereopairs. All the optical images under study were recorded under a cloudless sky.

Comparison of the estimates of the wave spectra obtained by different techniques was carried out in order to check the adequacy of the detection and measurement of variations in the wave spectra caused by the variability of the waves, associated both with natural processes and with anthropogenic influences.

The correlation between variations in the estimates of wave spectra obtained by different techniques was studied in [10]. We investigated the correlations of the main characteristics of the wave spectra obtained by three methods in two wavelength intervals: from 0.1 to 1 m and from 0.04 m to 0.4 m.

For the main characteristics to compare the wave spectra obtained by different techniques, we chose the rms elevation and the exponent of the power-law approximation of the wave spectrum in the indicated intervals.

The analysis of the obtained results showed that the correlation of estimates made by different techniques is quite high, and averages 0.8–0.9.

To check the adequacy of the developed techniques when working with the regions of the sea surface, where the wave spectrum is subject to anthropogenic impacts, an artificial anomaly of the spatial wave structure was created during the experiment using vegetable oil spilled onto the water.

The wave spectra obtained by different techniques in the area of the artificial anomaly were compared with similar spectra obtained earlier for the background sea surface.

The study of the background sea surface was carried out for two hours.

After that, an artificial anomaly was created by spilling about a liter of vegetable oil from the oceanographic platform. The creation of artificial films of surfactants has proven itself well as a method for studying anthropogenic pollution of the marine environment [22,27–29].

As a result of spilling the vegetable oil, an artificial slick was formed on the water. The area of this slick exceeded the water area under the platform. Special measurements of the thickness of the oil film on the water were not carried out. According to the estimates based on the volume of spilled vegetable oil and the approximate area of the slick, the film thickness was about ~10^{-6} m.

The artificial anomaly created with the spilled vegetable oil was investigated for about half an hour using three techniques.

Wave spectra measured by different methods were recalculated into the spectrum of elevations according to Equation (13).

According to our data, there is a noticeable damping of waves with lengths of 0.3–0.6 m, as well as shorter waves. Waves longer than one meter are not considered. The spectra in Figure 5 have single measurements with wavelengths over 0.5 m, for which the spectral power has decreased by about 2 times.

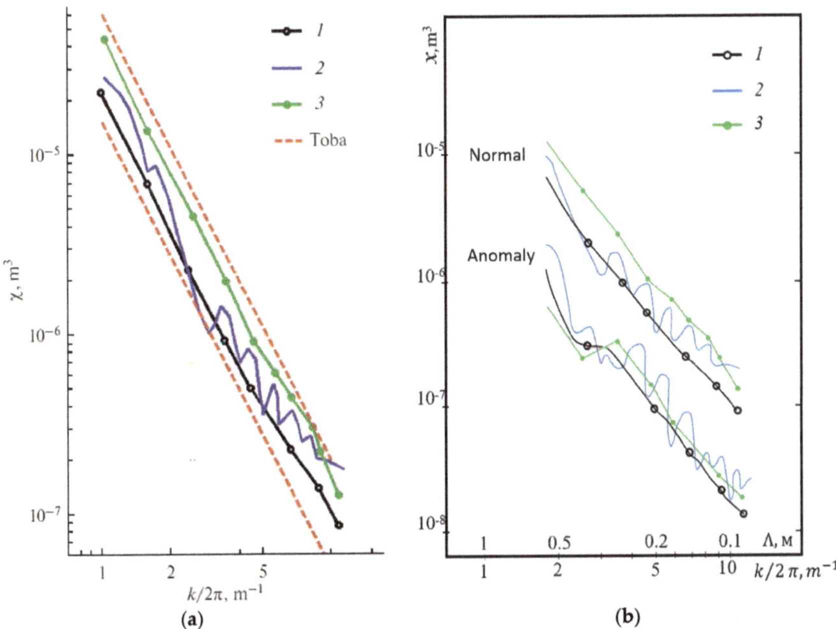

Figure 5. Spectra of sea surface elevations obtained by different methods [19]: (**a**) Average spectra during the experiment; (**b**) comparison of background spectra and spectra obtained in the presence of an anthropogenic anomaly. The figures indicate: 1—retrieving from images by a nonlinear multiposition method; 2—in situ measurements with a string wave recorder; 3—stereophotogrammetric measurements; red lines denote Toba approximation [26].

All the data presented can be divided into two classes—background and anomaly caused by oil on the water. If the background data are averaged over the results of all measurements of the data for all three methods, the rms elevation values given in Tables 1 and 2 will be obtained.

Table 1. Comparison of rms sea surface elevations measured by different techniques.

Class	Range λ, m	$\sigma_\zeta^{(R)}$, mm Non-Linear Retrieval	$\sigma_\zeta^{(CT)}$, mm Stereo-Photography	$\sigma_\zeta^{(K)}$, mm Wave-Recorder Measurements
Background	0.1–1.0	9.24	12.02	8.72
	0.04–0.4	2.9	7.1	4.1
Anomaly (Oil on water)	0.1–1.0	3.30	6.00	3.98
	0.04–0.4	2.0	2.4	2.9

Table 2. The ratio of characteristics in the background and anomalous areas.

Ratio	Range λ, m	Non-Linear Retrieval	Stereo-Photography	Wave-Recorder Measurement
Ratio of rms elevations	0.1–1.0	1.4	1.91	1.59
	0.04–0.4	1.65	2.5	1.37
Ratio of variances	0.1–1.0	1.96	3.64	2.51
	0.04–0.4	2.72	6.25	1.88

The tables also show the relationship between the background and the anomaly in the magnitude of the spectral power (squares of the root mean square elevation).

Thus, for the wavelength range from 0.1 to 1 m, the rms elevation in the area of the sea surface anomaly associated with the appearance of an oil slick decreases:

- 1.4-fold, according to the results of the developed technique;
- 1.91-fold, based on the results of processing stereopairs;
- 1.59-fold, according to the data of the in situ wave recorder.

Similarly, for the wavelength range from 0.04 m to 0.4 m, the rms elevation in the area of the oil slick decreases:

- 1.65-fold, according to the results of the developed technique;
- 2.5-fold, according to the results of processing stereopairs;
- 1.37-fold, according to the wave recorder data.

The obtained estimates of the spectral anomalies of the sea surface agree with the results of other researchers [22,27].

4. Discussion

The developed approach for retrieving the spatial spectra of slopes and elevations of surface waves from aerospace optical images takes into account the nonlinear modulation of the brightness field by the slopes of the sea surface and includes a number of computational operations, including the numerical synthesis of the slope fields of the sea surface with a given spatial spectrum and random phases, as well as modeling the brightness field, taking into account the conditions of illumination, reflection and light scattering.

In this case, parameterization of spatial-frequency filters using a set of parameters depending on the conditions of image registration can be used.

The conducted studies have demonstrated the adequacy of the application of the method for retrieving the spectra of sea wave slopes and elevations from space optical images of high spatial resolution, which takes into account the nonlinear nature of the modulation of the brightness field by slopes of the sea surface.

The possibility of using the developed method for studying the variability of sea waves in the coastal zone under conditions of limited fetch is shown.

According to the results of two different complex experiments, the spatial spectral characteristics of sea waves, estimated from remote sensing data, were compared with the corresponding characteristics measured by in situ means under controlled conditions. The data of the array of string wave recorders installed on the stationary oceanographic platform of the Black Sea hydrophysical sea truth test site, as well as the data of stereo photography from the same platform, were used as sea truth data for comparison. Comparison of the wave spectra and its statistical characteristics demonstrated the correspondence of the results obtained during the processing of satellite images of high spatial resolution and the results of processing data obtained by various ground-based means. The measure of the discrepancy between the background wave spectra obtained from remote and in situ data is about 0.07. At the same time, there is a good qualitative agreement between the spectra obtained by distance and in situ methods in the area of an artificially created anomaly associated with the damping of short waves by an oil film on water.

Thus, the adequacy of remote measurement of both the time-average spectra of the sea surface and the variations of these spectra associated with unsteady waves has been experimentally confirmed. The use of high spatial resolution images and retrieving operators, taking into account the nonlinear nature of the modulation of the optical signal by the surface slopes, makes it possible to remotely determine the characteristics of wave spectra in a wide range of wavelengths and over vast areas.

The developed methods will be used for remote monitoring of the state of sea waves, including for identifying natural and anthropogenic impacts on various water areas.

At the same time, the studies carried out have shown the prospects of further development of the elaborated approach to retrieve wave spectra from optical satellite images using retrieving operators that take into account the nonlinear modulation of the brightness field by surface slopes.

A practically important direction in the development of the approach is its combination with other methods to reconstruct the angular distribution of spectral energy for all directions. In particular, a possible integration of the proposed method with the method for estimating the wave spectrum from the statistical characteristics of solar flares [23], as well as with developing new methods based on fractal analysis of optical satellite images [30] looks promising. The development and validation of new approaches will require new complex experiments using synchronous satellite and sea truth measurements.

5. Conclusions

The research results presented in this article and in the cited works indicate that the developed method to retrieve spatial spectra of slopes and elevations of surface waves from aerospace optical images, taking into account the nonlinear modulation of the brightness field by slopes of the sea surface, can be used for aerospace monitoring sea areas.

The developed retrieving operator can be used under the changing conditions of aerospace surveys and wave generation due to the use of spatial-frequency filter parameterization using a set of parameters that depend on the conditions of image registration.

As shown in this work, the proposed approach can be used to identify and study the variability of the sea wave spectra associated with anomalous phenomena and processes, including those of anthropogenic origin.

Author Contributions: Conceptualization, A.M. and V.B.; methodology, A.M.; software, A.M.; validation, A.M.; formal analysis, V.B.; investigation, A.M.; resources, V.B.; data curation, V.B.; writing—original draft preparation, A.M.; writing—review and editing, V.B.; visualization, A.M.; supervision, V.B.; project administration, V.B.; funding acquisition, V.B. Both authors have read and agreed to the published version of the manuscript.

Funding: The work was carried out with the financial support of the project by the Russian Federation represented by the Ministry of Science and Higher Education of the Russian Federation, the Agreement No. 075-15-2020-776.

Institutional Review Board Statement: Not applicable. The study did not involve humans or animals.

Informed Consent Statement: Not applicable. The study did not involve humans.

Data Availability Statement: The study did not report or generate any data.

Acknowledgments: The authors are grateful to Vladimir A. Dulov for useful discussion and for cooperation in organization of experiments.

Conflicts of Interest: The authors declare no conflict of interest. The funders had no role in the design of the study; in the collection, analyses, or interpretation of data; in the writing of the manuscript, or in the decision to publish the results.

References

1. Bondur, V.G.; Grebenyuk, Y.V.; Sabinin, K.D. The spectral characteristics and kinematics of short-period internal waves on the Hawaiian shelf. *Izv. Atmos. Ocean. Phys.* **2009**, *45*, 598–607. [CrossRef]
2. Bondur, V.G.; Vorobjev, V.E.; Grebenjuk, Y.V.; Sabinin, K.D.; Serebryany, A.N. Study of fields of currents and pollution of the coastal waters on the Gelendzhik Shelf of the Black Sea with space data. *Izv. Atmos. Ocean. Phys.* **2013**, *49*, 886–896. [CrossRef]
3. Keeler, R.; Bondur, V.; Vithanage, D. Sea truth measurements for remote sensing of littoral water. *Sea Technol.* **2004**, *45*, 53–58.
4. Bondur, V.G. Satellite monitoring and mathematical modelling of deep runoff turbulent jets in coastal water areas. In *Waste Water—Evaluation and Management*; InTech: Rijeka, Croatia, 2011; pp. 155–180, ISBN 978-953-307-233-3. Available online: http://www.intechopen.com/articles/show/title/satellite-monitoring-and-mathematical-modelling-of-deep-runoff-turbulent-jets-in-coastal-water-areas (accessed on 28 July 2021). [CrossRef]
5. Bondur, V.G.; Zhurbas, V.M.; Grebenyuk, Y.V. Mathematical Modeling of Turbulent Jets of Deep-Water Sewage Discharge into Coastal Basins. *Oceanology* **2006**, *46*, 757–771. [CrossRef]
6. Phillips, O.M. *Dynamics of the Upper Ocean Layer*; Cambridge University Press: Cambridge, UK, 1977; 336p.
7. Collard, F.; Ardhuim, F.; Chapron, B. Routine monitoring and analysis of ocean swell fields using a spaceborne SAR. *Geophys. Res. Lett.* **2009**, *36*. [CrossRef]
8. Yurovskaya, M.V.; Dulov, V.A.; Chapron, B.; Kudryavtsev, V.N. Directional short wind wave spectra derived from the sea surface photography. *J. Geophys. Res.* **2013**, *118*, 4380–4394. [CrossRef]
9. Bondur, V.G.; Murynin, A.B. Methods for retrieval of sea wave spectra from aerospace image spectra. *Izv. Atmos. Ocean. Phys.* **2016**, *52*, 877–887. [CrossRef]
10. Bondur, V.G.; Dulov, V.A.; Murynin, A.B.; Yurovsky, Y.Y. A Study of Sea-Wave Spectra in a Wide Wavelength Range from Satellite and In-Situ Data. *Izv. Atmos. Ocean. Phys.* **2016**, *52*, 888–903. [CrossRef]
11. Bondur, V.G.; Dulov, V.A.; Murynin, A.B.; Ignatiev, V.Y. Retrieving Sea-Wave Spectra Using Satellite-Imagery Spectra in a Wide Range of Frequencies. *Izv. Atmos. Ocean. Phys.* **2016**, *52*, 637–648. [CrossRef]
12. Bondur, V.G.; Murynin, A.B. Measurement of Sea Wave Spatial Spectra from High-Resolution Optical Aerospace Imagery. In *Surface Waves*; Ebrahimi, F., Ed.; Intech: Rijeka, Croatia, 2018; pp. 71–88. Available online: https://www.intechopen.com/books/surface-waves-new-trends-and-developments/measurement-of-sea-wave-spatial-spectra-from-high-resolution-optical-aerospace-imagery (accessed on 28 July 2021). [CrossRef]
13. Bondur, V.G.; Murynin, A.B. Vosstanovlenie spektrov poverkhnostnogo volneniya po spektram izobrazheniy s uchetom nelineynoy modulyatsii polya yarkosti (Restoration of surface waves spectra from the spectra of images with the account for nonlinear modulation of the brightness field). *Opt. Atmos. Okeana* **1991**, *4*, 387–393. (In Russian)
14. Murynin, A.B. Restoration of the spatial spectrum of the sea surface from the optical images in a nonlinear model of brightness field. *Issled. Zemli Kosm.* **1990**, *6*, 60–70. (In Russian)
15. Stilwell, D. Directional energy spectra of the sea from photographs. *J. Geophys. Res.* **1969**, *74*, 1974–1986. [CrossRef]
16. Monaldo, F.M.; Kasevich, R.S. Daylight Imagery of Ocean Surface Waves for Wave Spectra. *J. Phys. Oceanogr.* **1981**, *11*, 272–283. [CrossRef]
17. Chapman, R.D.; Irani, G.B. Errors in estimating slope spectra from wave images. *Appl. Opt.* **1981**, *20*, 3645–3652. [CrossRef]
18. Murynin, A.B. Parameterization of filters retrieving the spatial spectra of sea surface slopes on the basis of optical imagery. *Issled. Zemli Kosm.* **1991**, *5*, 31–38. (In Russian)
19. Baranovskii, V.D.; Bondur, V.G.; Kulakov, V.V. Calibration of remote measurements of 2-D spatial wave spectra from optical images. *Issled. Zemli Kosm.* **1992**, *2*, 59–67. (In Russian)
20. Bondur, V.G.; Vorobyov, V.E.; Murynin, A.B. Retrieving Sea Wave Spectra Using High Resolution Satellite Imagery under Various Conditions of Wave Generation. *Izv. Atmos. Ocean. Phys.* **2020**, *56*, 887–897. [CrossRef]
21. Kosnik, M.V.; Dulov, V.A. Extraction of short wind wave spectra from stereo images of the sea surface. *Measur. Sci. Technol.* **2011**, *22*, 015504. [CrossRef]
22. Dulov, V.A.; Yurovskaya, M.V. Spectral Contrasts of Short Wind Waves in Artificial Slicks from the Sea Surface Photographs. *Phys. Oceanogr.* **2021**, *28*, 348–360. [CrossRef]
23. Kudryavtsev, V.; Yurovskaya, M.; Chapron, B.; Collard, F.; Donlon, C. Sun glitter imagery of ocean surface waves. Part 1: Directional spectrum retrieval and validation. *J. Geophys. Res. Ocean.* **2017**, *122*, 1369–1383. [CrossRef]
24. Yurovsky, Y.Y.; Dulov, V.A. MEMS-based wave buoy: Towards short wind-wave sensing. *Ocean Eng.* **2020**, *217*, 108043. [CrossRef]

25. Preetham, A.J.; Shirley, P.; Smits, B.E. A practical analytical model for daylight. In Proceedings of the 26th Annual Conference on Computer Graphics and Interactive Techniques, SIGGRAPH 1999, Los Angeles, CA, USA, 8–13 August 1999; pp. 91–100. [CrossRef]
26. Toba, J. Local balance in the air-sea boundary process. *Oceanogr. Soc. Jpn.* **1973**, *29*, 209–225. [CrossRef]
27. Ermakov, S.A.; Salashin, S.G.; Panchenko, A.R. Film slicks on the sea surface and some mechanisms of their formation. *Dyn. Atmos. Ocean.* **1992**, *16*, 279–304. [CrossRef]
28. Ermakov, S.A.; Sergievskaya, I.A.; Da Silva, J.C.B.; Kapustin, I.A.; Shomina, O.V.; Kupaev, A.V.; Molkov, A.A. Remote Sensing of Organic Films on the Water Surface Using Dual Co-Polarized Ship-Based X-/C-/S-Band Radar and TerraSAR-X. *Remote Sens.* **2018**, *10*, 1097. [CrossRef]
29. Ermakov, S.A. *Influence of Films on the Dynamics of Gravitational-Capillary Waves (Vliyanie Plenok na Dinamiku Gravitacionno-Kapillyarnyh Voln)*; Institute of Applied Physics, Russian Academy of Sciences: Nizhnij Novgorod, Russia, 2010; 165p.
30. Murynin, A.B.; Richter, A.A.; Shakhramanyan, M.A. Selection of Informative Features to Highlight Areas of Waste Disposal by High Spatial Resolution Satellite Imagery. *Izv. Atmos. Ocean. Phys.* **2019**, *55*, 1403–1415. [CrossRef]

Journal of
Marine Science and Engineering

Article

Characteristics of Three-Dimensional Sound Propagation in Western North Pacific Fronts

Jiaqi Liu [1,2,3], Shengchun Piao [1,2,3], Minghui Zhang [1,2,3], Shizhao Zhang [1,2,3], Junyuan Guo [1,2,3] and Lijia Gong [1,2,3,*]

1 Acoustic Science and Technology Laboratory, Harbin Engineering University, Harbin 150001, China; liujiaqi@hrbeu.edu.cn (J.L.); piaoshengchun@hrbeu.edu.cn (S.P.); zhangminghui@hrbeu.edu.cn (M.Z.); zhangshizhao@hrbeu.edu.cn (S.Z.); guojunyuan@hrbeu.edu.cn (J.G.)
2 Key Laboratory of Marine Information Acquisition and Security, Ministry of Industry and Information Technology, Harbin Engineering University, Harbin 150001, China
3 College of Underwater Acoustic Engineering, Harbin Engineering University, Harbin 150001, China
* Correspondence: lijia.gong@hrbeu.edu.cn; Tel.: +86-1510-456-2982

Abstract: Oceanic fronts involved by ocean currents led to strong gradients of temperature, density and salinity, which have significant effects on underwater sound propagation. This paper focuses on the impact of the oceanic front on three-dimensional underwater sound propagation. A joint experiment of ocean acoustic and physical oceanography at the western North Pacific fronts is introduced. The measurement data for sound waves passed through the oceanic front is processed. The results are analysed and compared with the numerical simulation. It was found that transmission loss presented some difference when the source was located in the front centre and sound waves propagated towards water mass on opposite sides of the front centre. And when the sound field is excited by the underwater explosion at a depth of 200 m, the effects of the horizontal refraction cannot be neglected. On the other hand, the transmission loss for sound pressure fell sharply and rose rapidly at the side of cold water masses.

Keywords: oceanic front; ray theory; horizontal refraction

Citation: Liu, J.; Piao, S.; Zhang M.; Zhang S.; Guo, J.; Gong L. Characteristics of Three-Dimensional Sound Propagation in Western North Pacific Fronts. *J. Mar. Sci. Eng.* **2021**, *9*, 1035. https://doi.org/10.3390/jmse9091035

Academic Editor: Grigory Ivanovich Dolgikh

Received: 30 August 2021
Accepted: 16 September 2021
Published: 19 September 2021

Publisher's Note: MDPI stays neutral with regard to jurisdictional claims in published maps and institutional affiliations.

1. Introduction

Mesoscale oceanic eddies and fronts, as complex ocean dynamic phenomenon, are energetic contributors to mixing. The oceanic fronts are the boundary between warm and cold water masses, with very significant differences in temperature, density and salinity [1]. There are different kinds of oceanic fronts, such as the large-scale meandering front, buoyancy forced coastal water mass fronts and the interface region between the deep-ocean and the coastal regions [2]. Fronts of Kuroshio is a type of oceanic front caused by intensified current led to a strong gradient of sound in the water column, which show a notable effect on sound propagation. The dynamic ocean sound speed leads to the change of long-range acoustic propagation travel time, the movement of the sound channel axis or/and convergence zone and the horizontal refraction of rays path [3,4].

Oceanic fronts are observed in the global ocean, such as the Antarctic gyre, the Gulf of Mexico, the region between Hawaii and Alaska and the Kuroshio extension area. Lots of work has been done in different parts of the ocean and indeed, one of the topics on the research of the front is the mechanism of the formation and extinction of the front. To survey the generation and to evaluate vertical motions of the near-surface front, Mahadevan analysed model simulations of the oceanic front [5]. Sokolov et al. [6] used the expendable bathythermograph (XBT) and satellite altimeter to identify the subtropical front near Antarctica and South Australia. The analysis of the front position helps to obtain variation information in the annual sea surface temperature. In the long-term observation of the temperature gradient of fronts in the western North Pacific, Levine et al. [7] summarised the influence of front on topography at different frontal zones. Yang et al. [8] generalized heat

125

and energy transfer of fronts near the continental shelf in the South China Sea, especially the mass exchange and energy distribution in the vertical direction. The other research topic is the seasonal variation in front. Nakamura et al. [9] identified the reasons for the seasonal variations in salinity of the oceanic front in the vertical mixing layer of the western North Pacific. Nakamura proved that this variation in winter is mainly caused by strong eddies and deep-ocean cyclonic. Also, the mixing layer energy enhances in the middle of oceanic fronts, due to the existence of near-inertial internal waves in winter. The westward movement of frontal current affected by the invasion of the Kuroshio in the Luzon Strait is analyzed by Liu et al. [10]. The generation of temperature fronts was observed in summer, using the satellite remote sensing processed data. Xie et al. [11] outline that the season has no evident impact on the horizontal position of the front while the vertical structure in the Southwest Pacific. By measurement and statistics, they also concluded that the depth of the oceanic fronts in the subtropical zone in winter was greater than that in summer. The oceanic fronts on the continental shelf are observed by Tan et al. [12] in both spring and summer. Also, they modelled the three-dimensional current field of the oceanic fronts and obtained its dynamic mechanism and formation reason.

The horizontal asymmetry character resulted in the complication of acoustic propagation in the area of frequently occurs current. Therefore, many studies and researches have been done on sound wave propagation through fronts. Seismic waves reflection has been observed at the boundary between two water masses with different thermodynamic characteristics. Nakamura et al. [13] analysed the multi-channel seismic exploring data, inversed sound speed of the horizontal variation of the temperature front. It showed sound wave is blocked by variations of horizontal temperature using mathematically sound propagation models. Neubert et al. [14] choose acoustic rays theory to deduce the average coherence intensity of sound wave multiple paths caused by random oceanic front. Tang et al. [12] also discussed the sound propagation characteristics responses to the intensity variability of oceanic front in the South China Sea. Acoustic shadow zones are formed when the sound waves pass through the warm side of the oceanic front. Liu et al. [15] made a preliminary summary of sound propagation of acoustic pressure resulted from typical mesoscale phenomena, using the parabolic model FOR3D [16] to simulate the impact of mesoscale eddies and oceanic fronts on the sea surface acoustic channel and converging zone. In the Yellow sea, coupled normal mode theory used as a numerical method for traveltime tomography, estimated the subsurface velocity structure and inversed the thermocline significant changes caused by current quantitatively [17]. Due to the limitation of experimental conditions, they study sound propagation without considering sound wave horizontal refraction. The three-dimensional thermocline inversion remains to be further explored. Jian et al. [18] provided the difference of propagation loss of acoustic pressure with and without oceanic front by parabolic method. Guo et al. [19] studied the travel time of sound waves and distribution of sound pressure in the ocean of Taiwan.

Furthermore, researchers have paid extensive attention to Kuroshio fronts and perform a big amount of numerical and experimental studies. Chen et al. [20] applied the ray theory to simulate the sound propagation process of front and found a remarkable change of sound propagation at the surface sound channel [21]. Rainville et al. [22] explained the gradient of temperature arouse the signal variety in the East China Sea and verified by joint experiment processed data in 2001. Shapiro investigated propagating characteristics along with different directions of oceanic front in different seasons of the same year.

Although scientists have carried out a relatively comprehensive study on the western North Pacific front by comparing of oceanic fronts exists or not and obtained certain conclusions in the aspects of sound pressure transmission loss and time arrival structure, there is a lack of discussion and analysis on the three-dimensional sound field of the front. This paper discusses the degree of sound wave horizontal deflection with several depths of source and receiver and reveals strong horizontal refraction because of the oceanic front, moreover explains the anomalous distribution of the acoustic field, which is helping to design vertical spacing in an array and positions of the acoustic stations of acoustic tomography. This paper analyses numerically and experimentally how oceanic fronts affect three-dimensional acoustic propagation in the western North Pacific. As follows in Section 2, a three-dimensional oceanic front environment was modelled and used to resolve the sound propagation and the forecasting process of the sound transmission based on the ray theory. And then, the sound propagation experiments in the Kuroshio is introduced, and the comparison between the measured results and numerical results are carried out in Section 3. In Section 4, the numerical results of the three-dimension acoustic propagation and the horizontal refraction process are described, while conclusions and discussions are drawn in Section 5.

2. Numerical Analysis of Sound Propagation

2.1. Distribution Model of Frontal Temperature and Salinity

The distribution of temperature and salinity of the oceanic front is presented based on their formation mechanism. Calado et al. [23] studied the characteristics of historical statistical data of oceanic front. Therefore the temperature characteristic equation of a large-scale front as follows Equation (1).

$$T(x,y,z) = [T_0(x) - T_b(x)]\Phi(x,z) + T_b + \alpha(x,z)\Gamma(y) \tag{1}$$

where

$T_0(x)$—surface temperature;
$T_b(x)$—bottom temperature;
$\Phi(x,z)$—non-dimensional temperature profile;
$\alpha(x,z)$—along-steam distribution of the cross-frontal slope of the temperature;
$\Gamma(y)$—cross-frontal distribution of the temperature gradient $\alpha(x,z)$.

Water fronts temperature feature model parameters are given in Figure 1, and the same equation and drawing can be used to describe the salinity distribution of the oceanic front. In the actual measurement process, the depths and positions of measured temperature and salinity data can not accommodate our needs regarding sound field simulation. In order to obtain a more accurate environmental distribution in the frontal area, the most frequently used method is spatial interpolation using limited field measurements and remotely sensed data. Considering the difference in the distance between the sampling location and the interpolation point, the interpolation method of inverse distance weighted should be adopted for temperature and salinity data based on the interpolation formula as Equation (2), where expression of T is temperatures or salinity vary with longitude and latitude.

$$T(x,y) = \sum_{i=1}^{N} \frac{w_i T(x_i, y_i)}{\sum_{i=1}^{N} w_i}, \quad \text{where } w_i = \frac{1}{d(x,y;x_i,y_i)^2} \tag{2}$$

where

(x,y)—coordinates of observation point;
(x_i, y_i)—coordinates of interpolated point;
d—distance from (x_i, y_i) to (x,y);
N—total number of observation points used in interpolation.

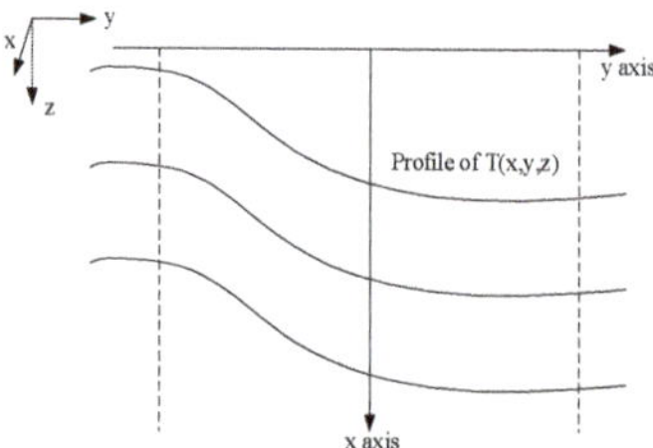

Figure 1. Vertical view of a simple velocity-based feature model of the temperature of front, *x* the dimensional along-front coordinate, *y* the cross-front coordinate with origin at the centre of the front and *z* the dimensional vertical coordinate (positive upward).

To obtain the three-dimensional speed field around the oceanic front, and precise the distribution of vertical structure for the temperature, salinity and sound speed effectively, the extrapolation depth of the environmental data should be executed until the mean depths between the ocean surface and the sediment of this area (nearly 5 km). Alan B. Coppen's [24] empirical formula as Equation (3) gives the relationship between the sound speed, temperature, salinity, and depth. Moreover, the limitation of temperature is 0 to 35 °C and for salinity is 0 to 45 parts per thousand. Then it is easy to calculate the sound speed profile for this frontal area.

$$
\begin{aligned}
c(x,y,z) &= c(x,y,0) + (16.23 + 0.253t)z + (0.213 - 0.1t)z^2 \\
&\quad + [0.016 + 0.0002(S - 35)](S - 35)tz, \quad \text{where} \\
c(x,y,0) &= 1449.05 + 45.7t - 5.21t^2 + 0.23t^3 + (1.333 - 0.126t + 0.009t^2)(S - 35)
\end{aligned}
\tag{3}
$$

where

t—temperature variable $t = T(x,y)/10$;
$T(x,y)$—interpolated temperature in degrees Celsius;
$S(x,y)$—interpolated salinity in parts per thousand;
z—depth in kilometres.

2.2. Sound Propagation Model

With the rapid development of underwater acoustics since the 1980s, the commonly used numerical methods include normal mode theory, ray theory, parabolic equation theory and finite-element modelling [25]. Among them, normal mode theory solves the Sturm–Liouville eigenvalue problem of the acoustic wave; parabolic equation theory and finite-element theory as numerical methods could accurately solve the three-dimensional underwater acoustic problem. Both the parabolic equation theory and the ray theory are effective methods for predicting underwater sound propagation to solve the range-dependent ocean [26], however parabolic equation theory reduce timeliness and intuitiveness compared with ray theory. Ray tracking theory is based on the generalized Snell's law [27], which adapts to study the sound deflection caused by the continuous variation of sound speed. Due to the different propagation paths of the acoustic ray from the source to the receiver position, the sound field at the receiving point is the superposition of the intensity of the ray path arrivals. The information including of each ray includes incident angle, soundtrack, intensity and phase. The horizontal deflection of the ray can be observed and verified intuitively by using the ray theory [28].

$$
\nabla^2 p(X) + \frac{\omega^2}{c^2(X)} p(X) = -\delta(X - X0)
\tag{4}
$$

where

ω—angular frequency of source signal;
X—receiver position;
$X0$—source position;
z—depth in kilometres.

Starting from the Helmholtz Equation (4), the sound pressure at a receiving position is actually a superposition of different acoustic rays (each ray is a normal to a wavefront), which can be expressed as,

$$p(X) = e^{i\omega\tau(X)} \sum_{j=1}^{\infty} \frac{A_j}{(i\omega)^j} \ \ (j = 1, 2, ...), \ \text{ where } \ \tau(s) = \tau(0) + \int_0^s c(s')ds' \tag{5}$$

where

X—ray trajectory $X = (x, y, z)$;
τ—travel times of the acoustic rays;
j—number of acoustic rays;
A_j—amplitude of j th acoustic ray;
s—arclength along the ray;
s'—ray trajectory;
c—sound speed at the position X.

Putting Equation (5) into Equation (4), equating terms of like order in angular frequency and referring to the ray theory section of [25] we obtained,

$$O(\omega^2) : |\nabla\tau(x, y, z))|^2 = \frac{1}{c^2(x, y, z)} \tag{6}$$

$$O(\omega^1) : 2\nabla\tau(x, y, z)\nabla A_0 + (\nabla^2\tau(x, y, z))A_0 = 0;$$
$$O(\omega^{1-j}) : 2\nabla\tau(x, y, z)\nabla A_j + 2(\nabla^2\tau(x, y, z))A_j = -\nabla^2 A_{j-1}, j = 1, 2, ...; \tag{7}$$

The equations of describing trajectories are indicated as "eikonal equation" Equation (6) and "transport equation" Equation (7), respectively. "Eikonal" regard as the change of phase of acoustic ray along with the ray trajectory. Since $\nabla\tau$ is vector perpendicular to the wavefronts, the relation of X and $\nabla\tau$ is written as $\frac{dX}{ds} = c\nabla\tau$. Putting Equation (6) into this differential equation we obtained: $\frac{d}{ds}\left(\frac{1}{c}\frac{dX}{ds}\right) = -\frac{1}{c^2}\nabla c$. In the Cartesian coordinate system, ray equations then are written in the following Equations (8) and (9):

$$\frac{d\xi}{ds} = -\frac{1}{c^2}\frac{dc}{dx}, \xi = \frac{1}{c}\frac{dx}{ds}$$
$$\frac{d\zeta}{ds} = -\frac{1}{c^2}\frac{dc}{dy}, \zeta = \frac{1}{c}\frac{dy}{ds}, \tag{8}$$
$$\frac{d\eta}{ds} = -\frac{1}{c^2}\frac{dc}{dz}, \eta = \frac{1}{c}\frac{dz}{ds}$$

where ξ, ζ and η are three auxiliary variables $k_0 = \omega/c_0$ is wave number calculated by default sound speed.

$$A_j(x, y, z)e^{ik_0 s(x,y,z)}\nabla^2 s(x,y,z) + \frac{2}{A_j(x, y, z)}\nabla A_j(x, y, z) \cdot \nabla s(x, y) = 0, j = 1, 2, ...; \tag{9}$$

$s = \sqrt{\xi^2 + \zeta^2 + \eta^2}$ is ray trajectories arclength. The two differential equations of the width and curvature of the acoustic beam are integrated to get the acoustic beam field near the sound rays inside the acoustic beam of the acoustic beam. Using Equations (8) and (9) the ray beam has been traced. Finally, using the fourth-order Runge-Kutta method we solve the initial value problem.

For the three-dimensional inhomogeneous distribution of sound speed caused by ocean dynamic effects such as oceanic fronts in the area, it is a challenge to simplify the sound speed profile into an analytical expression. Numerical calculation of integration equation of acoustic ray trajectory becomes complicated when the rays reach the sediment interface and reflect from the seabed, especially fluctuated seabed. Gauss beam [29] is selected to solve the shortcoming of the traditional geometric ray method in the acoustic shadow region and the infinite sound intensity in the caustic region. And sound intensity which is contributed by all the acoustic rays weighted meet the characteristics of the Gaussian distribution with the distance of the beam centre. The refraction phenomenon is revealed at the sudden change of sound speed position by the three-dimensional ray-tracing method. Tracing beam in two-dimensions (r, z), where $r(s)$ and $z(s)$ are the ray trajectory in the range–depth plane as a function of arclength s [28]. When it comes to the three-dimensions (x, y, z) case, we trace a set of rays, additionally, the rays form a fan over both azimuthal and elevation-declination angles.

Figure 2 shows a simplified shallow water coastal idealized front model as Lynch [30], the rigid seabed is flat at a depth of 100 m, and the boundaries of colder water (sound speed constant is 1488.5 m/s) and warmer water (sound speed constant is 1511.5 m/s) is the xz-plane as in Figure 2a. Interpretation of horizontal refraction by tracing the low grazing angles acoustic ray paths emanating from the source. As is exhibited in Figure 3a, Lloyd's mirror effect is observed through surface-reflected acoustic ray paths. On account of the change of sound speed in the horizontal plane, acoustic rays constantly refracted until the rays path curves upward and reached the receiver depth. With the increasing of horizontal grazing angles, the receivers gradually move further and further away from the point source. The fluctuation of the pressure field propagation difference (between a direct wave and a direct plus reflected wave) in Figure 2b demonstrates the acoustic rays superposition of interference from 11 km to 40 km. The enhancement of around 6 dB in the sound field is due to Lloyd's mirror reflection of a 200 Hz source nearby the coastal front.

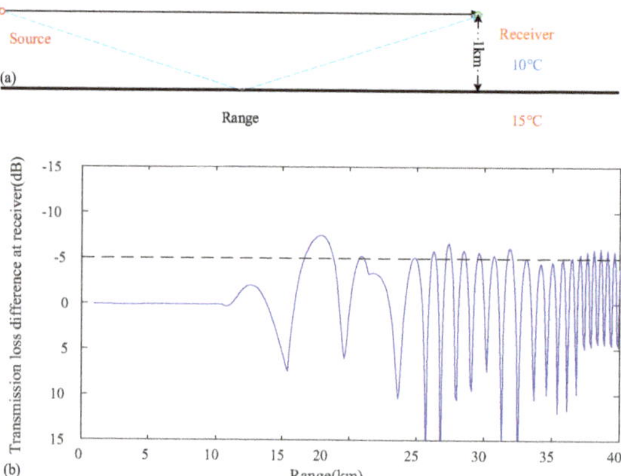

Figure 2. (**a**) A simplified model of front environment source and receiver located 1 km from xz-plane and (**b**) frontal reflection effects on the sound pressure field, the curve is the difference transmission loss between a direct wave and a direct plus horizontal reflected wave.

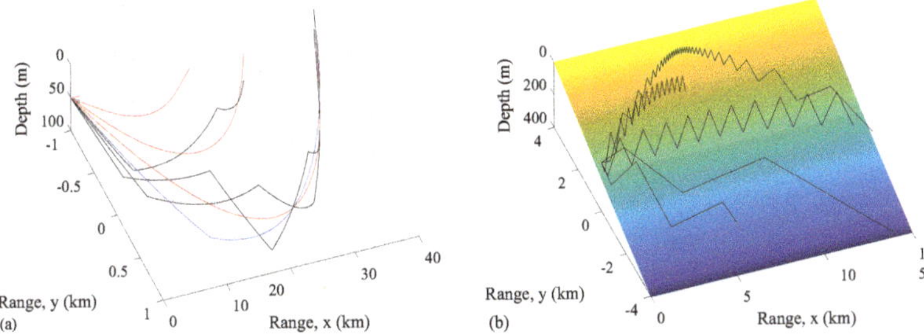

Figure 3. (**a**) Effect of a nearby oceanic front and (**b**) effect of a wedge seabed on horizontal refraction of the acoustic propagation.

The horizontal refraction effect of sound propagating standard wedge seabed is evident in comparison with that of nearby a front. Sound propagation experiences a faster rate of energy attenuation due to multiple seafloor reflections at steep angles and thus little to no deflection in the region (x [0–3 km], y [0–4 km]). However, the sound wave encounters the region (x [3–15 km], y [3–4 km]) of significant horizontal refraction, bending towards the direction of decreasing y. Due to the geometry of the wedge, the sound wave propagates longer distances before energy trending zero at (x [0–3 km], y [−4–0 km]). Using ray approximation, one could calculate rapidly and accurately to understand the propagation problem in the following sections. Furthermore, to satisfy the high-frequency approximation condition ($f > 10\, C_w/D_w$, where C_w is the average of water sound speed and D_w is the depth of water), the appropriate frequency is picked when studying the sound propagation problem. Thus, the ray-tracing method is chosen to study the refraction effect caused by the front and uneven seabed.

3. Comparison between Experimental and Numerical Simulation Results

The Kuroshio Current, one of the strongest surface oceanic currents in the whole ocean, intersects with the polar circulation and forms a large meander structure. The temperature and salinity distribution are extremely complex, especially in regions influenced by strong frontal currents. Nagai analysed the disturbance and temperature and salt distribution structure directly observed in the Kuroshio front sea area from the oceanographic perspective. Ocean energy dissipation is mainly influenced by the non-geotropic current and internal wave, with the changing of the temperature gradient [31]. The dynamic phenomena in the western North Pacific were identified and tracked, such as the size, life cycle and distribution of cyclonic and anticyclonic as respectively indicated in Figure 4a–d. The number of cold eddies and warm eddies in the western North Pacific are basically in the same magnitude, and the number of cyclonic is slightly larger.

From May to July 2019, a joint experiment of ocean acoustics and physical oceanography was carried out. The moving vessel profiler (MVP300) and conductivity-temperature-depth (CTD) measured the temperature/salinity gradients and sound speed profile. After four voyages along the ship trajectory, crossing the oceanic front, temperature and salinity data were collected continuously, which provided high-resolution environmental parameters for sound propagation study. Figure 5 shows the topography of the experimental area in the western North Pacific, the path of MVP300 (red lines) and the position of shipboard CTD cast (green rhomb) on 8 June. Figure 6 shows the sound speed of the ocean surface. The green rhomb and red lines are the locations of the CTD stations and MVP stations, respectively. Typical sound speed-depth curves were observed by underwater CTD in the western North Pacific, from 0 m to 2000 m in depth and deep oceans below 2000 m provided by remote sensing which is shown in Figure 7. Afterwards, Figure 8 illustrates four sound speed profiles in the water column collected by MVP.

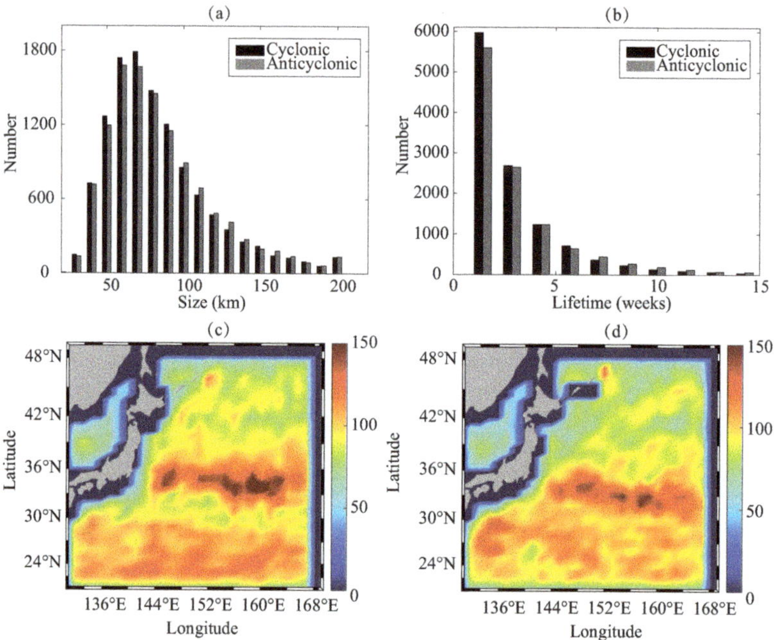

Figure 4. Statistical results of cyclonic and anticyclonic near Kuroshio obtained from 20 years of assimilation data from 1993 to 2014 (**a**) Kuroshio Current size (**b**) Kuroshio Current lifetime, (**c**) cyclonic size distribution and (**d**) anticyclonic size distribution.

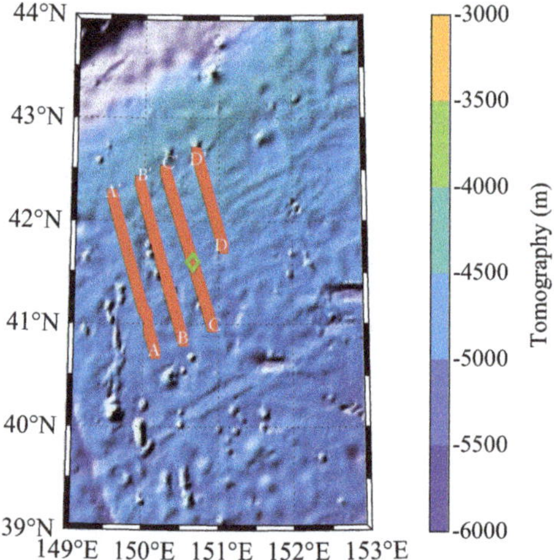

Figure 5. Topography of the western North Pacific. Colour shading indicates topographic relief from etopo1 data (global relief model of Earth's surface, for more information https://www.ngdc.noaa.gov/mgg/global accessed on 1 April 2021).

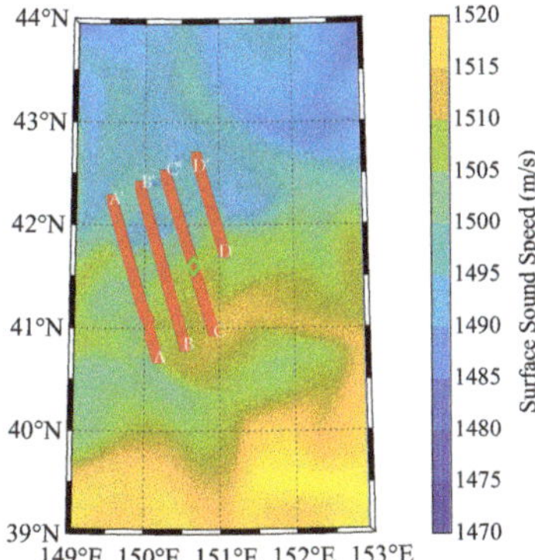

Figure 6. Sound speed of the ocean surface. Colour shading indicates sound speed fluctuation.

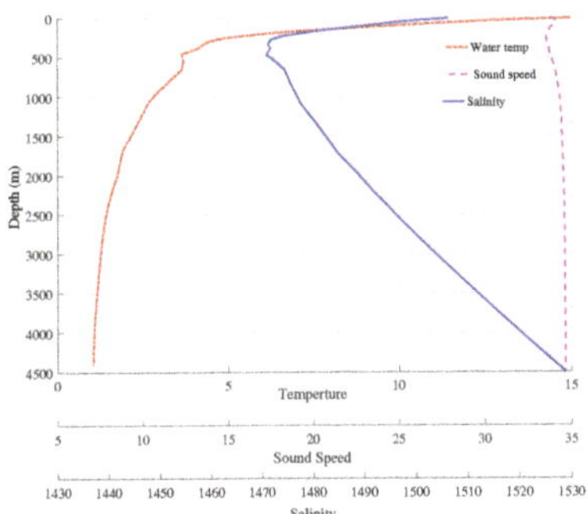

Figure 7. Distributions of sound speed profile (blue solid line), temperature (red dash-dotted line) and salinity (pink dashed line) for the depth from the ocean surface to bottom at the position of CTD.

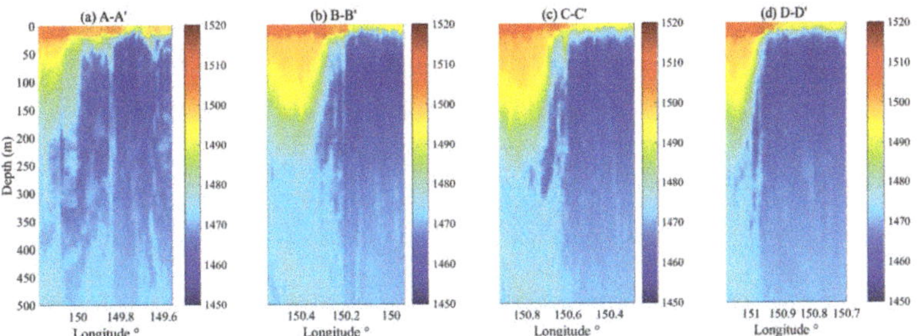

Figure 8. Measured sound speed profile from (**a–d**) is the survey line A–A′, B–B′, C–C′ and D–D′, respectively, which are marked in Figure 5.

We measured temperature variation in the horizontal direction, temperature and salinity consistently varied across the entire 3-D block, influenced by the warm Kuroshio current. The horizontal temperature gradient and salinity gradient between both sides of the oceanic front are large enough to detect the front, indicated in Figure 9a,b, respectively. The front has tracked to determine the placement of the receiving array for the following acoustic experiment. It is predicted that the frontal position located at 150.6° East longitude.

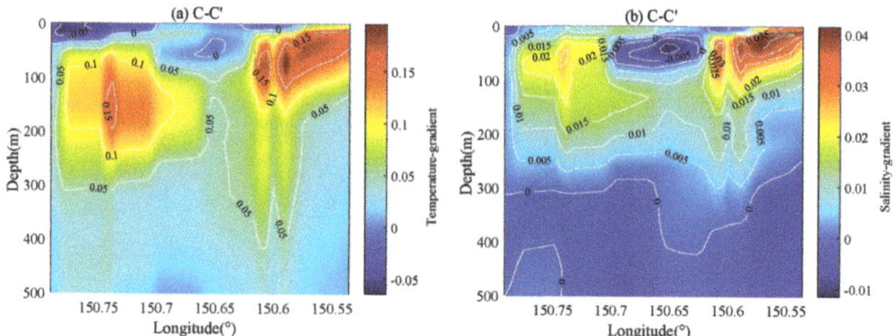

Figure 9. Distributions of temperature (**a**) and salinity (**b**) gradients.

Putting the relationship of depth, salinity and temperature into the Equation (3) and calculating the variation of sound speed profile, Figure 10a shows a layered structure of the sound velocity in the water column. Due to warm-water mass intrusion, sound velocity changes in both surface layer and thermocline. Figure 10b exhibits sound velocity profiles at different longitude in order to describe sound speed layer structure more clearly. There is a minimum sound speed in SSP that varies from 1470 m/s to 1455 m/s at a depth of roughly from 300 m to 100 m. Meanwhile, the maximum sound speed and the sound speed near the seabed reach up to 1547 m/s.

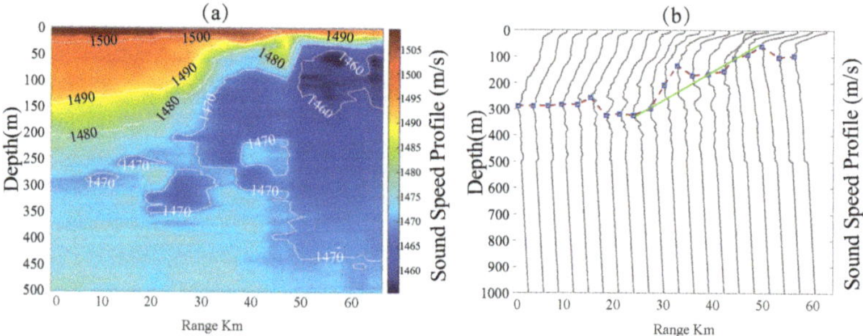

Figure 10. Distribution of sound speed profile on Longitude and depth, (**a**) vertical slice map and (**b**) Those sound speed profiles are equally divided into 20 slices in 60 km of range, green line shown the position of oceanic front and blue dot is the sound channel axis.

The horizontal distribution of sound speed was obtained through interpolation. The horizontal and vertical coordinates are marked with longitude and latitude. The sound velocity distribution at a depth of 250 m is enlarged in Figure 11. It is intuitively noticed that the change of sound velocity of the oceanic front change dramatically below the mixing layer. In Figure 11, the geodetic longitude and latitude coordinates are converted to a Cartesian coordinate system, and the position (149° E, 39° N, z) replaced by the point (0, 0, z). This figure shows the three-dimensional temperature field at positions $x = 10$ km, $x = 150$ km and $x = 340$ km with a depth of 200 m. The solid black lines in the middle are the isotherm distribution of the temperature on the section. Vertical sections are the interpreted temperature profile in the distribution of uniform distance, and three layers are shown with the map of interpreted temperature at 0, 100 m and 200 m depths in Figure 12.

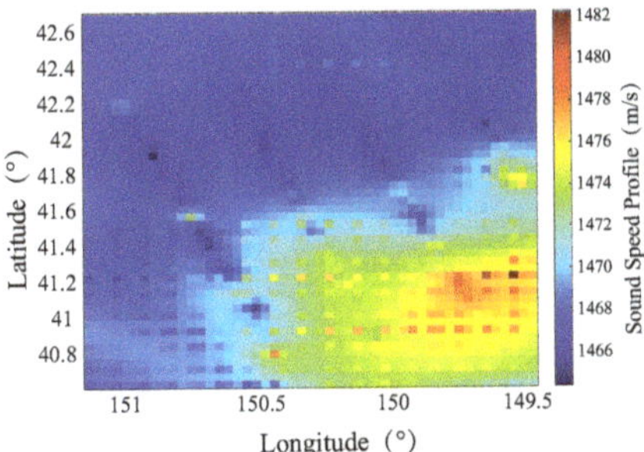

Figure 11. Regional sound speed profile for the depth of 250 m.

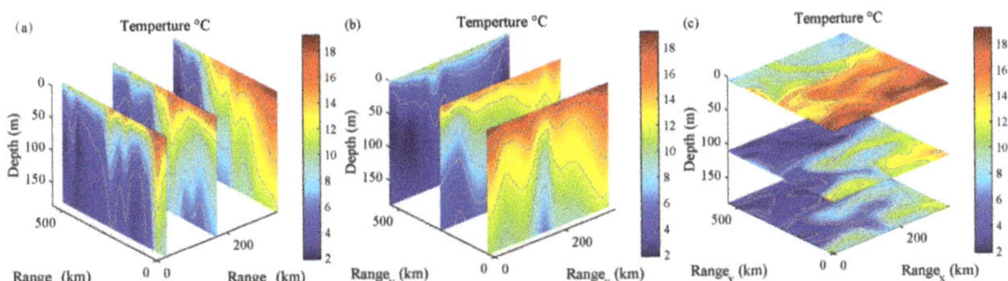

Figure 12. Three-dimensional temperature structure of the front transect. (**a–c**) are slices of the temperature in the yz plane, xz plane and xy plane, respectively

The following Figure 13 is a diagram of the sound propagation experiment configuration on the western North Pacific. The receiver system uses the ranges between a beacon fixed on the seabed and hydrophones on the submerged buoys aided by depthometers. There is a vessel in motion against the oceanic front. Melt cast explosives based on Trinitrotoluene (TNT) is used as high power and low-frequency underwater acoustic source. The transmission loss of sound pressure is acquired through theoretical calculations and experimental measurements. Two types of explosives are selected whose explosion depth is 100 m and 200 m. Moreover, more explosives occur on the side of the cold water as sound sources. The broadband explosives were thrown into the ocean every 1.7 km within 3 h from a vessel speed of 12 knots. The position of the blue rectangle is the vertical array of 10 self-contained underwater acoustic recorders, equally spaced from 50 m to 300 m. Figure 14 is one of the hydrophones used in this experiment, and the specifications of the hydrophones are shown in Table 1.

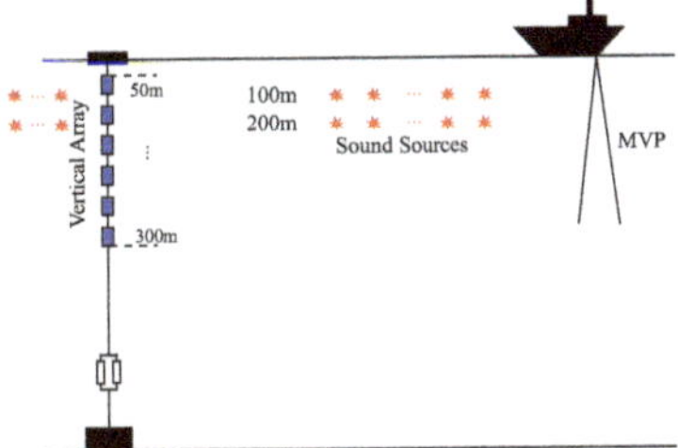

Figure 13. Experimental installation in the western North Pacific.

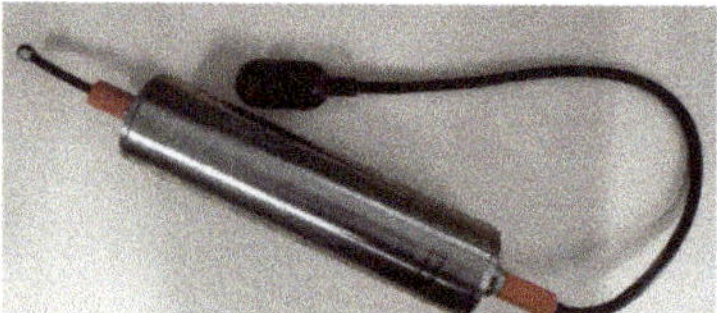

Figure 14. Hydrophone: signal conditioning, low-noise cable and data acquisition.

Table 1. Hydrophone specifications.

Parameter Type	Parameter Value (Range)
Frequency Range	10 Hz–20 kHz +3/−3 dB
Receiving Sensitivity	−203 dB re 1 V/μPa
Amplification Gain	20 dB
Filter Bandwidth	20 kHz
Sampling Rate	10 kHz 24 bit
Network Specific Interfaces	100 M/10 M Based TCP/IP
Supply Voltage and Current	19 V DC and 2 A

The actual depth of the vertical array influenced by ocean currents is not consistent with the preset values. After the depth of the vertical array was corrected, the receiving signals at the same depth were rearranged. The sampling frequency of the instrument is 10 kHz. To ensure the integrity of the received signal, the time window for intercepting the signal is selected according to the actual signal amplitude. The source signal spectrum was distributed from 5 Hz to 500 Hz, and the recorded signals were band-pass filtered with a Butterworth filter with a centre frequency of 120 Hz. The average coherence intensity of the sound wave is Equation (10). The pressure signal is also selected to have a peak frequency at 120 Hz and a 3 dB cut-off frequency from 106 Hz to 135 Hz. Ray theory is used to analyze the sound field and obtain broadband mean acoustic intensity by averaging the frequency points within the bandwidth in the interference field by Equation (10), where I is the sound intensity of the centre frequency of f_0, Δf is the half of frequency bandwidth.

$$I(f_0) = \frac{1}{2\Delta f} \int_{f_0-\Delta f}^{f_0+\Delta f} \frac{p^2(f)}{\rho c} df. \tag{10}$$

In this deep-sea sound propagation experiment, the acoustic reciprocity theorem was generalized for multiple sources and a fixed receiver array. In the sound field due to the source at explosive point, sound pressure received at any other hydrophones is the same as that would be produced at the explosive point if the source was placed at the depth of hydrophones. The results of signal processing received by different depth hydrophones are given by Figure 15. It can be seen from Figure 15a that there are no obvious differences for experimental results with different sound waves propagate direction (toward the cold side and the warm side), when the sound source was located in the mixed layer (75 m). However, it reflects the great differences that exist between the sound field propagated towards the warm and cold side of the ocean-front centre with three different depths of the acoustic sources located near the sound channel axis (150 m, 175 m and 200 m) is represented by Figure 15b–d. On the cold-water mass side, the distribution of the sound field is abnormal. There is a drop in the curves (after 3 km from the receiver array). On the contrary, transmission loss on the warm side of the oceanic front follows the law of reciprocal attenuation of the distance. These intuitive phenomena can be observed through the corrected experimental data and the sound pressure transmission loss curve drawn by simulation. Eventually, the agreement of experiment and modelling is satisfactory. At the bottom of the equipment, the float balls system is heavy enough for the vertical line hydrophone array to remain essentially in the experimental design. Although, the error may come from the fluctuation of the line array caused by the flow of the oceanic front. Figure 15 showed the calibrated depth of receiver hydrophone depends on the depth measurement. The standard deviation of explosion-signal receiver depths is within the allowed error range.

The whole depth of sound pressure transmission loss diagram is illustrated in Figure 16, simulated acoustic source located at depth of 175 m, the centre of the oceanic fronts is situated at the right cold side (red line in Figure 16). The warm water is on the right side of the front and the other side is cold. The sound field distribution on both sides of the oceanic front is discrepant. It is evident to observe that some acoustic rays turn downward until facing the centre of the oceanic front when the wave propagates to the right side, the other ray turning points. But on the other hand, there are no phenomena observed when the wave propagates to the left side. It can be concluded that the existence of the oceanic front does act as an acoustic lens that could strongly change the directions of acoustic propagation.

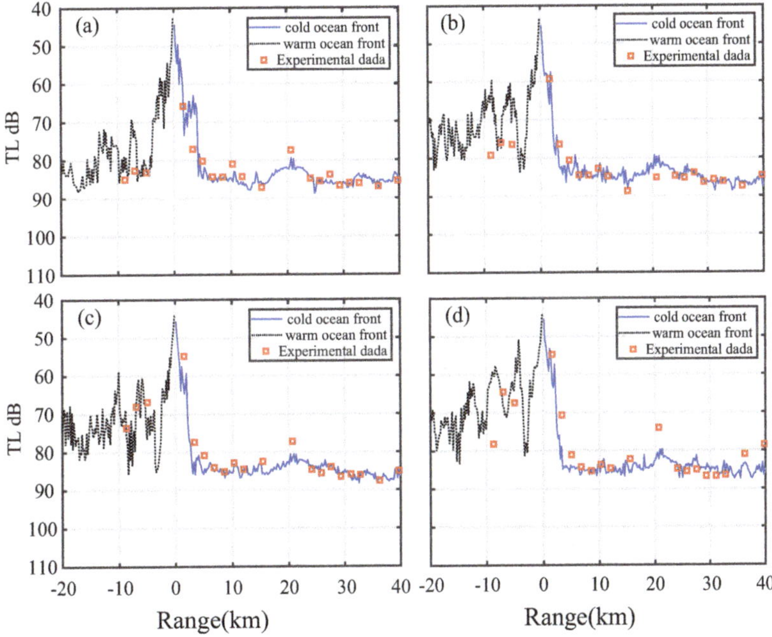

Figure 15. Transmission Loss in cold side (blue dotted line), warm side (black line) and experimental results with 100 m explosions (red square) in depths of 75 m (**a**), 150 m (**b**), 175 m (**c**) and 200 m (**d**).

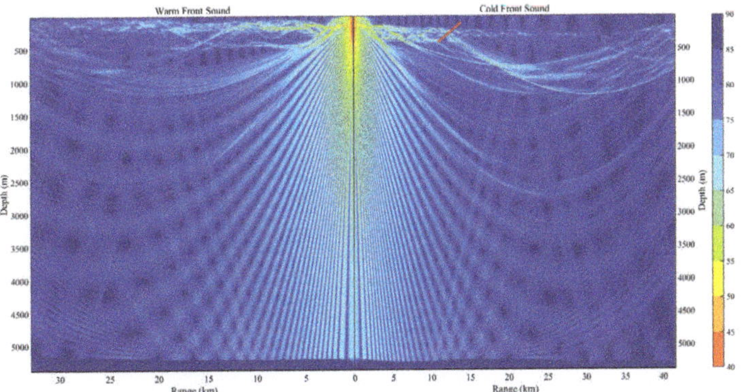

Figure 16. Transmission Loss in both warm and cold sides of oceanic front.

The data received on the hydrophone at 175 m depth were analysed, compared, and discussed here with the two-dimensional modelled result. Figure 17 indicates transmission loss generated by explosions at the depth of 200 m. The experimental data on the warm water side is in good agreement with the curve. Nevertheless, the 2-D simulation results at the side of cold water do not agree well with the experimental data. It is doubtful that the exciting sound pressure field at this depth is greatly affected by the horizontal refraction, which will be discussed in Section 4.

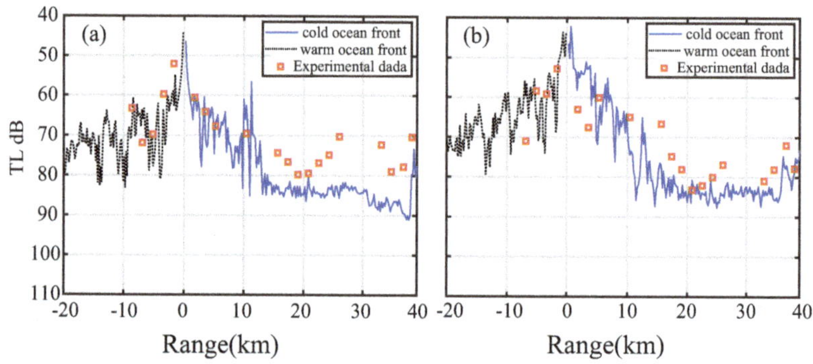

Figure 17. 2-D Transmission Loss in cold side (blue line), warm side (black dotted line) and experimental results with 200 m explosions (red square) in depths of 175 m (**a**) and 200 m (**b**).

Designing a band-pass filter with a centre frequency of 200 Hz, and a 3 dB bandwidth of 50 Hz, comparing the difference in transmission loss between the two depths of explosions, the abnormal distribution of the propagation loss at 4 km and 8 km from hydrophone arrays can be observed by Figure 18. The rapid fall was also measured, as a result of the 200 Hz filtered experimental data. According to the propagation loss curve generated by the explosions at depth of 200 m, it can also be noticed that some data at some certain distance from the receiving vertical array does not match well.

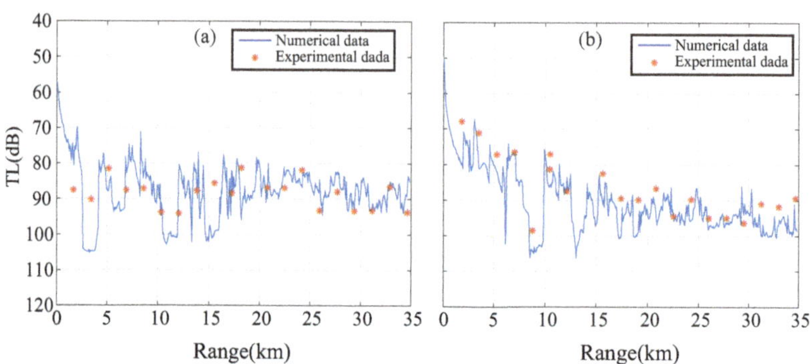

Figure 18. Comparison of Transmission Losses between numerical and experimental results for 175 m depth of hydrophone, (**a**) explosive source at depth 100 m and (**b**) at 200 m.

4. Discussion and Analysis Northwest Pacific Oceanic Front Sound Propagation and Horizontal Refraction Character

Both acoustic propagation direction and source position should be put into consideration when discussing the influence of ocean-front on the horizontal refraction of the acoustic field. In consideration of the deflection of acoustic rays in the two directions of the cold-water mass and the warm-water mass, the relationship between the maximum deflection angle of acoustic energy flow and the propagation distance is studied. The sound source propagates from the warm to the cold side, and more energy is confined to the mixed layer of the offshore surface. The horizontal refraction's angle is the angle between the direction of the signal arrival at the point of reception and the true direction (the sound energy flow) to the sound source in the horizontal plane. The horizontal refraction's angle symbol is also different from the thermocline, which is affected by the oceanic front. There is a misalignment between the direction of acoustic energy flow and the direction of the source to the receiver. Hence, positioning errors should be corrected when using horizontal energy flows to locate the source. This part is divided into three sub-problems for discussion. First of all, why are acoustic pressure transmission loss for the two sides of the oceanic front so different in the same receiver depth? Additionally, when the explosion source is located at 200 m depth, whether the three-dimensional simulation results are in better agreement with the experimental data than the two-dimensional simulation results? Eventually, sound wave propagation to the cold water side of the front is interpreted, using a 3D propagation model.

4.1. 3D Sound Propagation at Different Depths

The oceanic front is sloped, sound energy flow horizontal refraction is the most obvious of both sides of vertical direction with the strongest temperature gradient of the oceanic front. On the plane of the oceanic front, sound propagation away from front affected areas. According to the law of refraction, acoustic rays deflect to the area of the decrease of sound speed until the minimum of the temperature gradient occurs.

Topography is the controlled variable and is not allowed to change throughout the course of the simulation. The relationship between propagation attenuation and the oceanic front can quickly be established by changing the direction of propagation. The horizontal terrain with an average ocean depth of 5200 m is used for further simulation research. Figure 19a,b respectively show the sound speed top view from the sea surface at three depths and the sound pressure propagation loss in all directions which is excited by a sound source located near the centre of the front. To some extent, in Figure 19 green dotted line is the oceanic front, green solid line and the red solid line is regarded as the vertical directions of the oceanic front. In the Cartesian coordinate system with the sound source as the centre and the positive direction of the x-axis as 0 degrees, the direction of counterclockwise rotation of 72 degrees corresponds to the warm water side (red line), meanwhile the direction of anticlockwise rotation of 108 degrees corresponds to the cold water side (green line). To compare with Figure 19a,c, the oceanic front disturbs the distribution of the mixing layer. In the deep-sea mixed layer about 200 m as illustrated in Figure 19b, the sound pressure transmission loss at three depths illustrates an obvious difference. In the drawing, the green dotted line is the direction vertical to the oceanic front and the green solid line is the direction of the oceanic front. Figure 19 illustrates that with the effects of deep sound channel axis upward movement, when sound wave propagates to cold-water mass, more energy converges at the depth of 100 m under the sea surface; influenced by warm-water mass, sound channel axis is located under 200 m depth, and most of the acoustic rays skimmed over the sound channel axis region. With the increase of the transmission distance to the sound source, acoustic rays gathered in the deep-sea channel. Consequently, the cold side and the warm side are supposed to form concentric circles, a coherent circle of bright and dark with a similar distribution of sound speed appears.

Figure 20 shows the calculation results for 3D models at the depth of 175 m and 200 m, respectively. Comparing that with Figure 17 in the previous chapter, the tend of transmission loss curves basically agrees well with measurements. It's important to notice that the curve drawn by the 3-D ray theory method is in better agreement with experimental data than that by the 2-D ray theory method. We probably indicate that it is caused by the horizontal refraction of the acoustic rays, namely the three-dimensional effect of the sound field. In the horizontal direction of the sound field excited by the explosions whose depth is 100 m, the bending of the acoustic rays is not obvious enough to affect the distribution of the sound pressure transmission loss. However, the horizontal refraction excited by the explosions whose depth is 200 m cannot be neglected.

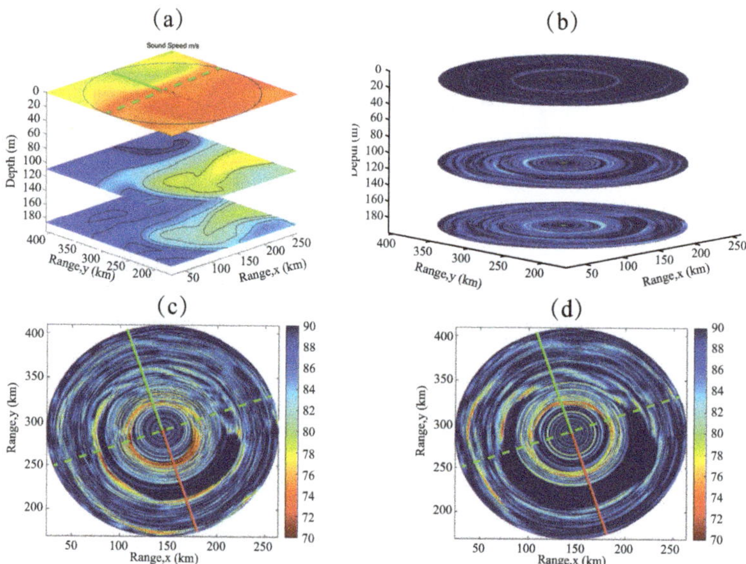

Figure 19. (**a**) Three dimensional structure of the oceanic front, (**b**) Transmission Losses 3-D for depths 0 m, 100 m and 200 m, (**c**,**d**) Transmission Loss slice for depth 100 m and 200 m.

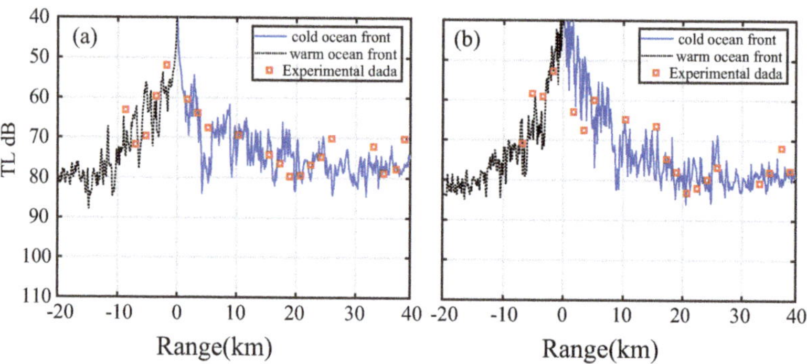

Figure 20. 3-D transmission loss in cold side (blue line), warm side (black dotted line) and experimental results with 200 m explosions (red square) in depths of 175 m (**a**), 200 m (**b**).

4.2. 3D Acoustic Ray Paths Explanations

There is no apparent difference between the curves of 2-D and 3-D simulation on the warm water side as shown by Figure 21a,b. The selective analysis is carried out by the changes of characteristic acoustic rays on the cold water side due to the influence of the oceanic front. Comparing the 2D model with the 3D model, there are more acoustic rays in the calculation process, and the sound pressure loses more energy with propagation illustrated in Figure 21. To observe the acoustic eigenrays, direct path arrival rays without surface-reflected and bottom-reflected rays represent more bending, which is the result of three-dimensional effects. At a distance of 27 km from the settled hydrophone receiver at the depth of 200 m, the 3D acoustic eigenrays have a unique convergence, and the unique convergence is not shown in the 2D simulation results, which also give evidence for Figures 17b and 20b curve distinctions at the position of 27 km.

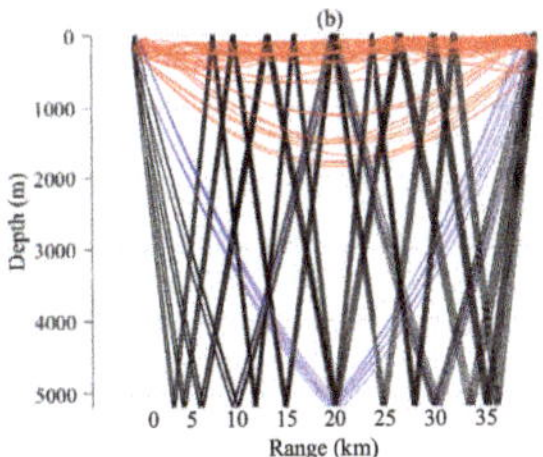

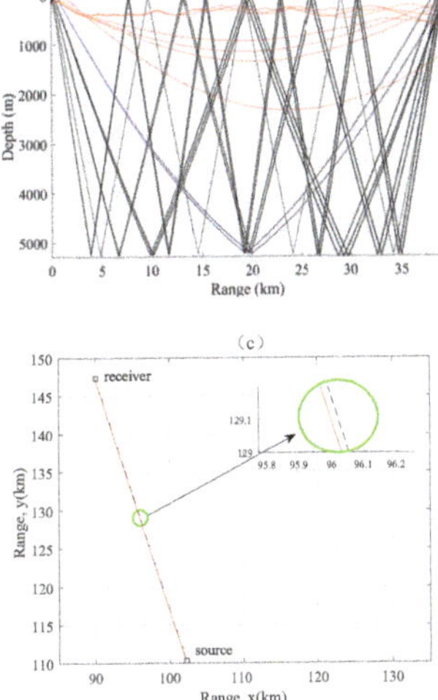

Figure 21. Comparison of sound rays between in (**a**) 2-D and (**b**) 3-D environment. Subfigure (**c**) is the top view of sound rays in 3-D simulation which showed horizontal refraction phenomenon. In the picture, red solid line is the acoustic ray and black dotted line is direction from source to the receiver.

4.3. Interpretation of Acoustic Anomaly Distribution Use 3D Propagation Model

The sound wave propagation to both sides of the oceanic front is discussed using the 3D simulation, and whether the transmission loss curve anomaly at the position of 8 km in Figure 18 is also shed light upon. Figure 22 shows the two-dimensional simulation of transmission loss distribution on the acoustic experiment's route. In Figure 23, the distribution of transmission loss is obtained from the three-dimensional simulation results of two excitation frequencies of sound sources. The red dotted line showing the sound pressure mutation is located exactly around 3 km away from the stationary as shown by the red circle in Figure 23a,b. The green dotted line showing the sound pressure mutation is located exactly around 8 km away from the stationary as shown by the green circle in Figure 23a,b. Whether the sound source frequency is 120 Hz or 200 Hz, the transmission loss at a depth of 100 m has no visible difference, while the transmission loss at 200 m depth displays a marked difference, especially at a long distance.

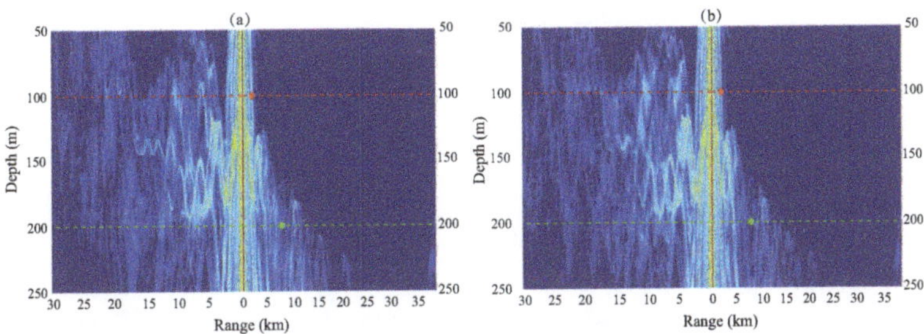

Figure 22. 2D simulation of sound pressure transmission loss at the frequency of (**a**) 120 Hz and (**b**) 200 Hz.

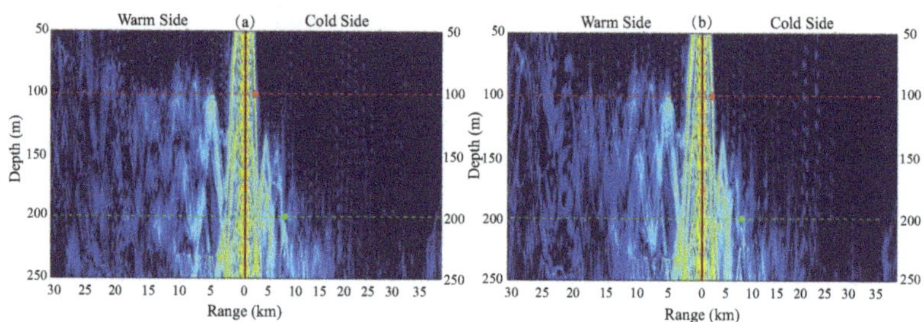

Figure 23. 3D simulation of sound pressure transmission loss at the frequency of (**a**) 120 Hz and (**b**) 200 Hz.

5. Conclusions

The Northwest Pacific oceanic front presented in this paper shows the significant effects of acoustic propagation in 2-D and 3-D models. As the common ocean dynamics phenomenon, it focuses the energy by horizontal and vertical plane refraction. The oceanic front changes the sound channel axis on both sides of the front centre. Especially on the cold water side, the sound channel is near the ocean surface which coincides with the depth interval of the mixing layer. The surface channel consists of a deep refracted path and several paths reflected from the surface and deeper upwards refraction. When the depth of hydrophone receiver is shallower than 150 m, transmission loss curves of the wave propagating to both sides of the oceanic front have the same tendency. However, when the depth of the hydrophone receivers is deeper than 150 m, there are significant differences between the curves in both directions. Comparison of diverse explosion depths by experiment expresses abnormal distribution of transmission loss, and yet anomaly position is uncertain. The decline of sonar action range owes to the oceanic front. More intuitively, the 3-D acoustic ray method is applied to study the acoustic ray paths which propagate from the explosion source towards both sides of the oceanic front. The comparison of the three-dimensional sound propagation between the experimental and numerical results illustrates that the three-dimension front environment should be considered when we discuss sound propagation and transmission loss at a depth of 200 m.

The horizontal refraction of acoustic rays carried out by front and undulated seabed are discussed. It is of interest to know how an oceanic front fell the amplitude of sound pressure at the cold water mass side. Furthermore, marine mammals nearby the sea surface could hardly be connected at some depth. Moreover, studying the effect of sound propagation could amend the inaccuracies of fish finders detection for the western North Pacific region. The additional advantage of studying horizontal refraction in the three-dimensional impact of the oceanic front is that the conclusion gives some idea and guides the placement of the vertical hydrophone array and tomography station. There are several alternative methods are available for solving these problems but spend much more time. Our further work is to overcome the slow computation and to expand our study to time-variant sound field characteristics with the existence of the oceanic front.

Author Contributions: Conceptualization, S.P.; methodology, J.L. and S.P.; software, J.L.; validation, J.L. and L.G.; formal analysis, J.L. and S.P.; investigation, J.L., L.G. and M.Z.; resources, S.P. and J.G.; writing—original draft preparation, J.L. and L.G.; writing—review and editing, J.L., S.P., L.G., S.Z., M.Z. and J.G.; visualization, S.P. and L.G.; supervision, S.P.; project administration, S.P.; funding acquisition, S.P. All authors have read and agreed to the published version of the manuscript.

Funding: This research was funded by National Natural Science Foundation of China, grant number 11904065.

Institutional Review Board Statement: Not applicable.

Informed Consent Statement: Not applicable.

Data Availability Statement: Not applicable.

Acknowledgments: The measuring of oceanic front data of the north-western ocean and the experiments of acoustic propagation were accomplished with the help of College of Meteorology and Oceanography in National University of Defense Technology.

Conflicts of Interest: The authors declare no conflict of interest.

References

1. Bowman, M.J.; Esaias, W.E. *Oceanic Fronts in Coastal Processes*; Springer: Berlin/Heidelberg, Germany, 2012.
2. Gangopadhyay, A.; Robinson, A.R. Feature-oriented regional modeling of oceanic fronts. *Dyn. Atmos. Ocean.* **2002**, *36*, 201–232. [CrossRef]
3. Henrick, R.; Siegmann, W.; Jacobson, M. General analysis of ocean eddy effects for sound transmission applications. *J. Acoust. Soc. Am.* **1977**, *62*, 860–870. [CrossRef]
4. Henrick, R.; Burkom, H. The effect of range dependence on acoustic propagation in a convergence zone environment. *J. Acoust. Soc. Am.* **1983**, *73*, 173–182. [CrossRef]
5. Mahadevan, A.; Tandon, A. An analysis of mechanisms for submesoscale vertical motion at ocean fronts. *Ocean Model.* **2006**, *14*, 241–256. [CrossRef]
6. Sokolov, S.; Rintoul, S.R. Structure of Southern Ocean fronts at 140 E. *J. Mar. Syst.* **2002**, *37*, 151–184. [CrossRef]
7. Levine, E.R.; White, W.B. Bathymetric influences upon the character of North Pacific fronts, 1976–1980. *J. Geophys. Res. Ocean.* **1983**, *88*, 9617–9625. [CrossRef]
8. Yang, W.; Li, B.; Gao, L.; Li, R.; Liu, C.; Ma, L. Ocean front locations in the southwest pacific. *Chin. J. Polar Res.* **2020**, *32*, 469–482. (In Chinese)
9. Nakamura, H.; Inoue, R.; Nishina, A.; Nakano, T. Seasonal variations in salinity of the North Pacific Intermediate Water and vertical mixing intensity over the Okinawa Trough. *J. Oceanogr.* **2021**, *77*, 199–213. [CrossRef]
10. Liu, Z.; Hou, Y. Kuroshio Front in the East China sea from satellite SST and remote sensing data. *IEEE Geosci. Remote Sens. Lett.* **2011**, *9*, 517–520. [CrossRef]
11. Xie, L.; Pallàs-Sanz, E.; Zheng, Q.; Zhang, S.; Zong, X.; Yi, X.; Li, M. Diagnosis of 3D vertical circulation in the upwelling and frontal zones east of Hainan Island, China. *J. Phys. Oceanogr.* **2017**, *47*, 755–774. [CrossRef]
12. Ying, K.; Linhui, P. Modal wave number spectrum for mesoscale eddies. *J. Ocean Univ. Qingdao* **2003**, *2*, 218–223. [CrossRef]
13. Nakamura, Y.; Noguchi, T.; Tsuji, T.; Itoh, S.; Niino, H.; Matsuoka, T. Simultaneous seismic reflection and physical oceanographic observations of oceanic fine structure in the Kuroshio extension front. *Geophys. Res. Lett.* **2006**, *33*, L23605. [CrossRef]
14. Neubert, J.A. Mean multipath intensity relation for sound propagation through a random ocean front. *J. Acoust. Soc. Am.* **1982**, *72*, 222–225. [CrossRef]
15. Liu, Q. Research on Sound Propagation in the Ocean Mesoscale Phenomenon. Ph.D. Thesis, Harbin Engineering University, Harbin, China, 2006.
16. Lee, D.; Saad, Y.; Schultz, M.H. *An Efficient Method for Solving the Three-Dimensional Wide Angle Wave Equation*; Technical Report; Yale University: New Haven, CT, USA, 1986.
17. Wang, N.; Liu, J.Z.; Gao, D.Z.; Gao, W.; Wang, H.Z. An overview of the 2005 YFIAE: Yellow Sea Oceanic Front and Internal Waves Acoustic Experiment. *J. Acoust. Soc. Am.* **2008**, *124*, 2444. [CrossRef]
18. Jian, Y.; Zhang, J.; Jia, Y. A Sound Speed Computation Model in Oceanic Front Area and Its Application in Studying the Effect on Sound Propagation. *Adv. Mar. Sci.* **2006**, *24*, 166–172.
19. Guo, T.; Gao, W. Phenomenon of ocean front and it impact on the sound propagation. *Mar. Forecasts* **2015**, *32*, 80–88. (In Chinese)
20. Chen, C.; Yang, K.; Duan, R.; Ma, Y. Acoustic propagation analysis with a sound speed feature model in the front area of Kuroshio Extension. *Appl. Ocean Res.* **2017**, *68*, 1–10. [CrossRef]
21. Zhang, Y.; Yang, K.; Xue, R.; Huang, C.; Chen, C. Convergence zone analysis for a source in the front area of Kuroshio Extension based on Argo data. In Proceedings of the OCEANS 2019—Marseille, Marseille, France, 17–20 June 2019; pp. 1–4.
22. Shapiro, G.; Chen, F.; Thain, R. The effect of ocean fronts on acoustic wave propagation in the Celtic Sea. *J. Mar. Syst.* **2014**, *139*, 217–226. [CrossRef]
23. Calado, L.; Gangopadhyay, A.; Da Silveira, I. Feature-oriented regional modeling and simulations (FORMS) for the western South Atlantic: Southeastern Brazil region. *Ocean Model.* **2008**, *25*, 48–64. [CrossRef]
24. Coppens, A.B. Simple equations for the speed of sound in Neptunian waters. *J. Acoust. Soc. Am.* **1981**, *69*, 862–863. [CrossRef]
25. Jensen, F.B.; Kuperman, W.A.; Porter, M.B.; Schmidt, H. *Computational Ocean Acoustics*; Springer: New York, NY, USA, 2011.
26. Hovem, J.M. *Ray Trace Modeling of Underwater Sound Propagation. Documentation and Use of the PlaneRay Model*; SINTEF: Trondheim, Norway, 2011.
27. Porter, M.B. *The Bellhop Manual and User's Guide: Preliminary Draft*; Technical report; Heat, Light, and Sound Research, Inc.: La Jolla, CA, USA, 2011; Volume 260.
28. Porter, M.B. Beam tracing for two- and three-dimensional problems in ocean acoustics. *J. Acoust. Soc. Am.* **2019**, *146*, 2016–2029. [CrossRef] [PubMed]

29. Červeny, V.; Pšenčík, I. Gaussian beams and paraxial ray approximation in three-dimensional elastic inhomogeneous media. *J. Geophys.* **1983**, *53*, 1–15.
30. Lynch, J.F.; Colosi, J.A.; Gawarkiewicz, G.G.; Duda, T.F.; Pierce, A.D.; Badiey, M.; Katsnelson, B.G.; Miller, J.E.; Siegmann, W.; Chiu, C.S.; et al. Consideration of fine-scale coastal oceanography and 3-D acoustics effects for the ESME sound exposure model. *IEEE J. Ocean. Eng.* **2006**, *31*, 33–48. [CrossRef]
31. Nagai, T.; Tandon, A.; Yamazaki, H.; Doubell, M.J.; Gallager, S. Direct observations of microscale turbulence and thermohaline structure in the Kuroshio Front. *J. Geophys. Res. Ocean.* **2012**, *117*, C08013. [CrossRef]

Journal of
Marine Science
and Engineering

Article

Scholte Wave Dispersion Modeling and Subsequent Application in Seabed Shear-Wave Velocity Profile Inversion

Yang Dong [1,2,3], Shengchun Piao [1,2,3], Lijia Gong [1,2,3], Guangxue Zheng [1,2,3], Kashif Iqbal [1,2,3], Shizhao Zhang [1,2,3] and Xiaohan Wang [1,2,3,*]

[1] Acoustic Science and Technology Laboratory, Harbin Engineering University, Harbin 150001, China; dong_yang@hrbeu.edu.cn (Y.D.); piaoshengchun@hrbeu.edu.cn (S.P.); lijia.gong@hrbeu.edu.cn (L.G.); zgx1057@hrbeu.edu.cn (G.Z.); kashifjamshed607@hrbeu.edu.cn (K.I.); zhangshizhao@hrbeu.edu.cn (S.Z.)
[2] Key Laboratory of Marine Information Acquisition and Security, Ministry of Industry and Information Technology, Harbin Engineering University, Harbin 150001, China
[3] College of Underwater Acoustic Engineering, Harbin Engineering University, Harbin 150001, China
* Correspondence: wangxiaohan@hrbeu.edu.cn

Abstract: Recent studies have illustrated that the Multichannel Analysis of Surface Waves (MASW) method is an effective geoacoustic parameter inversion tool. This particular tool employs the dispersion property of broadband Scholte-type surface wave signals, which propagate along the interface between the sea water and seafloor. It is of critical importance to establish the theoretical Scholte wave dispersion curve computation model. In this typical study, the stiffness matrix method is introduced to compute the phase speed of the Scholte wave in a layered ocean environment with an elastic bottom. By computing the phase velocity in environments with a typical complexly varying seabed, it is observed that the coupling phenomenon occurs among Scholte waves corresponding to the fundamental mode and the first higher-order mode for the model with a low shear-velocity layer. Afterwards, few differences are highlighted, which should be taken into consideration while applying the MASW method in the seabed. Finally, based on the ingeniously developed nonlinear Bayesian inversion theory, the seafloor shear wave velocity profile in the southern Yellow Sea of China is inverted by employing multi-order Scholte wave dispersion curves. These inversion results illustrate that the shear wave speed is below 700 m/s in the upper layers of bottom sediments. Due to the alternation of argillaceous layers and sandy layers in the experimental area, there are several low-shear-wave-velocity layers in the inversion profile.

Keywords: Scholte wave; theoretical dispersion curve; stiffness matrices; layered media

Citation: Dong, Y.; Piao, S.; Gong, L.; Zheng, G.; Iqbal, K.; Zhang, S.; Wang, X. Scholte Wave Dispersion Modeling and Subsequent Application in Seabed Shear-Wave Velocity Profile Inversion. *J. Mar. Sci. Eng.* **2021**, *9*, 840. https://doi.org/10.3390/jmse9080840

Academic Editors: Grigory Ivanovich Dolgikh, Kostas Belibassakis and Philippe Blondel

Received: 15 June 2021
Accepted: 30 July 2021
Published: 2 August 2021

Publisher's Note: MDPI stays neutral with regard to jurisdictional claims in published maps and institutional affiliations.

1. Introduction

The MASW method was initially developed as an inversion method in order to map the near-surface of Earth's structure, by recording and analyzing dispersion of Rayleigh-type surface waves. The method requires an accurately recorded Rayleigh wavefield to be analyzed for its dispersive properties. In addition, the method involves the inversion of the dispersion curve to be performed in order to yield the site properties capable of controlling the observed Rayleigh wave dispersion, i.e., mainly the shear wave velocity profile.

Contemporary studies have exhibited that the MASW method is an effective geoacoustic parameter inversion tool, which employs the dispersion curve of Scholte-type surface waves on the seafloor [1–6]. The applicable methodology of underwater MASW is explicitly illustrated in Figure 1. However, in contrast, a negligible amount of work on theoretical modeling of Scholte wave dispersion curves has been reported to date. Due to the varying excitation circumstances of the Rayleigh wave and Scholte wave, Rayleigh waves propagate along the free surface of a solid medium, whereas Scholte waves propagate at the fluid/solid interface. Although there are many methods for computing the dispersion

145

curve of the Rayleigh wave, they are not suitable for Scholte wave simulation. In this regard, it is necessary to establish the theoretical Scholte wave dispersion simulation model.

The computation of the surface wave dispersion curve not only serves for inversion, but also is the primary step of simulating the guided wave fields. The Thomson–Haskell matrix method [7,8] is an efficient method for dispersion curve computation at low frequencies. However, due to the exponentially rising term in the Thomson–Haskell matrix, the method is unstable at a relatively high frequency. In order to overcome the obstacle of a high frequency numerical value, varying attempts have been made in the literature [9–11]. Another common method for computing the dispersion curve is the reflection and transmission matrix method, which was proposed by Kennett [12,13]. Based on the dispersion forward model developed over the years, the dispersion characteristics of multi-order surface waves in complex environments are studied, which lays a foundation for its engineering application [14–16].

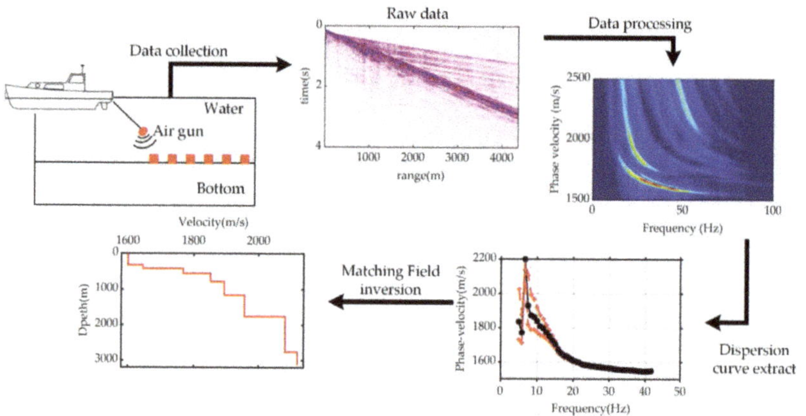

Figure 1. The applicable methodology of underwater MASW.

The stiffness matrix method is a relatively mature method to simulate the propagation of waves in layered media. It has been successfully applied to various problems [17–20] during the most recent few decades. The stiffness matrix method conjoins the displacements and stresses on both sides of the interface via the so-called stiffness matrix. The stiffness matrix is endowed with novel physical meaning, but its idea is analogous to the reflection and transmission matrix method. The stiffness matrix method has some characteristics as well as improvements, which can encourage the employment of the method in theory as well as in certain applications. Some of such characteristics are introduced and discussed in this particular study, including: (a) the forms of water layer stiffness matrices capable of analyzing fluid-elastic-coupled models; (b) the stiffness matrix of a vertically transverse isotropic elastic layer.

The shear wave velocity profile of the seafloor not only provides an important indicator for sedimentary rigidity, but also plays an important role in sedimentary characteristics, seismic detection, and assessment of earth hazards [21]. Furthermore, the variation of shear wave velocities provides an important mechanism for ocean acoustic propagation loss in environments with low shear velocities. This particular mechanism may be regarded as a major factor to be considered in establishing acoustic propagation models as well as in predicting sonar performance [22]. It is an efficient and convenient method to estimate the velocity gradient of the shearing wave in the seabed by inverting the dispersion curve of the seawater–seabed interface wave [6,23]. In recent years, Bayesian inversion methods have made great progress and played an important role in estimating seabed properties and uncertainties based on Bayesian formulas [24,25]. It can estimate the parameters of the maximum posteriori probability (MAP) model effectively, and then analyze the uncertainty of the inversion results qualitatively and quantitatively from the perspective of statistics. In our inversion work, high quality multi-channel seismic data are processed and good Scholte wave dispersion curves are obtained. Our Bayesian inversion produces unexpected new results. Previous inversion results of s-wave velocity profiles show that the s-wave velocity increases with the increase of depth. We found that under Jiao Lai basin sedimentary can be divided into argillaceous layers (clayey silt and silty sand) and the sand layer, two kinds of sediment appear alternately, caused the frequent alternation of high speed and low speed layer.

In the ensuing sections of the manuscript, we present the theoretical formulas of the stiffness matrix method and their root-searching schemes. In addition, the stability of the stiffness matrix method is analyzed. Similarly, the anomalies among the Rayleigh-type surface wave and Scholte type surface wave are illustrated. Furthermore, the impact of dispersion curves due to a vertically transverse isotropic layer is discussed. Finally, the fundamental and first higher-order mode Scholte waves excited by a large-volume air gun array near the sea surface are obtained by employing multi-channel seismic records. Similarly, ocean bottom shear wave profiles are obtained by phase velocity inversion.

2. Brief Description of Theory

2.1. Scholte Wave Dispersion Modeling

The stiffness matrix method is employed in this particular study to directly relate the forces at the layer interface to the displacements at the same position. The horizontal layered solid medium model is considered in this study, whose surface layer is a liquid layer, which can actually simulate the stratified marine earth model, as depicted in Figure 2. Considering a plane wave propagating in the model, the x-axis represents the wave propagation direction along the sea surface, whereas the z-axis is computed as a positive downward into the medium.

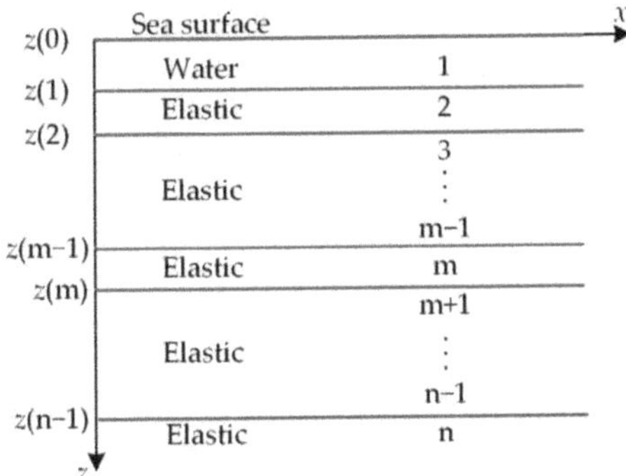

Figure 2. Layered ocean structure with horizontal interfaces.

In order to express the stiffness matrix of the elastic medium from the second layer to the $(n - 1)th$ layer, the reference [17] has derived it in detail. The external load vector on the upper and lower interfaces of the mth layer are expressed by the displacements at the interfaces as:

$$\begin{bmatrix} p_m(z_{m-1}) \\ p_m(z_m) \end{bmatrix} = \begin{bmatrix} K_{11}^m & K_{12}^m \\ K_{21}^m & K_{22}^m \end{bmatrix} \begin{bmatrix} d_m(z_{m-1}) \\ d_m(z_m) \end{bmatrix}, \tag{1}$$

or briefly,

$$P_m = K^m D_m, \tag{2}$$

where K^m is the stiffness matrix of the layer with 2×2 submatrix K_{ij}^m; $P_m = \begin{bmatrix} \sigma_m(z_m) & \tau_m(z_m) & -\sigma_m(z_{m-1}) & -\tau_m(z_{m-1}) \end{bmatrix}^T$ is the external load vector, the symbols σ and τ represent the normal and shear stresses at the interface, respectively;

and $D_m = \begin{bmatrix} u_m(z_{m-1}) & w_m(z_{m-1}) & u_m(z_m) & w_m(z_m) \end{bmatrix}^T$ is the displacement vector, the symbols u and w represent the horizontal and vertical displacements at the interface, respectively. It can be illustrated that K^m is symmetric.

For the half-space layer, which satisfies the infinite radiation condition, there is no upward propagating wave. The relationship among potential functions, displacements, and stresses at the upper interface is:

$$
\begin{bmatrix} u_n(z_{n-1}) \\ w_n(z_{n-1}) \\ \sigma_n(z_{n-1}) \\ \tau_n(z_{n-1}) \end{bmatrix} = \begin{bmatrix} -ik & -i\beta_n & ik & i\beta_n \\ -i\alpha_n & -ik & i\alpha_n & -ik \\ \mu_n a_n & 2i\mu_n k\beta_n & \mu_n a_n & -2i\mu_n k\beta_n \\ -2i\mu_n k\alpha_n & \mu_n a_n & 2i\mu_n k\alpha_n & \mu_n a_n \end{bmatrix} \begin{bmatrix} 0 \\ 0 \\ \varphi_n^D(z_{n-1}) \\ \psi_n^D(z_{n-1}) \end{bmatrix},
\tag{3}
$$

where $a_m = 2k^2 - \omega^2/(c_{sm}^2)$; $\alpha_n = \frac{\omega}{c}\sqrt{1 - (c/c_{pn})^2}$ and $\beta_n = \frac{\omega}{c}\sqrt{1 - (c/c_{sn})^2}$ are vertical wavenumbers; μ_n is the Lamé coefficients ($\mu_1 = 0$ for the water layer) and φ_n^D is the potential function of the downward propagating wave in the half-space layer. We transformed Equation (3) into the following form:

$$
\begin{bmatrix} d_n(z_{n-1}) \\ F_n(z_{n-1}) \end{bmatrix} = \begin{bmatrix} H_{12}^n \\ H_{22}^n \end{bmatrix} \Phi_n(z_{n-1}),
\tag{4}
$$

where $\Phi_n(z_{n-1}) = \begin{bmatrix} \varphi_n^D(z_{n-1}) & \psi_n^D(z_{n-1}) \end{bmatrix}^T$, $H_{12}^n = \begin{bmatrix} T_{13}^n & T_{14}^n \\ T_{23}^n & T_{24}^n \end{bmatrix}$, $d_n(z_{n-1}) = \begin{bmatrix} u_n(z_{n-1}) \\ w_n(z_{n-1}) \end{bmatrix}$, $F_n(z_{n-1}) = \begin{bmatrix} \sigma_n(z_{n-1}) \\ \tau_n(z_{n-1}) \end{bmatrix}$ and $H_{22}^n = \begin{bmatrix} T_{33}^n & T_{34}^n \\ T_{43}^n & T_{44}^n \end{bmatrix}$. According to the above equation, the potential function can be expressed as:

$$
\Phi_n(z_{n-1}) = (H_{12}^n)^{-1} d_n(z_{n-1}),
\tag{5}
$$

$$
\Phi_n(z_{n-1}) = (H_{22}^n)^{-1} F_n(z_{n-1}),
\tag{6}
$$

Then, the stress vector can be expressed by the displacement vector as:

$$
F_n(z_{n-1}) = H_{22}^n (H_{12}^n)^{-1} d_n(z_{n-1}),
\tag{7}
$$

Defining the external forces on upper interface z_{n-1} as $p_n(z_{n-1}) = F_n(z_{n-1})$, Equation (7) can be expressed in the form of stiffness matrix as in [26]:

$$
p_n(z_{n-1}) = K^n d_n(z_{n-1}),
\tag{8}
$$

There is no shear component in the fluid layer, and the displacement of the sea surface and seabed can be expressed by the potential function as:

$$
\begin{bmatrix} w_1(z_0) \\ w_1(z_1) \end{bmatrix} = \frac{\alpha_1}{\sinh \alpha_1 h_1} \begin{bmatrix} -\cosh \alpha_1 h_1 & 1 \\ -1 & \cosh \alpha_1 h_1 \end{bmatrix} \begin{bmatrix} \varphi_1(z_0) \\ \varphi_1(z_1) \end{bmatrix},
\tag{9}
$$

where w_1 is the vertical displacement and φ_1 is the velocity potential in water. In the background of the layer stiffness matrix method, we need to consider the volume perturbance as the flux condition at the interface, for example, the injection volume is expressed by the source in the form of the air gun at the interface. We employ a set of constitutive laws of $\bar{v}_1(z_0) = -j\omega w_1(z_0)$, $\bar{v}_1(z_1) = -j\omega w_1(z_1)$, $p_1(z_0) = -\sigma_1(z_0)$ and $p_1(z_1) = \sigma_1(z_1)$. Then, we acquire the following stiffness matrix [17] as:

$$
K^1 = \frac{\rho_1 \omega^2}{\alpha_1 \sinh \alpha_1 h_1} \begin{bmatrix} \cosh \alpha_1 h_1 & -1 \\ -1 & \cosh \alpha_1 h_1 \end{bmatrix},
\tag{10}
$$

The aforementioned stiffness matrices for the water and elastic layers are employed to construct the global stiffness matrix for a layered system. The global stiffness matrix is developed by conjoining the stiffness matrices of Equations (2), (8) and (10) as:

$$
K = \begin{bmatrix}
K_{11}^1 & K_{12}^1 & 0 & 0 & 0 & 0 & 0 & 0 \\
0 & \begin{bmatrix} 0 & 0 \\ 0 & K_{22}^1 \end{bmatrix} + K_{11}^2 & K_{12}^2 & 0 & 0 & 0 & 0 & 0 \\
K_{21}^1 & & & & & & & \\
0 & K_{21}^2 & K_{22}^2 + K_{11}^3 & K_{12}^3 & 0 & 0 & 0 & 0 \\
0 & 0 & K_{21}^3 & K_{22}^3 + K_{11}^4 & K_{12}^4 & 0 & 0 & 0 \\
0 & 0 & 0 & \ddots & \ddots & \ddots & 0 & 0 \\
0 & 0 & 0 & 0 & K_{21}^{n-3} & K_{22}^{n-3} + K_{11}^{n-2} & K_{12}^{n-2} & 0 \\
0 & 0 & 0 & 0 & 0 & K_{21}^{n-2} & K_{22}^{n-2} + K_{11}^{n-1} & K_{12}^{n-1} \\
0 & 0 & 0 & 0 & 0 & 0 & K_{21}^{n-1} & K_{22}^{n-1} + K^n
\end{bmatrix}, \quad (11)
$$

The global stiffness matrix for a system comprising of $(n-2)$ elastic layers, with the half-space layer, and overlaid by water, is a $(2n-1) \times (2n-1)$ symmetric matrix, which satisfies the following relationship as:

$$
p = Kd, \quad (12)
$$

The sea surface free boundary condition is given by:

$$
p_0 = 0, \quad (13)
$$

For any other interface due to the continuity of stress, all elements are equivalent to zero. Thus, the solution for Scholte waves must satisfy $Kd = 0$, and the dispersion function is therefore expressed as:

$$
\Delta(\omega, k) = |K| = 0, \quad (14)
$$

The overall stiffness matrix of the entire system can be combined by employing this particular method. The global load vector corresponds to the external forces at the interface of the layered system. By using the stiffness matrix method, the characteristic equation can be established to determine the mode shapes of the layered system. Thus, the theoretical dispersion curve of the surface wave propagating in the layered system can be ascertained.

2.2. Bayesian Inversion Theory

The solution of the inversion problem is transformed into the solution of parametric posteriori probability density (PPD) based on Bayesian theory in this particular study. Adaptive Simplex Simulated Annealing (ASSA), a nonlinear global optimization algorithm, is used to solve the maximum of a posterior probability (MAP) model, and then a Bayesian information criterion (BIC) is employed to select the model.

The experimentally measured data vector is set as **d** with element d_i, and **m** as the model vector composed of the bottom parameters m_i to be retrieved. Both d_i and m_i were random variables, which were correlated by Bayes' rule [27]

$$
P(\mathbf{m}|\mathbf{d}) = P(\mathbf{d}|\mathbf{m})P(\mathbf{m})/P(\mathbf{d}), \quad (15)
$$

where $P(\mathbf{m}|\mathbf{d})$ is the PPD; $P(\mathbf{m}|\mathbf{d})$ is the conditional probability density function (PDF) of **m** under measured data; $P(\mathbf{m})$ is the prior PDF of **m**, $P(\mathbf{d})$ is the PDF of **d**. As the $P(\mathbf{d})$ is independent of **m**, and the $P(\mathbf{d}|\mathbf{m})$ can be regarded as the likelihood function $L(\mathbf{d}|\mathbf{m})$ for the measured data [27], Equation (15) can be written as:

$$
P(\mathbf{m}|\mathbf{d}) \propto L(\mathbf{m})P(\mathbf{m}), \quad (16)
$$

The likelihood function is ascertained by the form of the data and the statistical distribution of the data errors. Regarding the difficulty of acquiring independent estimation of error statistics in pragmatic applications, the assumption of unbiased Gaussian error is adopted for processing, and the form of the likelihood function is:

$$L(\mathbf{m}) = P(\mathbf{d}\,|\mathbf{m}) \propto \exp[-E(\mathbf{m})], \tag{17}$$

where $E(\mathbf{m})$ is the cost function. Hence, the $P(\mathbf{m}\,|\mathbf{d})$ becomes:

$$P(\mathbf{m}\,|\mathbf{d}) \propto \exp[-E(\mathbf{m})]P(\mathbf{m}), \tag{18}$$

Normalizing the above equation, we have:

$$P(\mathbf{m}\,|\mathbf{d}) = \frac{\exp[-E(\mathbf{m})]P(\mathbf{m})}{\int \exp[-E(\mathbf{m}')]P(\mathbf{m}')d\mathbf{m}'}, \tag{19}$$

The domain of integration spans the M-dimensional parameter space, and the integrated components are distinguished by single quotation marks, where M is the number of parameters waiting for inversion.

In Bayesian theory, the posterior probability density (PPD) of $\mathbf{m}$ can be regarded as the inversion outcomes. To interpret the M-dimensional PPD requires estimating properties of the parameter value, uncertainties, and inter-relationships, i.e., the MAP model $\hat{\mathbf{m}}$, mean model $\overline{\mathbf{m}}$, model covariance matrix $\mathbf{C_m}$ and marginal probability distributions $P(m_i\,|\,\mathbf{d})$. These properties are respectively defined as:

$$\hat{\mathbf{m}} = \text{Arg}_{\max}\{P(\mathbf{m}\,|\mathbf{d})\}, \tag{20}$$

$$\overline{\mathbf{m}} = \int \mathbf{m}'\, P(\mathbf{m}'\big|\mathbf{d})d\mathbf{m}', \tag{21}$$

$$\mathbf{C_m} = \int (\mathbf{m}' - \overline{\mathbf{m}'}))(\mathbf{m}' - \overline{\mathbf{m}'})^T\, P(\mathbf{m}'|\mathbf{d})d\mathbf{m}', \tag{22}$$

$$P(m_i|\mathbf{d}) = \int \delta(m_i - m_i')P(\mathbf{m}'\big|\mathbf{d})d\mathbf{m}', \tag{23}$$

where T is the transpose symbol.

3. Dispersion Simulation of Layered Ocean Bottom

Theoretical dispersion curves can be obtained by solving Equation (14). There are four main techniques to ascertain the roots, i.e., the bisection searching technique [28], two dimensional searching technique [29], least absolute determinant [30], and phase-velocity scanning technique [31]. Since the singularity of a stiffness matrix can be better characterized by its minimum absolute eigenvalue, the method of least absolute determinant is employed for this particular study.

3.1. Fundamental Mode Dispersion

The method was analyzed for a number of varying cases, of which a couple are presented here. Table 1 illustrates the first model of three layers of an elastic medium covered with a liquid layer. The compressional wave velocity v_p, shear wave velocity v_s of elastic medium and the medium density increases with depth. The compressional wave velocity of water is 1500 m/s and the density is 1.0 g/cm^3.

An effort to ascertain the roots of the dispersion function of Model 1 at the frequency of 5, 10 and 20 Hz is conducted. The absolute determinant of stiffness matrix and their roots are illustrated in Figure 3. By comparing the top and bottom panels of Figure 3, the roots of high frequency dispersion function in the phase velocity domain are more similar to those of the low frequency dispersion function. In the bottom panel of Figure 3, the frequency is high, i.e., up to 20 Hz, and no numerical instability is found in the stiffness matrix method. We have computed the dispersion function of the Scholte wave, i.e., up to

50 Hz, and did not observe any numerical instabilities of high frequency. Furthermore, we acquired reliable and stable solutions by testing other models as well.

Table 1. Parameters of layered ocean Model 1.

Case 1	v_p (m/s)	v_s (m/s)	Density (g/cm^3)	Thickness (m)
Layer 1	1500	-	1.0	100
Layer 2	1800	500	1.5	70
Layer 3	2400	800	1.9	50
Layer 4	2800	1200	2.3	Infinite

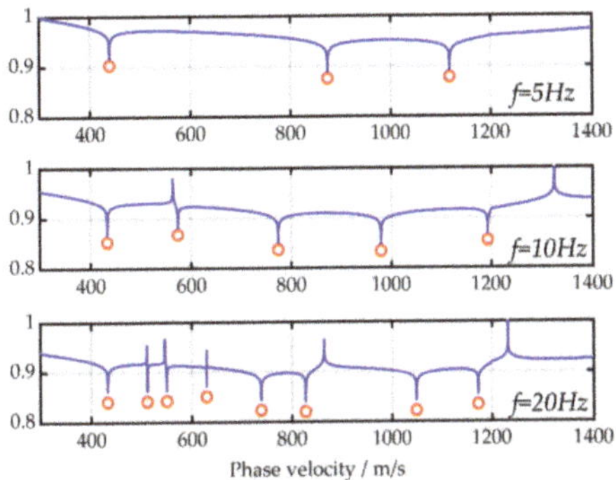

Figure 3. Normalized dispersion function of the Stiffness matrix method of model 1 at the frequencies of 5, 10, and 20 Hz, along with its roots (illustrated by circles).

By solving the dispersion function at the frequency 0.1–50 Hz, the first root at all frequencies constitutes the fundamental mode dispersion curve of the Scholte wave (red solid line in Figure 4a). In order to analyze the influence of the water layer, the fluid layer of Model 1 was removed, then the dispersion curve of the Rayleigh wave was computed (Figure 4a blue circles). At high frequencies, the phase velocities of both types of surface waves do not alter with frequency. The wave velocity is higher for the fundamental mode in case there is no liquid surface, the fundamental mode Scholte wave velocity is 432 m/s, while the Rayleigh wave velocity is 475 m/s. It is evident that regardless of the existence of the liquid layer, the fundamental mode of the dispersion curve of the surface wave has no cut-off frequency.

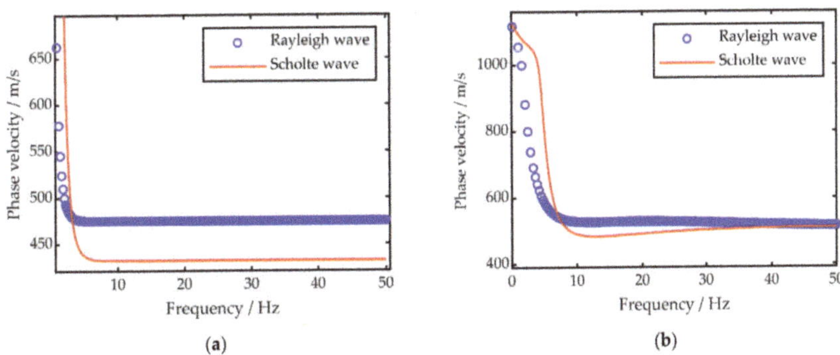

Figure 4. Fundamental mode dispersion curve of Model 1 (**a**) and Model 2 (**b**). The red solid line is the Scholte wave dispersion curve and the blue circles are the Rayleigh wave dispersion curve of the models removing the water layer.

Subsequently, Model 2, with a low shear-velocity layer, is analyzed. The low shear-velocity layer is defined such that the shear velocity of the layer is lower than that of its overburden layer. The model parameters are exhibited in Table 2. The third layer is a low shear-velocity layer with a thickness of 20 m. In addition, Model 2 is different from Model 1 in that its layer thickness is smaller and the medium parameters fluctuate severely. It is pertinent to mention that Model 2 employs the same root search scheme as Model 1.

Table 2. Parameters of layered ocean Model 2.

Case 2	v_p (m/s)	v_s (m/s)	Density (g/cm^3)	Thickness (m)
Layer 1	1500	-	1.0	20
Layer 2	1800	600	1.5	10
Layer 3	2400	500	1.9	20
Layer 4	2400	800	1.9	30
Layer 5	2800	1200	2.3	Infinite

The red solid line in Figure 4b represents the fundamental mode dispersion curve of Model 2, and the blue circles are the dispersion curve of the Rayleigh wave which is computed by removing the fluid layer of Model 2. By comparing the left and right panels of Figure 2, the dispersion curves of Model 2 exhibit larger phase-velocity perturbations than those of Model 1. The surface wave penetration depth in the elastic medium is roughly one-half of the Scholte wavelength [32]. When the thickness of layers is small, the penetration depth of the Scholte wave in the solid is larger when compared to the thickness of upper layers in the selected frequency band (Figure 5b). Consequently, the phase velocity will be affected by the layered model parameters. For the model with a large layer thickness, the penetration depth of the Scholte wave is smaller than the thickness of the first elastic layer (Figure 5a), and the dispersion curve is close to the case of the semi-infinite seabed. The bending of the dispersion curve of Model 1 requires very low frequency.

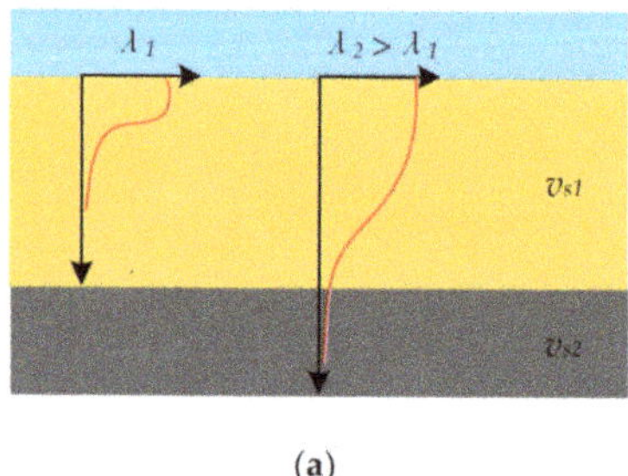

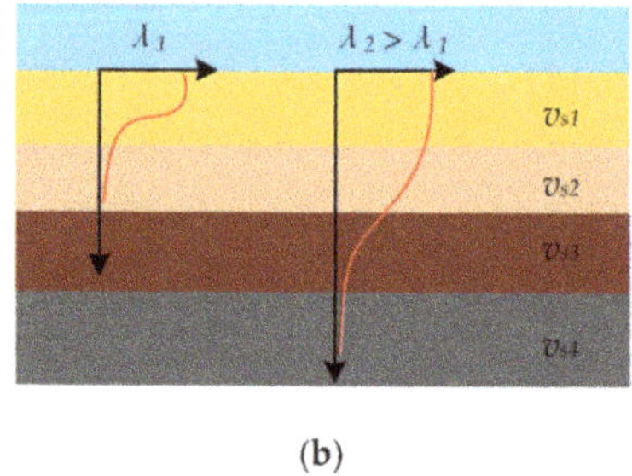

(a)   (b)

Figure 5. Fundamental mode Scholte wave penetration depth in the ocean bottom. (**a**) The model of large layer thickness; (**b**) the model of smaller layer thickness. In the figure represents the wavelength of Scholte waves (The subscript 1 represents the high frequency and 2 repre-sents the low frequency), and represents the s-wave velocity of the ith layer.

3.2. Fundamental and Higher Mode Dispersion

Figure 6a demonstrates the dispersion curves of the first nine modes, i.e., fundamental mode, as well as eight higher modes of Model 1. Higher-order modes have larger phase-velocity perturbations than the lower-order modes. All the higher modes approach the velocity of 500 m/s at high frequencies, which is the shear wave velocity in the first elastic layer of Model 1.

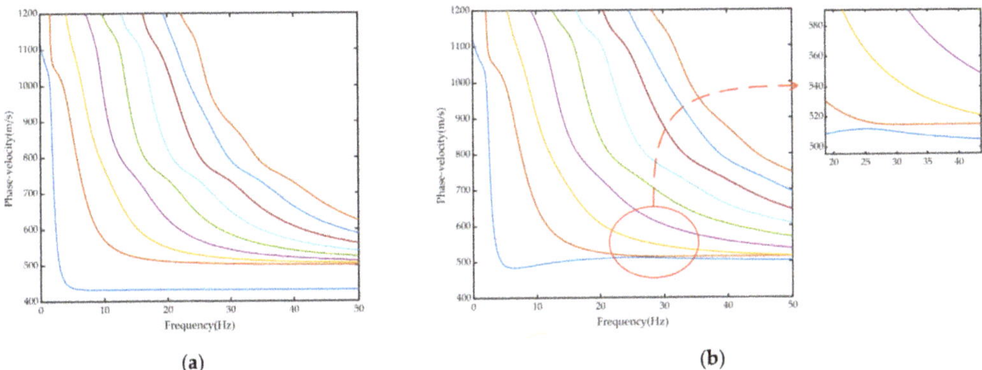

Figure 6. The dispersion curves of the first nine modes of (**a**) Model 1 and (**b**) Model 2.

Figure 6b represents the dispersion curves of the first nine modes, i.e., fundamental mode, as well as eight higher modes of Model 2. Due to the existence of a low shear-velocity layer, the first two mode curves get very close to each other at a frequency range of 25–30 Hz. The zoomed panel in Figure 6b illustrates that there is no point of intersection, and the two curves indeed do not intersect. However, the distance among two curves is negligibly small. Boaga, J. et al. [33] proved that the Rayleigh wave dispersion curve has the same rule when there is a low shear-velocity layer. They titled it, "osculation frequency", and proved that it exists in strong impedance contrast. Below this frequency, the energy of the vertical component is transferred from the fundamental to the first higher mode. In particular, the consequences of inverting the apparent dispersion curve are severe, and result from the combination of the two modes. Due to this combination, it is erroneously assumed to correspond to the fundamental mode. The errors in terms of depth and velocity of the bedrock can be large, and these parameters are of utmost interest in engineering surveys.

The higher mode dispersion curve in Figure 6a as well as in Figure 6b appear to be more sensitive to the layered model parameters as compared to the fundamental mode. Scholte wave amplitudes decrease exponentially with increasing distance from the interface, so it is confined to a narrow stratum in the proximity of the interface. For higher modes, the horizontal wave number $k = \omega / c_{phase}$. The higher the order, the smaller the incidence angle, and the larger the number of seabed layers are affected, as illustrated in Figure 7. Employing MASW for the inversion of ocean bottom parameters, including higher order modes, offers additional information leading to higher resolution and smaller uncertainties.

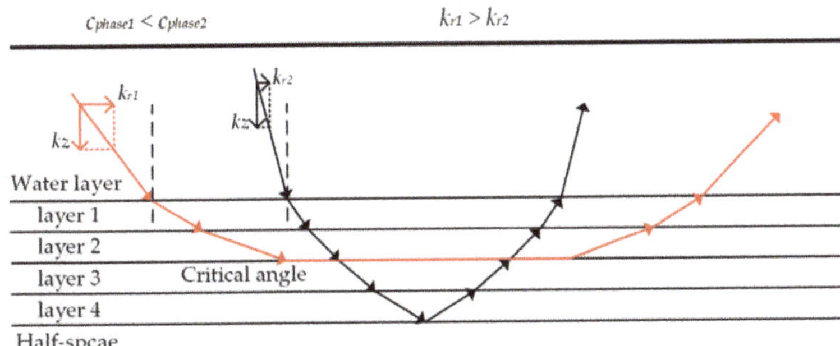

Figure 7. Plane waves with different incident angles propagating in the layered model. In the figure λ represents the wavelength of Scholte waves (The subscript 1 represents the high frequency and 2 represents the low frequency), and v_{si} represents the s-wave velocity of the ith layer.

4. Geoacoustic Inversion with Experimental Data

4.1. Experimental Configuration and Data Description

The multi-channel seismic exploration experiment was carried out in the southern Yellow Sea of China. The experimental area is exhibited in Figure 8a. A west–east-oriented survey with water depth ranging from 20 to 80 m was conducted. The topographic perturbations along the survey line are depicted in Figure 8b. The sound source was comprised of an air gun array with a total capacity of 2940 in^3 and was deployed at a depth of 8 m. Marine towing cable was employed for the acquisition purposes, cable sinking depth remained at about 12 m, and the receiving array had a spacing of 12.5 m. The shot spacing is 37.5 m.

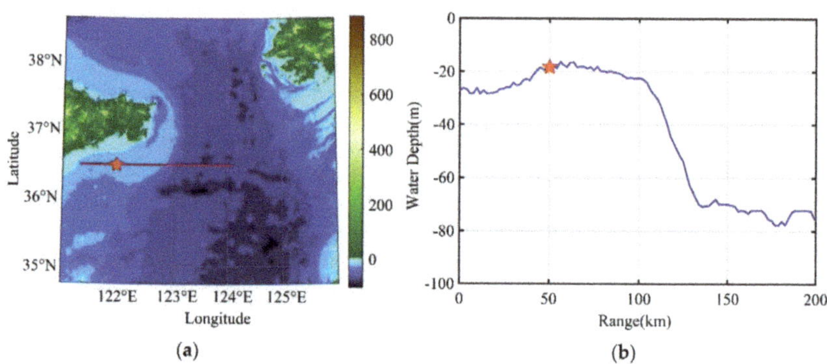

(a)   (b)

Figure 8. (**a**) The experimental area with survey line (red line). (**b**) The topographic perturbations along the survey line.

The temporary signal recording was of good quality. Figure 9 illustrates the time domain signals recorded by a single shot. The relative position of the shot point is indicated by a star in Figure 8. In order to improve the signal-to-noise ratio of the interface wave, the data were passed through a 1–12 Hz bandpass filter. The very-low-frequency arrival signal marked by the arrow is identified as the Scholte wave with a group velocity of about 150 m/s. After filtering, the time-frequency analysis method was employed to extract dispersion curves, as described below.

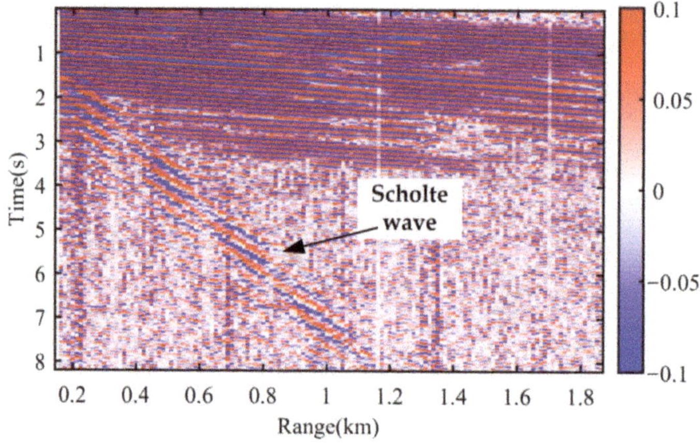

Figure 9. Signals recorded by a single shot.

Dispersion curves extracted by time-frequency analysis are usually divided into two categories: single-sensor method and multi-sensor method [34]. The single sensor method estimates the group velocity dispersion of one trace. In this paper, the forward model is used to solve the dispersion characteristics of phase velocity, and the cost function employed in the subsequent inversion process is the difference value of phase velocity. Furthermore, the most obvious phase velocity dispersion curve images are usually obtained by employing the multi-sensor approach [35]. A couple of multi-sensor processing methods are usually employed to extract phase velocity dispersion curves: frequency wavenumber (f-k) spectrum and slowness frequency (p-ω) transform. In this study, the slowness frequency (p-ω) transform method is employed to extract the dispersion data. Firstly, we transform the signal from the space-time (x-t) domain to the (τ-p) domain by linear Radon transform:

$$m(p,\tau) = \int_{-\infty}^{\infty} d(x, t = \tau + px)dx, \tag{24}$$

where τ is the reduced time; p is slowness. The integral across x in Equation (24) is done at constant τ, which is a slanting line in the (x,t)-plane. Secondly, take the Fourier transform of data in the (τ-p) domain along the time direction to obtain the slowness frequency (p-ω) data. Finally, the dispersion energy spectrum of the surface wave can be extracted by converting the data from the (p-ω) domain to the phase velocity frequency (v-f) domain by employing the relation $p = 1/v$. The dispersion diagram of the Scholte wave shown in Figure 9 is plotted in Figure 10. A couple of modes are extracted for the Scholte wave: the fundamental mode and the first higher-order modes. The modes are relatively well-separated. The fundamental mode between 1.6 and 3.2 Hz and the first higher-order mode between 2.4 and 3.0 Hz are obvious, with sharp peaks and good continuity. The black curves are the extracted dispersion data, which will be used for Bayesian inversion.

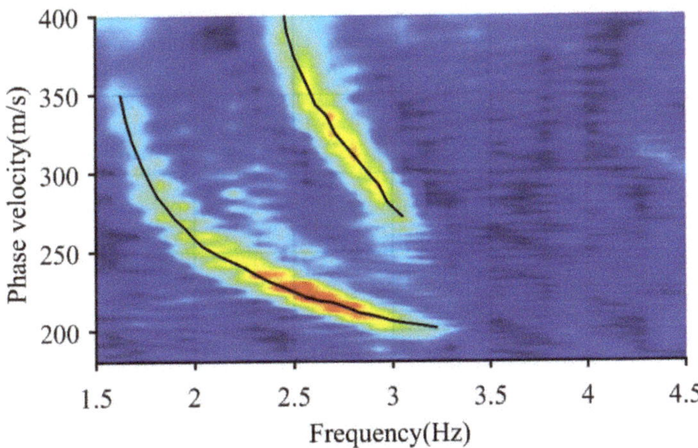

Figure 10. Scholte wave phase velocity dispersion diagram. The extracted fundamental mode and first higher-order mode dispersion data are represented by black curves.

4.2. Analysis of Inversion Result

The initial model parameters are illustrated in Table 3. The dispersion of the Scholte wave is mainly affected by the bottom shear wave velocity structure, but is relatively less sensitive to the other parameters. The initial P-wave velocity structure and density structure are determined by the empirical function [36]:

$$v_p = 1.511 + 1.304d - 0.714d^2 + 0.257d^3, \tag{25}$$

$$v_p = 2.3304 - 0.257\rho + 0.4877\rho^2, \tag{26}$$

where v_p is P-wave velocity, d is depth and ρ is density. The initial shear-wave velocity profile is determined from dispersion data, and its semi-infinite layer parameters (maximum) and seabed surface parameters (minimum) are determined from the range of phase velocities based on the empirical relationship [37]:

$$v_s = c_R / 0.88, \tag{27}$$

where c_R is the phase velocity of surface wave. The shear-wave velocity of other layers is linearly distributed by depth. In our inversion, the number of layers in the shear wave velocity profile is determined by minimizing the Bayesian information criterion (BIC) [5]. Here, a model of 13 layers was chosen, with each layer having a constant thickness of 50 m.

Table 3. Parameters of the initial model.

layer	v_p (m/s)	v_s (m/s)	Density (g/cm^3)	Thickness (m)
1	1500	-	1.0	20
2	1537	238	1.49	50
3	1602	266	1.54	50
4	1667	294	1.60	50
5	1732	322	1.65	50
6	1797	350	1.69	50
7	1863	378	1.74	50
8	1928	405	1.78	50
9	1993	433	1.83	50
10	2057	461	1.87	50
11	2123	489	1.91	50
12	2189	517	1.95	50
13	2254	545	1.99	Inf

The final inversion results are illustrated in Figure 11a, which estimate the shear wave velocity profile along the estimated standard deviation and express the standard deviation as an error bar. A total of 23 unknown parameters were estimated, including 12 shear wave velocities and 11-layer thickness. The sea bottom depth alters from 20 to 700 m, and the shear wave sound velocity ranges from 181–653 m/s. From shallow to deep, the standard deviation is about 50 to 150 m/s. The upper sediments at the bottom of the ocean are extremely soft and have very small shear wave velocities. In addition, low S-wave velocity layers appear at the depth of 200 and 500 m. The actual sedimentary structure under the seabed can be divided into argillaceous layer (clayey silt, silt) and sandy layer. The two kinds of sediments alternate, resulting in frequent alternation of high-speed and low-speed layers. The same conclusion is obtained from the seismic profile of CSDP-2 acoustic well logging in the near sea area in reference [32]. It can be seen from the velocity curve given in the literature that although the overall velocity increases with the increase of depth, it shows the characteristics of alternating change under the background of the overall trend. The theoretical dispersion curves are in strong agreement with the experimental values, as illustrated in Figure 11b.

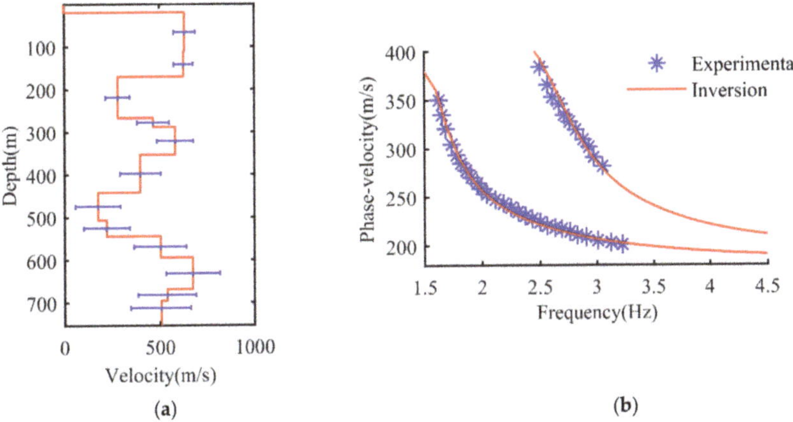

Figure 11. (**a**) Inversion results of shear wave velocity profile with an error bar. (**b**) Comparison between experimental and theoretical dispersion curves.

5. Conclusions

Stiffness matrix methods are introduced in this study for the fundamental and higher mode of Scholte wave dispersion curve computations. It is tested in a vertically transverse isotropic elastic layer, and remains stable for high frequencies in the unlimited range. The dispersion curves of the fundamental mode of the Scholte wave and the fundamental mode of the Rayleigh wave are compared without a fluid layer, and the wave velocity is higher for the fundamental mode when there is no liquid surface. By computing the dispersion curves of different ocean environment models, it is found that when there is a low shear-velocity layer, the coupling phenomenon between the fundamental mode and the first higher-order mode occurs. Due to the large incident angle of high-order modes, it is difficult to reach the critical angle in the layered seabed, i.e., the incident depth is larger. The higher mode dispersion curve appears to be more sensitive to the layered model parameters as compared to the fundamental mode.

In our multi-channel seismic exploration experiment, the Scholte wave generated by the near-sea air gun array is acquired. The fundamental and first-higher mode Scholte wave dispersion curves are extracted from the experimental data, and ocean bottom shear wave profiles are obtained by phase velocity inversion. The thickness and parameters of the topmost layer are estimated with relatively negligible errors. The maximum depth of penetration is nearly 700 m below the sea floor, and the sensitivity is highest at the top layers. The very low shear wave velocities near the sea floor explain the weak excitation of Scholte waves at relatively low frequencies. The inversion results show that the sedimentary structure in the experimental sea area can be divided into a muddy layer and sandy layer, and the two kinds of sediments appear to alternate. As a result, this shear wave velocity structure will be capable of providing significant information regarding the study of seismic processing, offshore field response analysis, and underwater acoustic transmission loss computation in this particular area.

Author Contributions: Conceptualization, S.P.; methodology, Y.D. and S.P.; software, Y.D.; validation, Y.D. and L.G.; formal analysis, Y.D. and S.P.; investigation, Y.D. and G.Z.; resources, S.P.; writing—original draft preparation, Y.D., S.Z. and K.I.; writing—review and editing, Y.D., S.P., L.G., G.Z., S.Z. and K.I.; visualization, S.P. and L.G.; supervision, S.P. and X.W.; project administration, S.P.; funding acquisition, S.P. and X.W. All authors have read and agreed to the published version of the manuscript.

Funding: This research was funded by National Natural Science Foundation of China, grant number 11904065.

Institutional Review Board Statement: Not applicable.

Informed Consent Statement: Not applicable.

Conflicts of Interest: The authors declare no conflict of interest.

References

1. Kaufmann, R.D.; Xia, J.; Benson, R.C.; Yuhr, L.B.; Casto, D.W.; Park, C.B. Evaluation of MASW data acquired with a hydrophone streamer in a shallow marine environment. *J. Environ. Eng. Geophys.* **2005**, *10*, 87–98. [CrossRef]
2. Kugler, S.; Bohlen, T.; Bussat, S.; Klein, G. Variability of scholte-wave dispersion in shallow-water marine sediments. *J. Environ. Eng. Geophys.* **2005**, *10*, 203–218. [CrossRef]
3. Park, C.B.; Miller, R.D.; Xia, J.; Ivanov, J.; Sonnichsen, G.V.; Hunter, J.A.; Good, R.L.; Burns, R.A.; Christian, H. Underwater MASW to evaluate stiffness of water-bottom sediments. *Lead. Edge* **2005**, *24*, 724–728. [CrossRef]
4. Xuan, N.N.; Dahm, T.; Grevemeyer, I. Inversion of Scholte wave dispersion and waveform modeling for shallow structure of the Ninetyeast Ridge. *J. Seismol.* **2009**, *13*, 543–559. [CrossRef]
5. Li, C.; Dosso, S.E.; Dong, H.; Yu, D.; Liu, L. Bayesian inversion of multimode interface-wave dispersion from ambient noise. *IEEE J. Ocean. Eng.* **2012**, *37*, 407–416.
6. Du, S.; Cao, J.; Zhou, S.; Qi, Y.; Jiang, L.; Zhang, Y.; Qiao, C. Observation and inversion of very-low-frequency seismo-acoustic fields in the South China Sea. *J. Acoust. Soc. Am.* **2020**, *148*, 3992. [CrossRef] [PubMed]
7. Thomson, W.T. Transmission of elastic waves through a stratified solid medium. *J. Appl. Phys.* **1950**, *21*, 89–93. [CrossRef]
8. Haskell, N.A. The dispersion of surface waves on multilayered media. *Bull. Seismol. Soc. Am.* **1953**, *43*, 17–34. [CrossRef]
9. Buchen, P.W.; Ben-Hador, R. Free-mode surface-wave computations. *Geophys. J. Int.* **1996**, *124*, 869–887. [CrossRef]
10. Dunkin, J.W. Computation of modal solutions in layered, elastic media at high frequencies. *Bull. Seismol. Soc. Am.* **1965**, *55*, 335–358. [CrossRef]
11. Watson, T.H. A note on fast computation of Rayleigh wave dispersion in the multilayered half-space. *Bull. Seismol. Soc. Am.* **1970**, *60*, 161–166.
12. Kennett, B. Reflections, rays and reverberations. *Bull. Seismol. Soc. Am.* **1974**, *64*, 1685–1696.
13. Chen, X. A systematic and efficient method of computing normal modes for multilayered half-space. *Geophys. J. Int.* **2010**, *115*, 391–409. [CrossRef]
14. Hu, S.; Zhao, Y.; Wu, J.; Ge, S. Forward modeling of multichannel Rayleigh wave dispersion curve in laterally inhomogeneous media. *Chin. J. Geophys.* **2021**, *64*, 1699–1709.
15. Han, C.; Liu, Z.P.; Zhang, R.H.; Zhang, H.L.; Yang, L.I.; Huang, Y.Y.; University, S.J. Numerical simulation of the Rayleigh surface wave two-dimensional physical models. *Prog. Geophys.* **2017**, *32*, 357–362.
16. Yang, Z.; Chen, X.; Pan, L.; Wang, J.; Xu, J.; Zhang, D. Multi-channel analysis of Rayleigh waves based on Vector Wavenumber Tansformation Method (VWTM). In *Geophysical Research Abstracts*; EBSCO Industries, Inc.: Birmingham, AL, USA, 2019.
17. Kausel, E.; Roësset, J.M. Stiffness matrices for layered soils. *Bull. Seismol. Soc. Am.* **1981**, *71*, 1743–1761. [CrossRef]
18. Wang, Y.; Rajapakse, R. An exact stiffness method for elastodynamics of a layered orthotropic half-plane. *J. Appl. Mech.* **1994**, *61*, 339–348. [CrossRef]
19. Rokhlin, S.I.; Wang, L. Stable recursive algorithm for elastic wave propagation in layered anisotropic media: Stiffness matrix method. *J. Acoust. Soc. Am.* **2002**, *112*, 822–834. [CrossRef] [PubMed]
20. Tan, E.L. Stiffness matrix method with improved efficiency for elastic wave propagation in layered anisotropic media. *J. Acoust. Soc. Am.* **2005**, *118*, 3400–3403. [CrossRef]
21. El Allouche, N.; van der Neut, J.; Drijkoningen, G. Separation of PS-wave reflections from ultra-shallow water OBC data using elastic wavefield decomposition. In Proceedings of the 75th EAGE Conference & Exhibition incorporating SPE EUROPEC 2013, London, UK, 10–14 June 2013; p. cp-348-00797.
22. Richardson, M.D.; Briggs, K.B.; Bentley, S.J.; Walter, D.J.; Orsi, T.H. The effects of biological and hydrodynamic processes on physical and acoustic properties of sediments off the Eel River, California. *Mar. Geol.* **2002**, *182*, 121–139. [CrossRef]
23. Isse, T.; Kawakatsu, H.; Yoshizawa, K.; Takeo, A.; Shiobara, H.; Sugioka, H.; Ito, A.; Suetsugu, D.; Reymond, D.J.E.; Letters, P.S. Surface wave tomography for the Pacific Ocean incorporating seafloor seismic observations and plate thermal evolution. *Earth Planet. Sci. Lett.* **2019**, *510*, 116–130. [CrossRef]
24. Zheng, G.; Zhu, H.; Wang, X.; Khan, S.; Li, N.; Xue, Y. Bayesian inversion for geoacoustic parameters in shallow Sea. *Sensors* **2020**, *20*, 2150. [CrossRef] [PubMed]
25. Bonnel, J.; Dosso, S.E.; Eleftherakis, D.; Chapman, N.R. Trans-dimensional inversion of modal dispersion data on the new england mud patch. *IEEE J. Ocean. Eng.* **2020**, *45*, 116–130. [CrossRef]
26. Lin, S. *Advancements in Active Surface Wave Methods: Modeling, Testing, and Inversion*; Iowa State University: Ames, IA, USA, 2014.
27. Dosso, S.E.; Dettmer, J. Bayesian matched-field geoacoustic inversion. *Inverse Probl.* **2011**, *27*, 055009. [CrossRef]
28. Nazarian, S.; Stokoe, K.H., II. In Situ Deyermination of Elastic Moduli of Pavement Syatems by Spectral-Analysis-of-Surface-Waves Method: Practical Aspcts. 1985. Available online: https://trid.trb.org/view/268865 (accessed on 29 July 2021).
29. Lowe, M. Matrix techniques for modeling ultrasonic waves in multilayered media. *IEEE Trans. Ultrason. Ferroelectr. Freq. Control* **1995**, *42*, 525–542. [CrossRef]

30. Supranata, Y.E. *Improving the Uniqueness of Shear Wave Velocity Profiles Derived from the Inversion of Multiple-Mode Surface Wave Dispersion Data*; University of Kentucky: Lexington, Kentucky, 2006; Volume 67.
31. Ryden, N.; Park, C.B. Fast simulated annealing inversion of surface waves on pavement using phase-velocity spectra. *Geophysics* **2006**, *71*, R49–R58. [CrossRef]
32. Pico Vila, R. *Waves in Fluids and Solids*; IntechOpen: London, UK, 2011.
33. Boaga, J.; Cassiani, G.; Strobbia, C.L.; Vignoli, G. Mode misidentification in Rayleigh waves: Ellipticity as a cause and a cure. *Geophysics* **2013**, *78*, EN17–EN28. [CrossRef]
34. Dong, H.; Hovem, J.M.; Frivik, S.A. Estimation of shear wave velocity in seafloor sediment by seismo-acoustic interface waves: A case study for geotechnical application. In *Theoretical And Computational Acoustics 2005: (With CD-ROM)*; World Scientific: Singapore, 2006; pp. 33–43.
35. Lang, S.; Kurkjian, A.; McClellan, J.; Morris, C.; Parks, T. Estimating slowness dispersion from arrays of sonic logging waveforms. *Geophysics* **1987**, *52*, 530–544. [CrossRef]
36. Hamilton, E.L. Vp/Vs and Poisson's ratios in marine sediments and rocks. *J. Acous. Soc. Am.* **1979**, *66*, 1093–1101. [CrossRef]
37. Xia, J.; Miller, R.D.; Park, C.B. Estimation of near-surface shear-wave velocity by inversion of Rayleigh waves. *Geophysics* **1999**, *64*, 1390–1395. [CrossRef]

Journal of
Marine Science and Engineering

Express Image and Video Analysis Technology QAVIS: Application in System for Video Monitoring of Peter the Great Bay (Sea of Japan/East Sea)

Vitaly K. Fischenko [1,*], Anna A. Goncharova [1], Grigory I. Dolgikh [1], Petr S. Zimin [1], Aleksey E. Subote [1], Nelly A. Klescheva [2] and Andrey V. Golik [1]

1   V.I. Il'ichev Pacific Oceanological Institute, Far Eastern Branch of the Russian Academy of Sciences, 690041 Vladivostok, Russia; goncharova@poi.dvo.ru (A.A.G.); dolgikh@poi.dvo.ru (G.I.D.); zimin@poi.dvo.ru (P.S.Z.); subote.ae@poi.dvo.ru (A.E.S.); golik@poi.dvo.ru (A.V.G.)
2   School of Natural Sciences, Far Eastern Federal University, 690922 Vladivostok, Russia; klenel@mail.ru
*   Correspondence: fischenko@poi.dvo.ru

Abstract: The article describes the technology of express analysis of images and videos, recorded by coastal video monitoring systems, developed by the authors. Its main feature is its ability to measure or evaluate in real time the signals of sea waves, sea level fluctuations, variations of underwater currents, etc., on video recordings or streaming video from coastal cameras. The real-time mode is achieved due to processing video information read not from files, but from the graphic memory of the screen. Measurements of sea signals can be carried out continuously for a long time, up to several days, with high sampling rate, up to 16 Hz, at several points of the observed water area simultaneously. This potentially allows studying the entire spectrum of wave movements, from short waves with periods of 0.3–0.5 s to multi-day fluctuations at the sea level of a synoptic scale. The paper provides examples of the use of this technology for analyzing images and videos obtained in the network of scientific video monitoring of the Peter the Great Bay (Sea of Japan/East Sea), deployed by the authors.

Keywords: coastal video monitoring; streaming video; image and video processing; real-time mode; subpixel resolution; wind waves; swell; seiches; underwater currents; microseisms

Citation: Fischenko, V.K.; Goncharova, A.A.; Dolgikh, G.I.; Zimin, P.S.; Subote, A.E.; Klescheva, N.A.; Golik, A.V. Express Image and Video Analysis Technology QAVIS: Application in System for Video Monitoring of Peter the Great Bay (Sea of Japan/East Sea). *J. Mar. Sci. Eng.* **2021**, *9*, 1073. https://doi.org/10.3390/jmse9101073

Academic Editor: Francesca De Serio

Received: 2 July 2021
Accepted: 20 September 2021
Published: 1 October 2021

Publisher's Note: MDPI stays neutral with regard to jurisdictional claims in published maps and institutional affiliations.

1. Introduction

Systems for long-term video surveillance of coastal waters have been actively deployed around the world since the early 2000s. A great contribution to this process was made by the activities of the scientific group of R. Holman from the Coastal Imaging Lab (CIL) at Oregon State University, which was engaged in the development and application of photo and video equipment for studying the morphodynamics of sea coasts since the early 1980s [1,2]. In 1992, they developed and tested a standard coastal video surveillance system, ARGUS [3]. Information from 5–6 cameras was used for long-term monitoring of dynamic processes in the coastal zone of the sea, which affect the coastline. Quantitative methods were developed for describing hydrodynamic and morphodynamic processes in the surf zone based on the analysis of ARGUS video data [4–6]. Later, more than 30 such complexes were deployed in several countries around the world as part of the international CoastView project [7]. Data from all stations were promptly sent to the Internet network, for example, on the CIL website (http://cil-www.coas.oregonstate.edu/, accessed on 26 July 2021). It helped to popularize the project and stimulated the deployment of remote scientific video surveillance systems for coastal waters in different countries, including those not covered by the ARGUS network.

Largely under the influence of CoastView, in 2007, it was decided to start deploying a network of integrated scientific video surveillance systems of one of the most important water areas of the Far Eastern region of Russia—Peter the Great Bay—located in the southern part of the Primorsky Krai. The network was supposed to be an important part of the Scientific Monitoring System (SMS) of the Bay, launched in the Pacific Oceanological Institute of the Far Eastern Branch of the Russian Academy of Sciences (POI FEB RAS) in 2005. The developers of QAVIS-technology were engaged in the deployment of the SMS "cyberinfrastructure"—a set of technical and software tools for the quick delivery, processing, and presenting to scientists of data from several monitoring experiments conducted by scientific groups of POI FEB RAS. The data of the coastal video surveillance, organized at key points of the Bay, was to significantly supplement the SMS databases with information that would be useful in interpreting all other scientific data collected in the system. This determined the main tasks of the video monitoring system (VMS): 1—collecting, storing, and providing scientists with operational and historical video data on the state of the Bay's waters, 2—developing and applying in research video-based methods for quantitative assessment of the parameters of sea processes—currents, sea-level fluctuations, and others.

The QAVIS program (Quick Video and Image Analysis for Scientists) has proven useful in performing the second group of tasks. It allows one to measure, in several local points of the observed water area, wave signals $h(t)$ or other signals, for example, the signal of the sea surface brightness $B(t)$, carrying information about the properties of $h(t)$, from sea video recordings or directly from video streams. Signal measurements can be made with a frequency of up to 16 Hz, continuously, for a long time, up to several days. This potentially makes it possible to analyze the entire spectrum of wave processes in the water area, from short wind waves to synoptic sea-level fluctuations. QAVIS is complemented by Stitcher and OceanSP. The first one allows stitching the signals received in QAVIS with other geophysical signals into a single multi-channel file. OceanSP provides a set of tools for analyzing such multi-channel signals. The complex of programs presented in this paper, together with the methods of their application, is proposed as QAVIS-technology for studying sea processes based on coastal videos.

The article will provide information about Peter the Great Bay VMS, QAVIS technology, and its capabilities in studying wave processes and sea-level fluctuations based on VMS data.

This article examines several examples of the use of QAVIS in measuring the characteristics of wave processes and evaluating their frequency properties. In some cases, these measurements were compared with data from alternative surveillance tools, which, in general, confirmed the objectivity and effectiveness of the corresponding QAVIS measurement techniques. In other cases, it was not possible to obtain data from alternative measuring instruments for various reasons. In these cases, it cannot be stated that the accuracy and efficiency of QAVIS measurements are fully proven. The presentation of such techniques in the article is justified since they determine the directions for future research of promising measuring techniques using QAVIS and professional oceanographic instruments.

2. Material and Methods

2.1. Peter the Great Bay Video Monitoring System

As of 1 January 2018, the VMS consisted of twelve surface and three underwater cameras installed in 7 geographical points on the coast and islands of Peter the Great Bay and on two small research vessels (see the locations of installation sites in Figure 1).

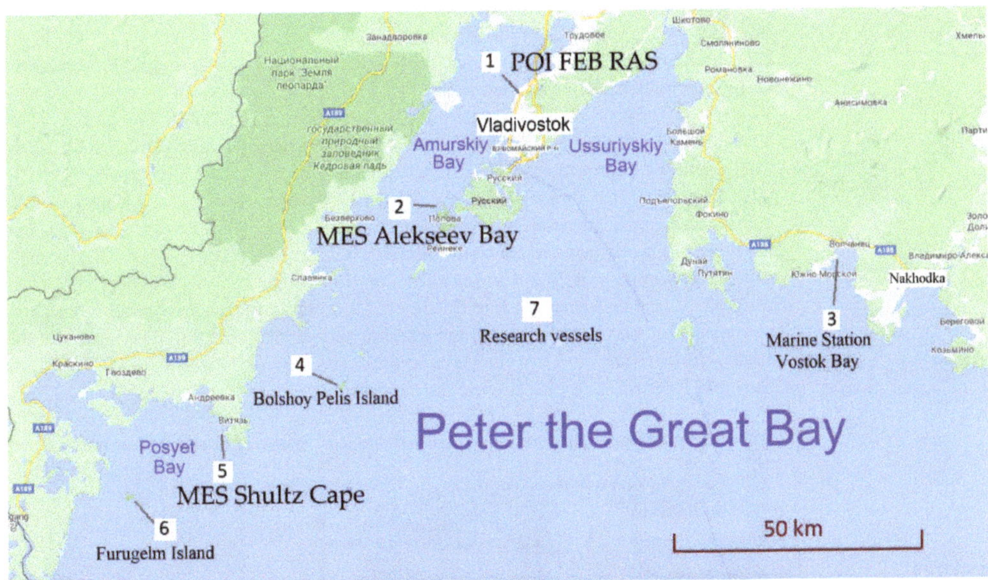

Figure 1. Locations of surface and underwater video surveillance camera installation (Google Map service used).

At point 1 (the roof of the POI FEB RAS building), up to three stationary and PTZ cameras with a view of the waters of Amurskiy Bay worked at different times. PTZ cameras, under the control of the operator or program, could be rotated horizontally (Pan) and vertically (Tilt) and change the multiplicity of optical zoom (Zoom) by increasing or decreasing the focal length of the lens. At point 2 (Marine Experimental Station (MES) "Alekseev Bay" on Popov Island), two surface and two underwater cameras were used. PTZ camera Axis-233d provided a 360° view of the waters around the island. The stationary surface camera in Alekseev Bay was used as part of the "videowavemeter". Underwater cameras were located on the bottom of Alekseev Bay, at 100 and 200 m from the shore. At point 3 (Marine Station of the Institute of Marine Biology FEB RAS in Vostok Bay), a stationary IP camera was used to view the nearest water area from 2014 to 2018. At point 4 (scientific cordon of the Far Eastern Marine Reserve on the Bolshoy Pelis Island), from 2014 to 2020 there was an IP camera with a view of nearby water areas. At point 5 (Marine Station "Cape Schultz") in the Posyet Bay, from 1 to 3 surface cameras and one underwater camera were used at different times. At point 6 (Marine Reserve cordon on the Furugelm Island), a stationary camera was installed, with a view of the Posyet Bay water area in the direction of Cape Schultz. Point 7 indicates the location of two small research vessels of the Institute, on which one stationary camera and one PTZ camera were installed.

Information from all cameras, along with data from other remote scientific observations, was sent, on the set schedule, to the SMS databases (DB). From each stationary camera, one snapshot per minute and one-minute-long video per hour were entered into the DB. Day-long animations were compiled from the snapshots and were also saved in the SMS database. PTZ cameras additionally make a 360° view of the water areas every hour and send snapshots to form panoramic images on the SMS computing servers. Every day, 10–20 GB of video data were stored in the DB. About 30 Terabytes of video data were collected between 2009 and 2021. To request video data, the Institute's researchers used the corporate web service of SMS inaccessible from the outside. A separate web service had been developed for open access to data from several cameras (https://poi.dvo.ru/live/, accessed on 25 September 2021). The state of the VMS by 2011 is described in [8].

Institute staff analyzed images and videos from SMS databases using professional programs, such as Adobe Photoshop and Adobe Premiere Pro, and also the QAVIS program. The latter can analyze not only previously recorded video files, but also live video streams simultaneously from several cameras in real time. Figure 2 shows the process of analyzing "live video" from 7 cameras installed in 6 points of the bay. Three computers were used: 4 streams were processed on the first one, 2 on the second, and 1 on the third computer. The number of camera installation locations is indicated on each stream, in accordance with Figure 1. Estimates of the wave signal (for cameras 1, 2, 4) or signals that carry information about the frequency properties of the wave signal (for cameras 3, 5, 5′, 6 were measured. Continuous measurements with a sampling frequency of 2 Hz were carried out in the summer of 2017 for more than 40 days. Swell waves arriving in the bay from the central and northern parts of the Sea of Japan were recorded several times during the measurement period. They were recorded at all points simultaneously and had the same base periods from 5 to 12 s. Wind wave spectra varied significantly at different points, as they are determined by local wind conditions.

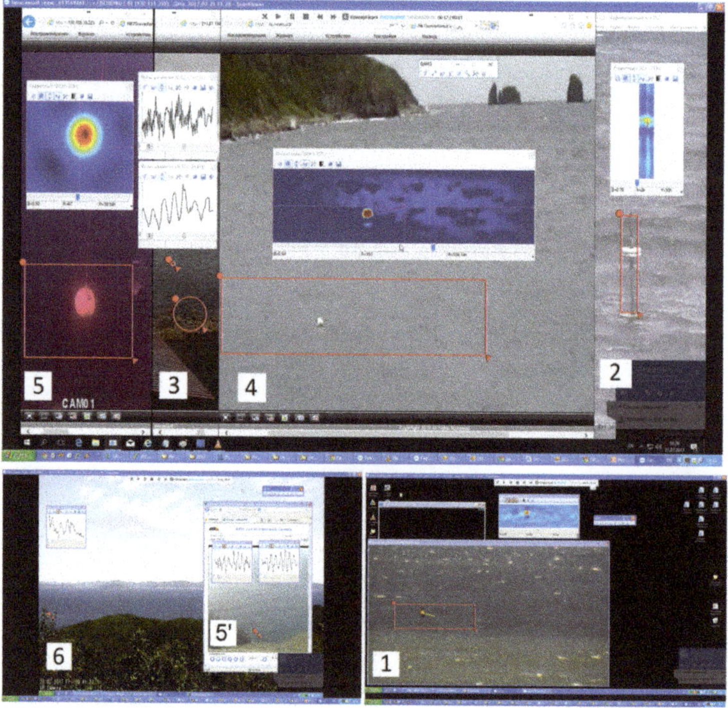

Figure 2. Synchronous measurements in real time of wave signals and their assessments at 6 points in Peter the Great Bay: (5) Cape Schultz, an underwater camera in Vityaz Bay, horizontal displacements of the underwater marker under the action of fluid movements caused by surface waves is measured; (3) Vostok Bay, the sea surface brightness signal, modulated by the passing waves, and the brightness signal of the region of the wave ashore are measured; (4) Bolshoi Pelis island, the signal of vertical movements of the float on the sea surface is measured; (2) Alekseev Bay on Popov Island, the wave signal is measured on base a "videowavemeter"; (6) Furugelm Island, the brightness signal in the breaking zone of waves emerging on a small rocky island is measured; (5′) Cape Schultz, view towards the Sea of Japan, the signal of the sea surface brightness modulation by passing waves is measured; (1) roof of the POI FEB RAS, view of the Amurskiy Bay, the signal of vertical movements of the navigational buoy under the action of waves is measured.

2.2. Sources of Geophysical Data, Used for Interpretation of QAVIS Video Analysis Data

The validation of the QAVIS methods for the analysis of coastal wave processes was carried out using synchronous geophysical observations performed in SMS or freely available on the Internet data from other organizations. Two automatic weather stations were located in points 2 and 5 (numbers as in Figure 1). The most important meteorological parameters for checking QAVIS methods are wind direction, wind speed, and atmospheric pressure. Sea level measurement stations, included in the Global Sea Level Observing System (GLOSS), were installed in Vladivostok, Posyet Bay, and Nakhodka (http://www.ioc-sealevelmonitoring.org/station.php?code=vlad, accessed on 25 September 2021). Measurements were carried out once per minute with an accuracy of 1 cm.

In 2008, specialists of the Kamchatka branch of the Unified Geophysical Service of the Russian Academy of Sciences installed the CMG-3ESPB broadband seismic station in point 5 on Cape Schultz. Its unique feature is its proximity to the coastline (distance—150 m), facing the open part of the Sea of Japan. As previous QAVIS observations show, swell waves entering the bay (typical periods of 5–14 s) can usually be detected simultaneously in wave signals and primary microseisms [9] of the seismic station's BHN channel. This channel registers the earth's surface vibration velocity along the "North-South" direction, facing the open part of the Sea of Japan. Primary microseisms are caused by swell waves running up to the coast of the cape, so their period coincides with the swell period. In addition to primary microseisms, the seismic signal contains more powerful secondary microseisms, the frequency of which is equal to twice the frequency of the swell waves. They arise due to the fields of standing waves caused by the superposition of direct and reflected from the cape waves. These standing waves send pressure pulses towards the seabed with a frequency twice that of the original swell wave. This causes secondary microseisms that propagate along the seabed in the form of Rayleigh waves in all directions, including towards the seismic station at Cape Shultz. Swell is a large-scale process; when it comes from the central part of the Sea of Japan to Peter the Great Bay, it manifests itself in all its local waters. Therefore, seismic data were used in the most difficult cases, for example, when studying the ability of QAVIS to register sea waves at long distances from the camera, up to 5 km.

In the same place, at Cape Schulz, two instruments were installed: a laser-interference complex, consisting of two laser strainmeters measuring oscillations of the earth's surface in the frequency range up to 1000 Hz in the North–South and East–West directions, and a laser nanobarograph, measuring fluctuations in atmospheric pressure [10,11]. They were used for comparison with QAVIS data in the range of wind and infragravity waves. Moreover, the POI FEB RAS has developed several designs of laser-interference meters for hydrospheric pressure variations–laser hydrophones [12,13]. They were installed at various points in Peter the Great Bay and registered variations in the hydrosphere pressure in the frequency band from 0.00001 to 1000 Hz. They were used for comparison with QAVIS data in the range of surface waves and seiche sea-level fluctuations.

Furthermore, data from satellite observations of the waters of the Sea of Japan, available on the Internet, were used to verify the results of QAVIS measurements. The most commonly used images of the sea surface are in the visible and radar ranges from the Sentinel-2 satellite, provided by the Sentinel-Hub EO Browser of the European Space Agency (EO Browser, Sinergise Ltd., https://apps.sentinel-hub.com/eo-browser/, accessed on 25 September 2021). In turn, specialists in satellite methods used data from QAVIS measurements of wave processes to interpret satellite information.

2.3. QAVIS Technology Software

The software consists of three programs: QAVIS, Stitcher, and OceanSP. The first one analyzes images or videos and, if necessary, stores the output spatial or temporal signals in text or binary files. Stitcher stitches QAVIS signals and any other time-synchronized geophysical signals into single multi-channel signals. The text format of the output files is simple; they can easily be processed in any software system for scientific calculations, for example, in Matlab. In this paper, another program, OceanSP, is presented. It implements many signal processing algorithms and works with the QAVIS data formats or data in text files.

Let's now take a closer look at the main program-QAVIS (Quick Analysis of Videos and Images for Scientists). It was developed as an easy-to-use program for scientists from any field of research that deals with images or videos. When working at a computer, scientists can notice spatial or temporal patterns in pictures and videos. They may have ideas on how these patterns can be quantitatively described and applied in their professional activities. With the help of QAVIS, they can very quickly go to the practical investigation of assumptions and ideas directly on the observed image or video. Another important aspect of QAVIS quickness is that video processing is done in real time. Analysis of marine video surveillance data is one of the possible areas of application for QAVIS.

The program is compact; it consists of three files with a total volume of about 2 MB. It works on ordinary personal computers, running Windows XP, Windows 7, Windows 10. It requires the Microsoft.NET Framework package installed on the system, no lower than version 4.1. It is usually pre-installed with the operating system or can be downloaded from the Internet. QAVIS, like OceanSP, uses the fast Fourier transform library FFTW (http://www.fftw.org/, accessed on 25 September 2021). The QAVIS software, along with OceanSP and Stitcher, can be downloaded from the web page (http://oias.poi.dvo.ru/qavis/, accessed on 25 September 2021).

The main feature of the program is that it does not work with image and video files, but with information that is currently displayed on the computer screen. QAVIS reads static images and video frames from the screen's graphics memory. If visualization of information from some source file on the computer screen is performed at the scale of 1:1 (this can always be controlled), then the array of pixels $x(i,j)$, $i = \overline{1, N}$, $j = \overline{1, M}$ read from a rectangular area of the screen of size $N \times M$, is in most cases completely identical to the array of pixels in the file. In both cases, one pixel is encoded with three bytes that carry information about the intensity of the red (R), green (G), and blue (B) components of its color. If the scale is different, then the user decides how critical this is for analysis. When working with video in the QAVIS program, the user sets the frame processing frequency (FPS—frames per second) from the range 1–64. It defines the time interval between frame processing cycles: $T = 1/\text{FPS}$. If one frame is processed in a timeframe less than T, the real-time mode is achieved, and the results will be the same as when processing video files. Thus, replacing work with files with work with screen memory does not lead to information losses. The user's convenience is obvious. They do not have to look for ways to download a file image of a video of interest from the Internet, deal with video data storage formats, or look for suitable software for processing. The price of such convenience is that many of the functions supported in professional image and video processing programs are not implemented in QAVIS. However, as the experience of using the program shows, the functions available in it are enough to solve many practical tasks of sea video monitoring.

When loading QAVIS, a small window appears that is always displayed on top of all other windows, and is opened by the user (Figure 3).

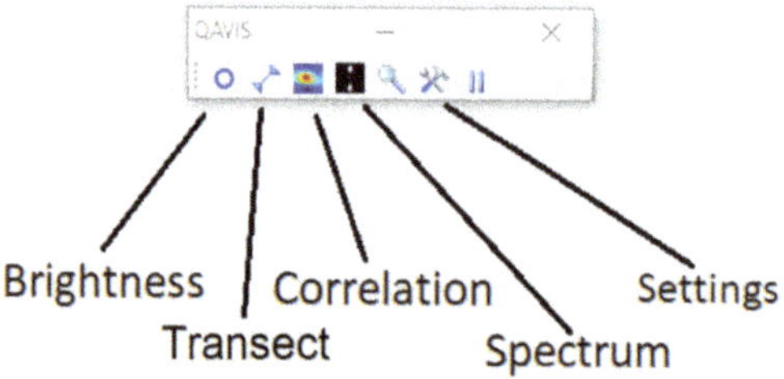

Figure 3. QAVIS Main Menu.

If necessary, important parameters of the program can be modified in the Settings section. QAVIS works only with scalar images, by default, grayscale (gray). Each pixel is assigned a scalar value, composed of three-color components according to the rule $x = (R + G + B)/3$. In the settings, users can select the mode of operation with only one color component: R, G, or B. Moreover, the frequency of data processing cycles is set in the settings, FPS. Several options are offered: 1, 2, 5, 9, 16, 21, 32, 64. For sea video monitoring, the FPS = 5 option is usually sufficient. In this case, the program allows one to analyze wave processes with periods of 0.4 s or more. In the settings, the user can set a scale by specifying the linear size of one pixel. Another important parameter is enabling/disabling the display mode of auxiliary markers on the image. Such markers (vectors of movement of objects on the water surface, vectors of wave velocity, schematic representations of wavefronts, etc.) are used for visual control of the correct operation of QAVIS tools, but sometimes interfere with a visual observation of the scene itself.

In general, there is one technical problem associated with the display of auxiliary markers on top of the image, which users should take into account. Due to the peculiarities of operating systems, such displays that do not distort the main image under the markers cannot be implemented in Windows 10. QAVIS recognizes this situation and makes auxiliary displays in separate information windows.

To analyze the images and videos observed on the computer screen, QAVIS presents four tools: Brightness, Transect, Correlation, and Spectrum, which are called from the initial window (Figure 4). Each tool has its own "selector" with which the user specifies the area of analysis on the computer screen. At the beginning of each clock cycle, the program reads the information under each selector from the screen, performs the appropriate processing, and displays the results in small information windows. Often, this information is sufficient for analyzing video data. In other cases, the user can record the signals produced by the tools to files that can be analyzed, after the recording is completed, by the user's programs, or by using OceanSP. In Figure 4, all four types of instruments are installed in the sea surface video viewer window after a ship passes from left to right. The operation of each tool in this example is explained below.

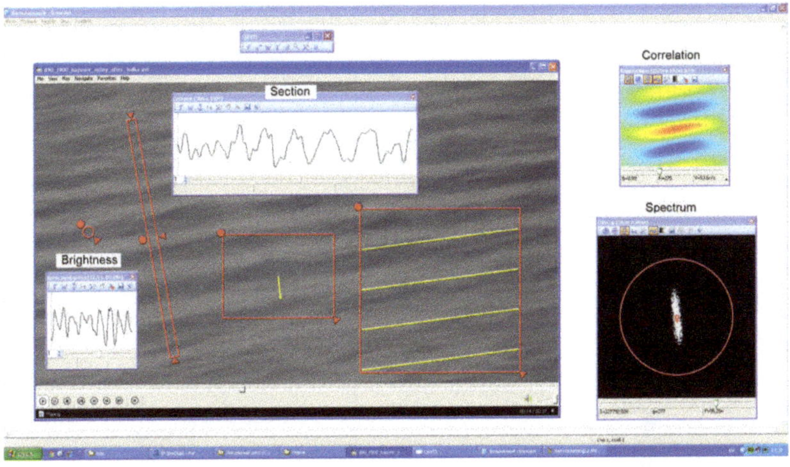

Figure 4. Basic QAVIS tools. The red selectors of the Brightness, Transect, Correlation, and Spectrum tools are arranged in order from left to right. The current results of the tools are displayed in the corresponding information windows.

2.3.1. Brightness Tool

The tool calculates the average pixel brightness value at each cycle under the round selector. The size (diameter) of the selector can be any, starting from 1; in this case, it is equal to 19 pixels. Below the selector, its information window is shown; it displays the last 200 samples of the brightness signal in the form of a graph, running from left to right, with the last time sample on the right. In the settings, if necessary, the user can change the number of displayed points. The tool is used to study the properties of the wave signal $h(t)$ at the selector location. The brightness value $B(t)$, at a given time t, strictly speaking cannot be called an estimate of the height of the water surface $h(t)$. However, a brightness signal, measured over a certain time interval Δt, will visually resemble a true wave signal $h(t)$. Analyzing a sufficiently long recording of the brightness signal under favorable conditions of visibility of the light flux modulation by wave slopes, the user can use spectral analysis methods to detect the presence of one or more wave systems (wind waves, swell, ship waves) in the water area, and can determine their main frequencies, and the dynamics of these frequencies over time. By installing several brightness tools in the water area, one will be able to estimate the direction and speed of movement of the detected wave systems, based on the time delays between markers, measured using cross-correlation analysis.

Scenes with a pronounced modulation of the light flux by waves (as in Figure 4) are rare. Therefore, other schemes can be used to evaluate the properties of wave processes. Thus, in (Figure 5a), three instruments record a signal of brightness changes at the points where swell waves exit to the Tarantsev Islands (Vityaz Bay). The brightness variations that are visible in the three information windows are obviously caused by the process of waves breaking on the shore.

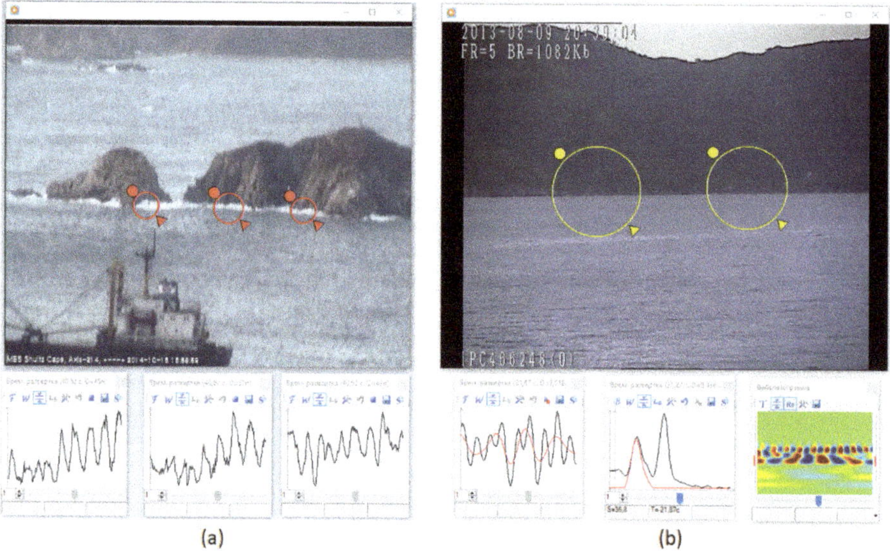

Figure 5. Two methods of detecting a wave signal with Brightness instrument: (**a**) registration of the processes of swell waves breaking when they come ashore; (**b**) registration of apparent movements of the shore during video surveillance from a ship, standing at the pier.

Figure 5b analyzes a one-minute video recording of 2013 from a PTZ camera, installed on a small research vessel. The ship swayed in waves; the camera directed at the opposite shore of the bay showed the swaying of the observed shoreline scene. Two instruments measure the average brightness at the "light sea–dark coast" boundary. At the bottom, in the left information window, the current waveform of the brightness signal of the left marker is displayed. On the right, for demonstration purposes, the display mode of the Fourier spectrum is set. Two peaks are visible in it, corresponding to periods of 5.4 s (its most likely cause is swell waves) and 2.4 s (wind waves). More to the right, the operation of the "Waveletogram" service is demonstrated. It displays the Waveletogram (scalogram of Morlet Continuous Wavelet Transform) of the first brightness signal. It shows two bands of alternating extremes, corresponding to the same periods of swell and wind waves. The user selected the lower band of extremes for wavelet filtering of the swell components; they are displayed in red in the information windows. The procedures for calculating, visualizing, and filtering Waveletograms implement the algorithms described in [14].

2.3.2. Transect Tool

When calling up a tool, the user is presented with a Transect selector in the form of a horizontally oriented elongated rectangle (Figure 6a). The selector can be stretched, widened, rotated, and positioned in the desired location on the screen (Figure 6c). The program calculates the signal of brightness change along the length of the selector $S(i)$, $i = \overline{1, N_s}$ at each QAVIS processing cycle. Here, N_s is the transect length in pixels. The calculation scheme is explained in (Figure 6b). At each point i of the transect, pixels are averaged over selector width (along the gray segment). The received signal is displayed as a waveform in the information window of the instrument (Figure 6d). Instead of the waveform, one can enable the display of the Fourier spectrum, or the correlation function, which is calculated for transect pairs that are separated in time by a user-specified number of QAVIS processing cycles. The information window displays information about the position of the correlation maximum and an estimate of the speed of movement of surface structures along the transect, measured in pix/sec. What is also useful is the display mode of the transect Waveletogram (Figure 6e). On it, the user can see the movement of structural elements of different scales (spatial periods), perform their wavelet-filtering, and then estimate their movement speeds in the mode of calculating the cross-correlation function.

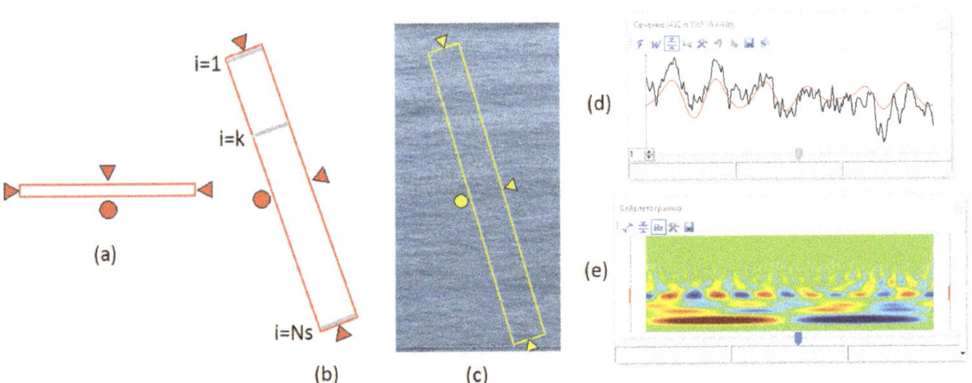

Figure 6. Transect tool: (**a**) basic selector; (**b**) stretched selector with a transect calculation scheme $S(i)$, $i = \overline{1, N_s}$; (**c**) example of transect installation on a sea scene; (**d**) waveform of transect; (**e**) Waveletogram of transect.

2.3.3. Correlation Tool

When the tool is requested, a rectangular selector appears (see Figure 4). If necessary, the user changes its size and puts it in the appropriate place on the screen. The QAVIS tool calculates and displays two-dimensional correlation functions between pairs of video frames in the information window at each time step. It works in two modes.

The fixed first frame mode is relevant for measuring the motion signals of certain objects, for example, the vibrations of a float on the sea surface. The user, by clicking on the corresponding button in the information window, fixes the initial frame with the object, after which the correlation of the current frame with the initial one is calculated at each cycle. The coordinates of the correlation maximum determine the displacement vector (in pixels) of the object on the current frame, relative to its original position. The correlation function is calculated on the basis of a two-dimensional discrete Fourier transform-DFT, implemented using the fast Fourier transform algorithm from the FFTW library. The convenience of this library is that it is very fast and at the same time does not rigidly require that the side lengths in pixels are powers of two. The program first calculates the DFT of both frames, multiplies of the DFT of the first frame by the complex conjugate DFT of the second frame, and then performs the inverse DFT. The result is a correlation function. It is possible to additionally configure the filter mask that will be applied in the frequency domain. This can improve the quality of tracking the movements of the object. If the visually controlled tracking quality satisfies the user, they can enable the recording of the maximum coordinates in a file. This will be a time signal of the object's movements, digitized at the frequency of FPS Hertz. If the object is a light marker on the sea surface, the vertical component is the wave signal $h(t)$, measured in pixels. The signal values can be presented in metric scales if the actual size of the object is known, and the pixel size can be estimated.

A fixed delay mode calculates the correlation between the current frame and a frame delayed by a set number of cycles. It is useful for estimating the directions and velocities of movement of some structures, such as waves. Figure 4 shows exactly this mode of operation of the tool. The time delay between frames is 0.19 s. The yellow vector indicates the direction of wave movement under the selector. The length of the vector is proportional to the speed of the waves.

2.3.4. Spectrum Tool

The user adjusts the size of the rectangular selector $[M \times N]$ and sets it to the desired location on the screen (see Figure 4). At each clock cycle, the program reads an image from the selector $x(m,n)$, $m = \overline{1,M}$, $n = \overline{1,N}$, then calculates and displays its quadratic Fourier spectrum in the information window: $S(k,l) = |X(k,l)|^2$, $k = -\frac{M}{2}, \frac{M}{2}-1$, $l = -\frac{N}{2}, \frac{N}{2}-1$, where X is the two-dimensional DFT of the image x. Visually noticeable bright localized regions in the spectrum indicate periodic structures in the image, such as sea waves. The location of these regions (k, l) relative to the center of the spectrum determines the orientation and spatial periods of the structures. The user can get these characteristics by pointing the mouse at a localized area in the spectrum. Users can also enable the auto-search mode for spectrum maxima in the specified frequency regions, whether rectangular or circular. In the version of QAVIS for Windows XP and Windows 7, it is possible to enable the mode of displaying the fronts of selected waves on the image itself. For Windows 10, the fronts are drawn in the information window. When analyzing a video, one can see how the fronts move synchronously with the waves. In the example in Figure 4, an annular zone of auto-search is defined, the maximum point found is marked with a cross, and the screen displays fronts consistent with the main wave structure.

The Spectrum tool can be used to analyze satellite images of the sea surface. Figure 7a shows a fragment of a radar image from a Sentinel-2 satellite of the water area around MES "Cape Schultz" dated 25 September 2020, 21:22 UTC. Using the Transect tool and Google Earth, the distance between two reference points was measured in pixels and meters, and then a scale factor of 14.1 m/pixels was determined. The Spectrum tool selector is set at the bottom right, and its information window with the calculated two-dimensional Fourier spectrum is shown at the top left. One bright region is visible in it (the second is only a symmetrical copy of it, this is a feature of 2d spectra). The user has selected an annular frequency range, and the program found the maximum in it, marking it with a cross. The text line at the bottom of the window shows the characteristics of the found wave structure, its orientation relative to the horizontal axis of 157°, and the spatial period or wavelength of 128.5 m.

Visual examination of the contrasted source image with and without overlapping wavefronts (Figure 7b) confirms that the fronts are superimposed correctly. Therefore, the found parameters objectively describe the wave process. Under the assumption that the observed waves are gravity waves, using dispersion relations (see, e.g., [15]) their period T at the known length λ can be calculated by the formula:

Where $g \approx 9.81$ m/s^2 is the acceleration of gravity, and h is the average sea depth in the observation area. In the area under consideration, the depth is approximately 50 m. Taking into account the estimated using QAVIS wavelength of 128.5 m, the period is obtained $T \approx 9.1$ s. It is typical for swell waves that are occasionally recorded in Peter the Great Bay. Moreover, in this case, it is very close to the period of primary microseisms of 8.9 s, which were recorded at the same time by the seismic station at Cape Schultz. This proves that the satellite image does show gravity swell waves, and in general, satellite methods can be effectively used to analyze wave processes in the Bay. On the other hand, this is an argument in favor of the fact that coastal seismic stations can be used as another useful tool for recording and describing the frequency properties of swell waves that come ashore at the station installation site.

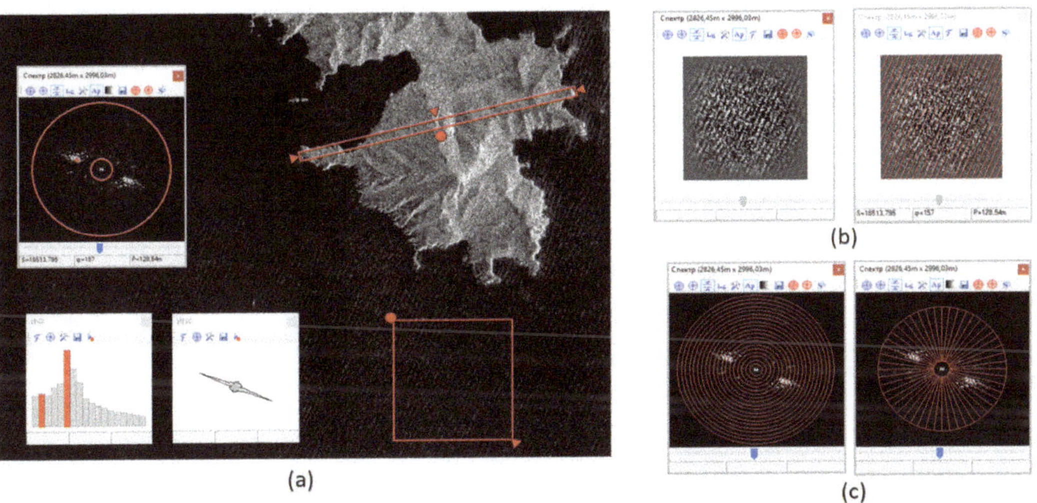

Figure 7. Spectrum tool. Analysis of the satellite (Sentinel-2) radar image, 25 September 2020 (Sentinel-Hub EO Browser used). (**a**) the image with selectors, the spectrum, IFC and ISC plots; (**b**) a fragment of source image without and with overlapping wavefronts; (**c**) IFC and ISC calculation schemes.

Note that in addition to the spectra, the program can calculate and display the so-called integral frequency characteristics (IFC) and integral spatial characteristics (ISC) (see the two windows at the bottom of the satellite image). The first is the distribution of spectral energy in a system of several concentric rings in the frequency domain. It characterizes a certain "average size distribution" of the image's structural elements. ISC is a distribution of spectral energy over a system of angular sectors. It characterizes the isotropic/anisotropic properties of an image, the presence and degree of expression of certain "anisotropy axes" in it. Note that these two systems of spectral characteristics were proposed for use as features in problems of recognition and classification of optical images by G.G. Lendaris and G. L. Stanley in 1970 in their pioneering work [16]. QAVIS allows the user to independently configure the IFC and ISC calculation schemes (Figure 7c).

The example considered in Figure 7 was intended primarily to demonstrate how to work with the Spectrum tool of the QAVIS program. The study of the possibility of evaluating the spatial and temporal periodicities of the swell waves using radar satellite images cannot be considered as finished. Additional research is needed to obtain a useful measurement technique.

3. Applications of QAVIS in the Peter the Great Bay Scientific Monitoring System

This section will review the results of several studies that were performed either only on the basis of QAVIS-measures and through using data from other geophysical observations.

3.1. Measurement of the Wave Signals, Based on Tracking Objects in the Sea at Long Distances

As noted above, direct estimates of $h(t)$ at a given point in the sea area can be obtained by measuring the signals of vertical movements of natural markers on the water surface (buoys, moored vessels, etc.) Measurements in pixels can be converted, for example, to centimeters if it is possible to estimate the scale factor K (cm/pixel), equal to the linear size of a pixel in the observation plane. Figure 8 on the left shows a diagram of video surveillance of an anchored marker and on the right a diagram of measuring its vertical movements in the observation plane tied to the marker.

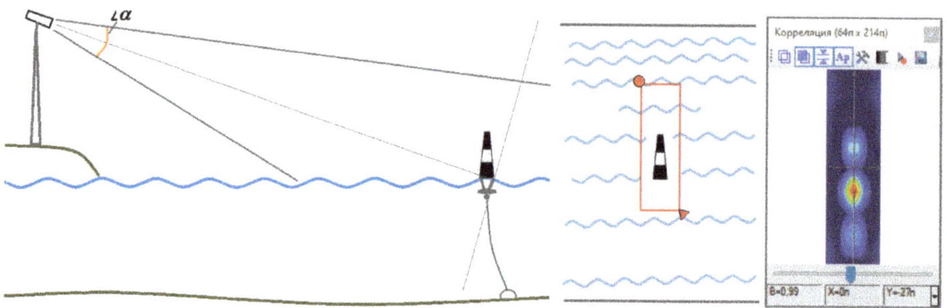

Figure 8. Measurement of the wave signal $h(t)$: the observation scheme of the marker object and the measurement scheme, based on video analysis.

The height of the marker in the field of the video frame in pixels $H_{pix} = 75$ and its real height $H_{real} = 300$ cm determine the pixel size $K = 4$ cm. It can be said that the system measures the wave signal with an accuracy of 4 cm. Given that the measured QAVIS signal can be represented as the sum of the true signal $h(t)$ and the error ε having a uniform distribution in the interval $\left[-\frac{K}{2}, \frac{K}{2}\right]$, it can be said that the measurements are made with a standard error $\sigma = \frac{K}{\sqrt{12}} \approx 0.288$, $K = 1.15$ cm.

It can be noted that, strictly speaking, the actual size of the pixel K in the observation plane, associated with the marker, is not constant, but depends on its location on the frame. Knowing the calibration parameters of the camera and the height of its position above sea level, the distance to the marker object and the signal of vertical displacements of the marker can be measured more precisely. However, this would most likely not allow QAVIS to process a video in real time. If necessary, based on the known calibration data, the previously recorded signal can be corrected. The noted problem with the variable pixel size is especially critical for cameras with wide-angle lenses (vertical viewing angle $\alpha = 20° - 40°$) but less critical for narrow-angle ones $\alpha = 2° - 5°$, which are mainly used for tracking remote marker objects.

An important question for applications is the distance at which wave processes can be measured using coastal cameras. Consider two examples of measurements made in May–June 2021 in Amurskiy Bay (point 1 in Figure 1) using two PTZ cameras installed in one of the rooms on the top floor and the roof of the building of the POI FEB RAS. Figure 9 shows a map of the measurement area; the yellow segment indicates the route from the Hikvision DS-AL camera (optical zoom of 1–30×) to the navigational buoy near Skrebtsov Island (distance of 1.4 km).

Figure 9. Measurements of waves at long distances in Amurskiy Bay: map of the area (Google Earth service used).

Figure 10 shows a live video stream from this camera (30× zoom), using the Ffplay program from the FFmpeg package (http://www.fftw.org/, accessed on 25 September 2021). Synchronously with it, the wave signal was measured using the Correlation tool (FPS = 5) of the QAVIS program. The height of the buoy above the water was 300 cm (the standard for buoys of this type); the height measured using QAVIS was 70 pixels. The scale factor was K = 4.3 cm/pix.

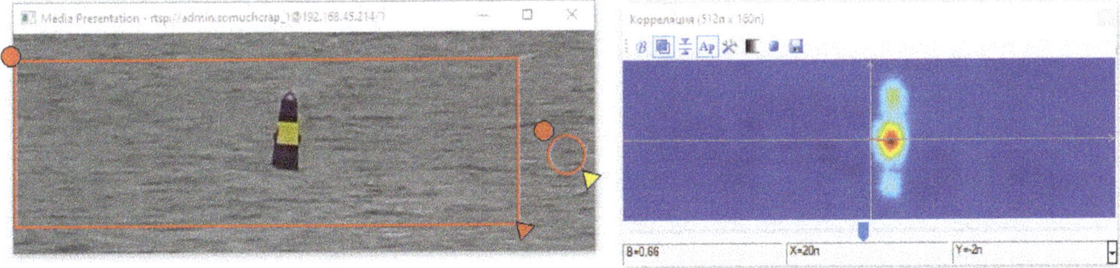

Figure 10. Registration of buoy movements: Correlation tool, Hikvision camera, optical zoom 30×.

Figure 11a shows a waveform of a 30-s fragment of the signal of vertical movements of the buoy to demonstrate the features of the measurement process. Despite the rough quantization step (4 cm), the signal displayed wave structures well. Figure 11b,c shows a waveform and a spectrogram (short-time Fourier transform) [17] of a three-day signal (11–14 June 2021). The gaps in the data correspond to nighttime. The spectrogram very clearly displays the temporal dynamics of the frequency composition of the waves. The swell response is clearly visible. On the first day, its frequency was stable near the value of 0.12 Hz (period of 8.3 s). In the next two days, it gradually increased to 2 Hz (5 s). Wind waves (frequency 0.3–0.5 Hz) were insignificant in the morning; their intensity increased in the afternoon. In general, it can be concluded that Hikvision-class cameras, in combination with QAVIS, were able to track the frequency dynamics of surface waves at a distance of 1.4 km.

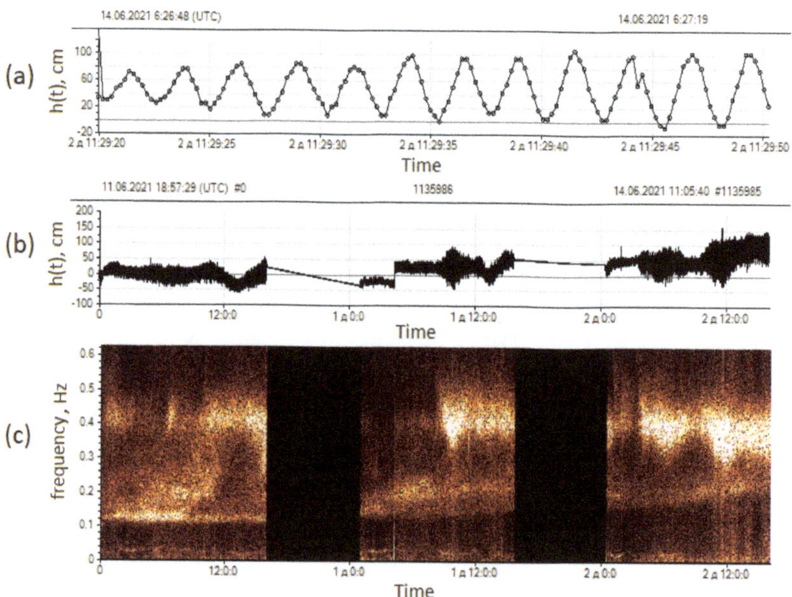

Figure 11. Results of 3-day (11–13 June 2021) measurement of the wave signals in Amurskiy Bay, based on the buoy verticals movement tracking with the Hikvision camera: (**a**) waveform of a 30-s fragment of the wave signal; (**b**,**c**) waveform and spectrogram of the entire three-day signal, respectively.

In addition to the frequency composition of the surface wave, it is possible to estimate the significant wave height (SWH) for different wave systems. Below is an explanation of the technology. Based on viewing the spectrogram, the time interval was determined, in which the wave system of interest to the user was visible and stationary. For example, for the swell in Figure 11c, it is logical to choose the first half of the first day. Figure 12a,b shows a three-hour recording of the $h(t)$ signal from this area and its spectrogram. It is evident that the swell component, which corresponds to a continuous bright stripe in the lower part of the spectrogram, was stationary. In the Fourier spectrum of the $h(t)$ signal, the leftmost peak corresponded to the swell waves, as shown in the upper half of Figure 12c. The blue curve shows a band-pass filter applied for frequency filtering of the swell component $hf(t)$ from the total signal $h(t)$. The spectrum shown in the lower half of Figure 12c confirmed that the filtration was correct. Figure 12d shows a 30-s fragment of the $hf(t)$ signal. Using the downward zero-crossing algorithm [18], individual waves were selected from $hf(t)$ (one of them is shaded in brown in Figure 12d); for each wave, two characteristics were determined—height (the difference between the maximum and minimum values) and duration (period). Figure 12e shows an OceanSP information window with the results of the statistical analysis of the extracted wave set. One of the statistics is SWH, the average height of 1/3 of the highest waves; in this case, SWH = 28 cm. In addition to statistics, the information window displays histograms of wave distribution by their heights and durations (periods).

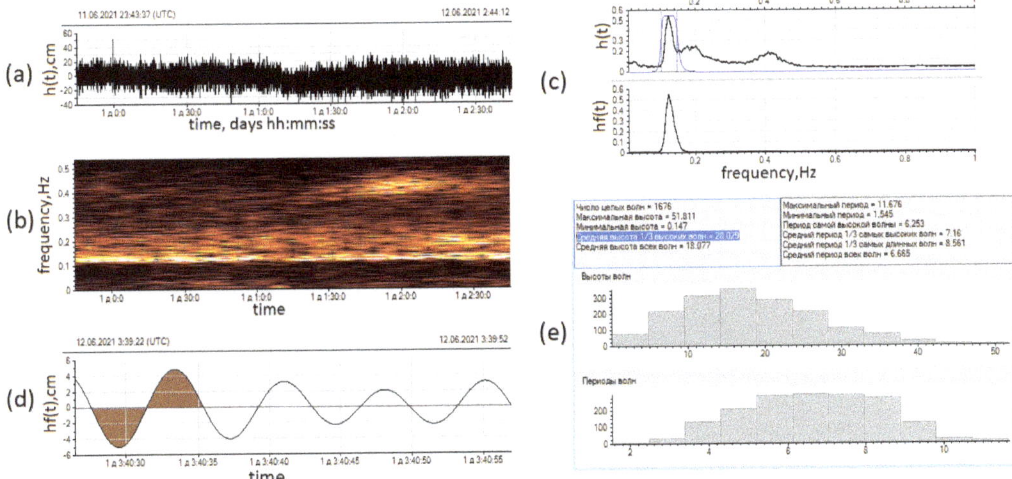

Figure 12. Methodology for calculating the significant wave height (SWH) implemented in OceanSP: (**a**) initial three-hour signal $h(t)$; (**b**) its spectrogram, the swell response is a bright band in the low-frequency region; (**c**) at the top is the spectrum of the signal $h(t)$ (black graph) and the frequency response of the band-pass (blue graph); at the bottom is the spectrum of the filtered signal $hf(t)$; (**d**) signal fragment $hf(t)$, one of the waves separated by the downward zero crossing algorithm is shaded in brown; (**e**) OceanSP information window with the results of statistical analysis of the set of selected waves.

The same analysis was performed to estimate the SWH of intense wind waves observed at the end of the third day of the study discussed above. SWH, in that case, was equal to 88 cm. Note that the described method can give overestimated SWH estimates if the frequency peak of the wave system of interest is observed in the spectrum against the background of powerful frequency components caused by noise or other wave systems close in frequency.

Experimental Study of One Simple Scheme for Obtaining Subpixel Resolution When Tracking Objects

On 11 June 2021, a small experimental study was carried out to practically investigate the possibility of achieving subpixel resolution when tracking objects. This problem is constantly in the focus of attention of specialists dealing with coastal video monitoring. Thus, in the work [19], it is considered in connection with the problem of the most accurate positioning of reference points, used for the georeferencing of video scenes, which must be observed continuously for several years. Under the influence of various factors, these points begin to shift, and it is important to track these changes with pixel accuracy, for example, with the accuracy $\Delta x_{new} = 0.2 \cdot \Delta x_{old}$, where Δx_{old} is the pixel size in the plane of reference points. Rodriguez-Padilla [19] describes a simple and effective method based on image subpixel cross-correlation that has improved the accuracy of positioning of the reference points. In principle, almost all approaches are based on specially developed procedures for interpolating source images or video frames into more dense sampling grids. The task is to find such procedures that would minimize a selected quality criterion that is important for a particular field, taking into account additional restrictions. However, it can be assumed that simpler interpolation schemes, such as those used in CCTV cameras Digital Zoom mode, can also produce subpixel resolution when tracking very distant small objects.

The following study was conducted to evaluate the subpixel resolution capabilities of bicubic interpolation using two PTZ cameras—OMNY 2030-IR and the previously described Hikvision DS-AELW. On 11 June 2021, simultaneously with Hikvision, an OMNY camera located on the roof of the institute streamed a navigational buoy scene with an optical zoom of 2.5×. The buoy looked like a very small object with a size of about 3 × 7 pixels; the height of one pixel in the observation plane of the buoy was approximately 43 cm. The streamings were carried out using the FFplay program simultaneously in two windows, the first—with the original resolution, the second—with 12-fold bicubic interpolation; the total zoom was 2.5 × 12 = 30×. Figure 13 shows the process of measuring the vertical movements of the buoy in both windows.

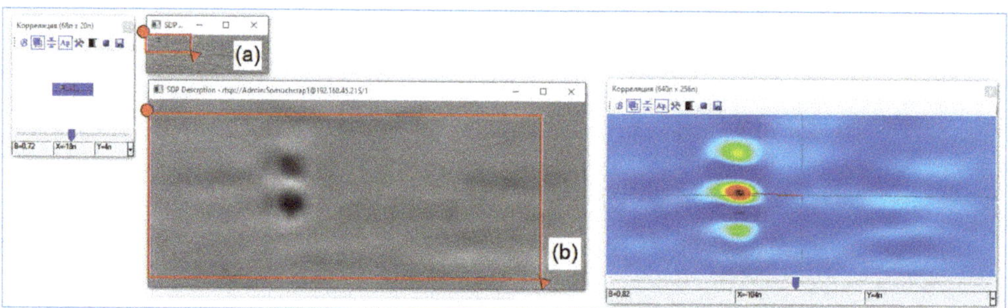

Figure 13. Recording of the buoy movements on two streams from the OMNY camera: (**a**) with optical zoom of 2.5×, (**b**) when using additional 12-fold bicubic interpolation.

At the same time, the Hikvision camera streamed with the maximum optical zoom of 30×. The signal of vertical movements of the buoy, measured on it, can be considered the reference signal. The main goal was to find out whether bicubic interpolation leads to an increase in the accuracy of tracking the movements of the buoy, i.e., to the effect of subpixel resolution. Let the names of the received buoy movement signals be associated with the streamings: h1-OMNY (2.5×), h2-OMNY (2.5 × 12 = 30×), h3-Hikvision (30×). All three signals are measured in pixels. Another signal used in the study was the BHN seismic signal, synchronously recorded at Cape Schultz (94 km from the buoy). As mentioned above, the BHN signal is used to monitor the occurrence of swell entering the bay. The BHN signal is presented on integer conventional units (CU). For it convert to actual velocities signal it is necessary to use a scale factor of 4.54×10^{-9} m/s.

Figure 14 shows the results of the study. Figure 14a shows the waveforms of 12 h recordings of all channels. The h1 signal does not look much like ordinary signals, as it has only three gradations of the level. There are also long sections of constant zero values. The h2 signal, obtained using bicubic interpolation of the first stream, looks more like a signal with continuous values due to having more gradations of the level. In the spectra of three QAVIS signals h1, h2, h3 (see Figure 14b), a swell response with a period of about 8 s (0.13 Hz) is noticeable, which is confirmed in the spectrum of the BHN seismic signal. In the spectra of the interpolated and reference signals, h2 and h3, more wave systems with periods of 5.4 s (0.18 Hz) and 2.5 s (0.4 Hz) are synchronously manifested. The analysis of the spectrograms in Figure 14c further confirms the positive effect of bicubic interpolation use. The spectrogram of the non-interpolated signal h1 contains only two small sections with swell responses. The h2 spectrogram conveys the temporal dynamics of wave systems almost as well as the spectrogram of the reference signal h3. Figure 14d–f considers the unique case when there are no changes at all in the h1 signal for 44 min; the buoy does not shift by a single pixel in the field of view of the OMNY camera. At the same time, the movements of the buoy are recorded in the interpolated h2 signal; there is a peak in their spectrum at the swell frequency, the same as in the h3 and BHN spectra. In addition, the h2 signal is significantly correlated with the reference signal h3 (see Figure 14f).

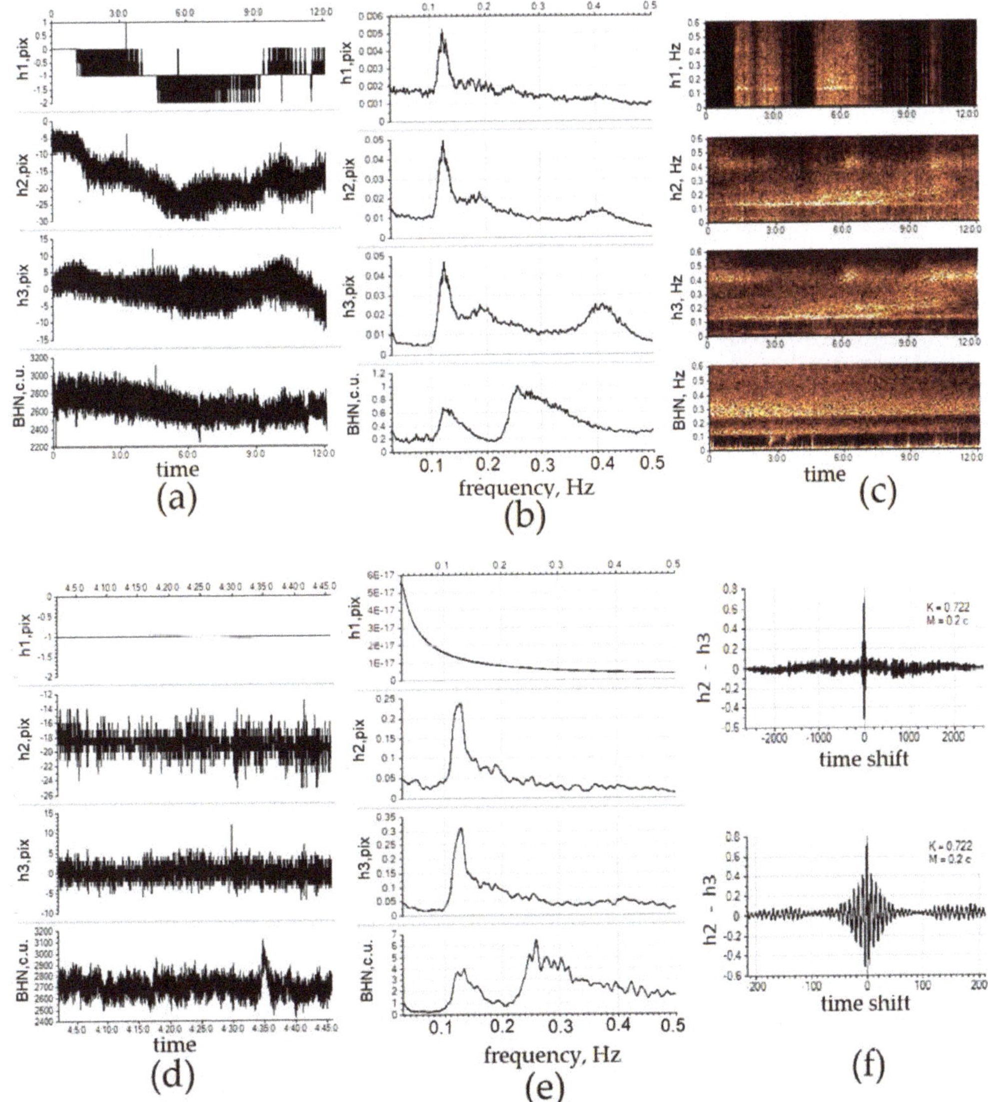

Figure 14. Results of the experimental study to investigate the possibility of using bicubic interpolation of a video stream for tracking the movements of anchored objects on the sea surface with the QAVIS program with subpixel accuracy. Oscillograms (**a**), spectra (**b**), spectrograms (**c**) of QAVIS signals h1, h2, h3, and seismic signal BHN, recording duration 12 h. Oscillograms (**d**) and spectra (**e**) of the same signals at a 44 min time interval, when the signal h1 was strictly constant. Correlation function between signals h2 and h3 over the entire interval of time shifts and on its central part (**f**).

Thus, a simple bicubic interpolation of a video fragment with a tracked marker can significantly improve the quality of tracking. Video streaming programs usually support the bicubic interpolation operation; many, IP cameras support the Digital Zoom with bicubic interpolation. This is important for the QAVIS-technology, since it will be possible to measure wave processes at significantly large distances in real time.

Let us return to the problem of studying the capabilities of the program for evaluating wave signals at ultra-long distances. In Figure 9, the green segment indicates the route from the OMNY PTZ camera on the roof of the POI FEB RAS building to the central spans of the low-water bridge over Amurskiy Bay (distance 5.4 km) (https://eng.gpsm.ru/activities/low-water-bridge-vladivostok/, accessed on 25 September 2021). From the autumn of 2020 to the present, several observation cycles have been performed to develop QAVIS-methods for estimating the parameters of sea processes at such large distances. Figure 15 shows the results of one of these studies.

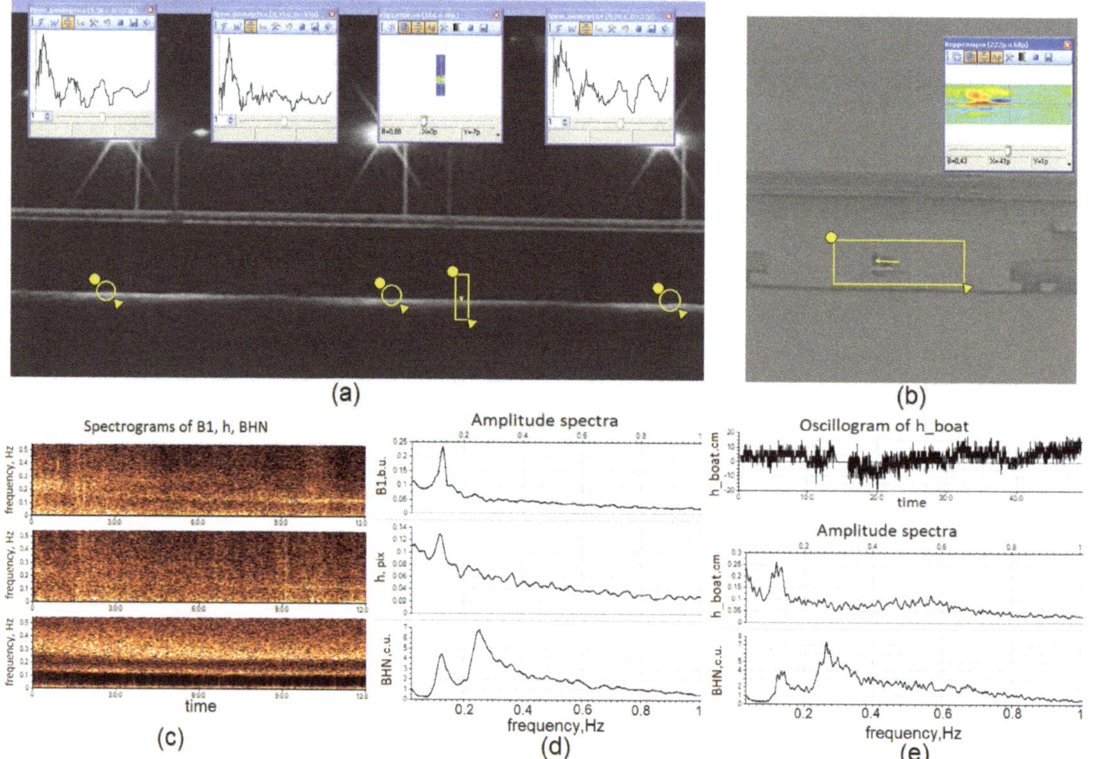

Figure 15. Night QAVIS-measurements of wave characteristics near the low-water bridge (**a**), daytime measurements of a fishing boat vibrations on the waves; (**b**), results of measurement analysis (**c–e**).

Figure 15a demonstrates the process of measuring fluctuations in the water of light spots from the bridge lights with instruments Brightness and Correlation. The measurements were carried out for 12 h during nighttime from 31 October to 1 November 2020. Figure 15c shows the spectrograms of signals B1 (brightness tool, left selector, measured in brightness units (b.u.) in the range from 0 to 255), h (correlation tool measured the vertical displacements in pixels of a strip of light on the water surface), BHN (a seismic signal, synchronously recorded at Cape Schultz, 98 km from the bridge). In the spectrograms of the "optical signals" near the end of the recording, light horizontal stripes are visible at a frequency of about 0.12 Hz. The same band is reproduced more clearly in the spectrogram of the BHN signal. Such frequency bands are characteristic of swell waves. In the spectra (Figure 15d) of 1-h fragments, the peaks corresponding to the swell periodicity of 8.3 s (frequency 0.12 Hz) are synchronously present for all three signals. Thus, the techniques of night measurements at very long distances tested in this example may well be applied to detect swell waves and describe their frequency properties.

Figure 15b,e demonstrates another small experiment, typical of the QAVIS technology. On the same day, 1 November, a fishing boat was accidentally discovered in one of the bridge spans on the main video broadcasts. In order not to overload the processing with a large number of measuring instruments, another video stream was launched on another computer; QAVIS was called to measure the movement of the boat bobbing on the waves (Figure 15b). Before the sailors left the camera's field of view, it was possible to record a 49-min signal of the vertical movements of the boat h_boat. Figure 15e shows the waveform of the h_boat signal, its spectrum, and the spectrum of the BHN seismic signal at Cape Schultz. The spectra show synchronously low-frequency peaks of the same swell near 8.3 s. Taking into account the known height of the low-water bridge, the scale factor $K = 3.66$ cm/pix was calculated and applied. Then, the swell component was isolated using a bandpass filter, and its significant wave height SWH = 7.1 cm was estimated.

Figure 16 shows the results of observations near the low-water bridge on the night of 14–15 October 2020, in which wind waves with a period of 2.1 s were detected and described.

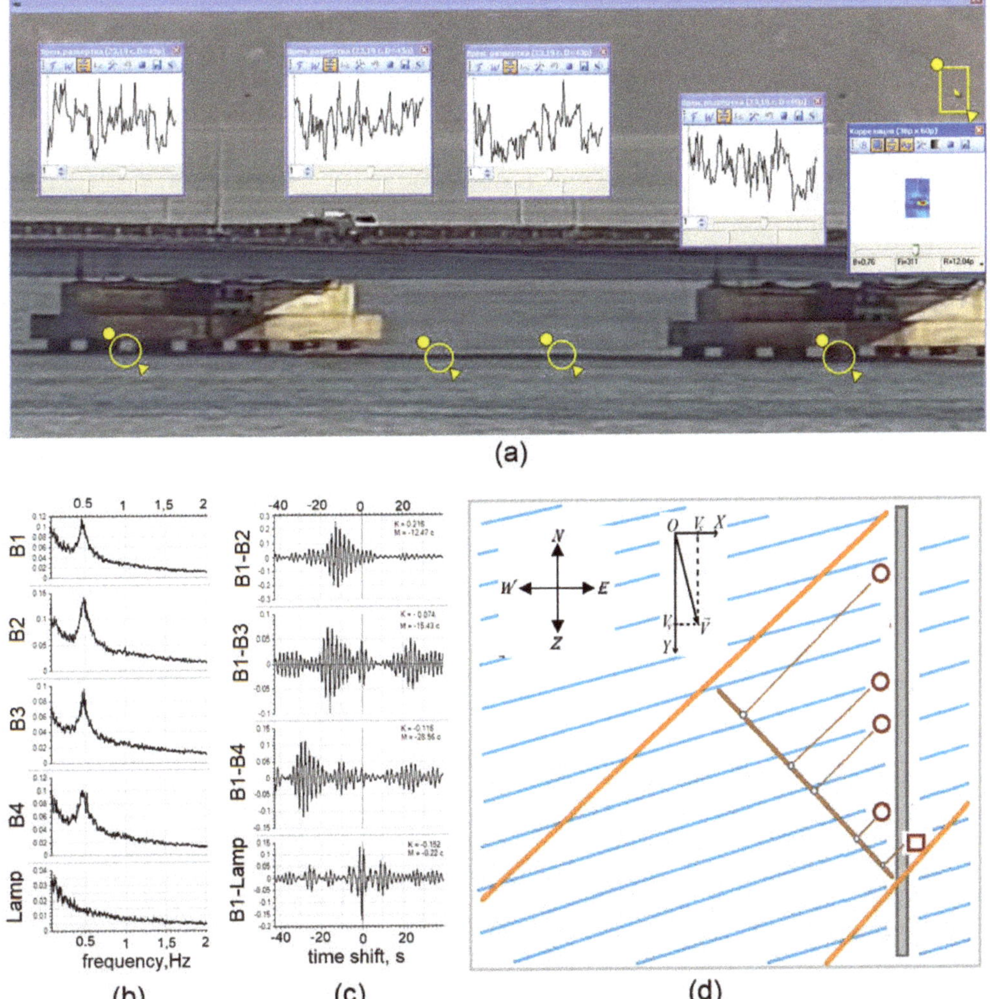

Figure 16. Registration of wind waves near the low-water bridge on 14–15 October 2020: (**a**) placement of meters on the video scene; (**b**) Fourier spectra of two-hour signal records; (**c**) mutual correlation functions between signals; (**d**) schematic map of the study.

Figure 16a shows the measurement tools set-up that was used in nighttime measurements. Four round selectors of the brightness instrument measured the signals of fluctuations in the brightness of light spots on the water at night: B1, B2, B3, B4. The correlation tool (rectangular selector at the top right) measured the movement of a street lamp on a bridge. Figure 16b shows the Fourier spectra of two-hour signal records. In the spectra of all brightness signals, a peak corresponding to a period of 2.1 s is noticeable. There is no such peak in the movement of the street lamp signal.

Then, the components of wind waves were extracted from all signals using a band-pass filter (the bandwidth is 0.4–0.6 Hz). Then, the correlation functions were calculated between the signal component B1 and the components of the remaining signals. In all correlation functions, the peaks of the correlation maxima were visually noticeable (see Figure 16c). In this case, the values of the maxima themselves were not large, from 0.08 to 0.23. It is noticeable on the graphs how the correlation maxima of the brightness signals successively move away from the point of zero shift with an increase in the distance between the measuring instruments. The values of time shifts between signals, determined by the program automatically according to the position of the maximum of the correlation module, were as follows: M12 = −12.47 s, M13 = −15.43 s, M14 = −28.56 s. The negative sign of the time delay means that the first signal is ahead of the second. The sequence of delays is logical if we take into account the uneven installation of the meters in Figure 16a.

A possible scenario for nighttime observations is presented on the schematic map shown in Figure 16d. The map shows the location of the bridge (gray vertical bar on the right), location of measuring instruments (brown circles) along the bridge, and conditional lines coming from the camera, limiting the field of observation of the scene to the left and right in Figure 16a, observation plane (brown line), orthogonal to the optical axis of the camera, the projection of the positions of the measuring instruments on the observation plane. The top left shows the wind rose. The direction of the bridge, and hence the line of placement of the meters along it, are oriented almost exactly along the "North-South" direction. This fact can be checked from Figure 9 and in the Google Earth service. The blue lines schematically represent wind waves moving from the direction of North-North-West. Usually, in October, north-westerly winds prevail over the Amur Bay. They could well have caused the wind waves shown in the diagram. The diagram also shows the velocity vector of the waves; its vertical component Vy can be estimated taking into account the known technical information about the dimensions of the bridge elements (https://eng.gpsm.ru/activities/low-water-bridge-vladivostok/, accessed on 25 September 2021). The distance D between the bridge supports is 63 m. Figure 16a shows that the distance between gauges 1 and 4 is slightly larger, D14 = 70 m. Thus, the projection of the wind waves velocity along the bridge can be estimated Vy = D14/M14 = 2.45 m/s.

The correlation between B1 and Lamp signals should be discussed. The correlation maximum is significant; it noticeably exceeds the average level of the surrounding correlations (see Figure 16c) but is reached at a very small time shift of 0.22 s, almost zero. When the distance between the meters is more than 70 m, the correlation with zero delay cannot be due to marine processes. The most likely cause is camera shake, which appears synchronously in the signals of all meters and therefore contributes to the correlation functions at zero time delay. With a strong jitter, situations are possible when the correlation of jitter components will exceed the correlation of real physical signals recorded by the meters. This can cause difficulties, and even errors, in the interpretation of observation results. In the previously considered case with the registration of swell waves near the bridge, significant peaks in the cross-correlation functions were observed for almost all pairs of signals. But only for two pairs of signals from closely spaced meters, these peaks were achieved with a nonzero delay. Apparently, at large distances, the correlation of real processes weakens, and the correlation of the jitter components begins to prevail.

Thus, in this section, it was shown that using QAVIS measuring techniques and OMNY 2030-IR class IP cameras, it is possible to detect swell and wind wave systems, and also to describe their frequency properties at distances of up to 5 km from a camera.

Strictly speaking, for wind waves, this statement is not supported by alternative field measurements. The seismic station at Cape Schultz does not register responses of wind waves in water areas remote from the Cape. The spectra and spectrograms of the buoy movement signal (Figure 14) contain powerful components in the range of 0.3–0.5 Hz, which usually reflect wind waves. However, the natural vibrations of the buoy structure may be manifested in the same frequency range. Therefore, it is advisable to conduct additional research using wave meters located next to the buoy. The fact of registration of wind waves near the low-water bridge (Figure 14) looks very reasonable. The data of all four wave meters are consistent with each other and satisfy one of the most realistic scenarios of the wave process for this part of Amur Bay. But even in this case, there were no data from alternative measuring systems.

3.2. Videowavemeters

Measurements of the wave signal $h(t)$, based on tracking anchored marker objects (buoys, ships), contain an uncontrollable error associated with horizontal movements of the marker on the sea surface. They can make a small indefinite contribution to the vertical displacements of the object in the field of the video frame. Usually, horizontal movements are relatively slow and practically do not affect the height measurements of faster surface waves. However, such errors may be unacceptable for estimating the parameters of tidal and seiche fluctuations in sea level. Therefore, several methods have been tested, using specially placed structures in the sea, in which the horizontal movements of markers are limited.

One of the most successful designs of a stationary videowavemeter was first tested in 2014 in Alekseev Bay (point 2 in Figure 1). A six-meter-long pipe was rigidly driven into the seabed at a depth of about four meters. The distance from the shore was 100 m (110 m from the camera). A free-sliding light marker structure was mounted on its end, protruding above the water. The installation of the marker on the stand ensured its visibility in case of a significant wave. A reflective tape was glued on the marker, illuminated at night from the shore with a narrow beam LED projector, which provided the possibility of round-the-clock measurements. The measurement accuracy is determined by the linear size of one pixel 0.48 cm. The observations made it possible to describe several seiche periodicities of Alekseev Bay previously noted in the works of [20], in addition to describing surface wave systems. Figure 17 shows the Alekseev Bay on the Google Earth map and the process of measuring the wave signal in the Alekseev Bay during the day and at night, using a videowavemeter.

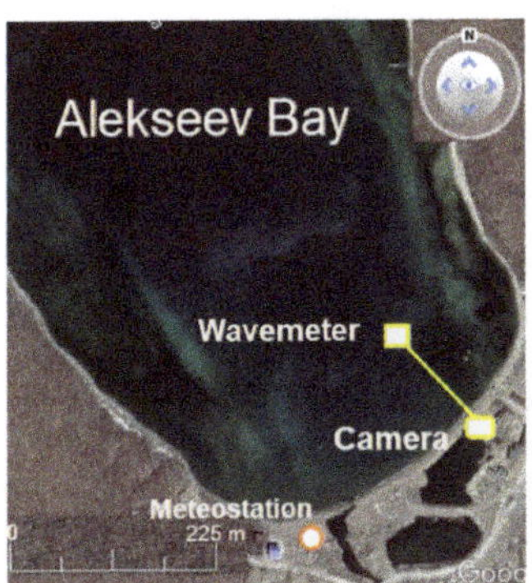

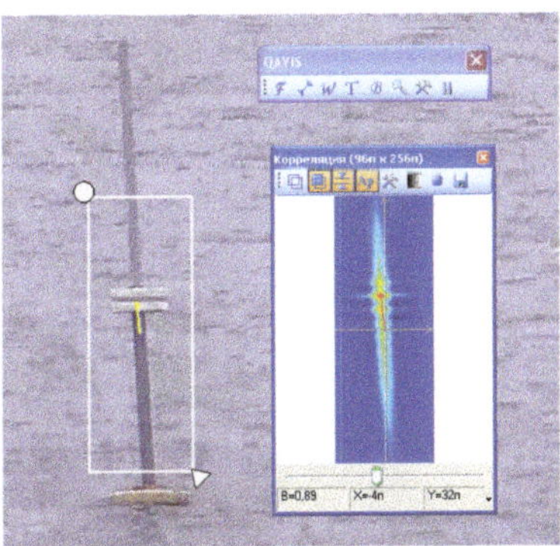

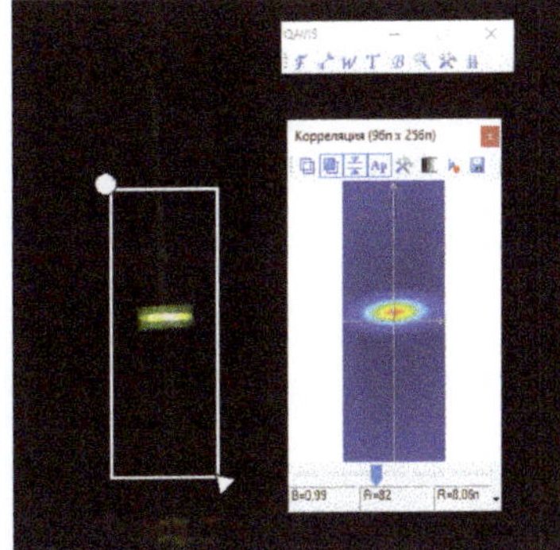

Figure 17. Videowavemeter in Alekseev Bay. Shown from left to right: its location (Google Map used); measuring process of wave signals in the daytime and at night, respectively.

Figure 18 shows an example of analysis of a seven-hour recording of the wave signal $h(t)$ from 22 April 2015. Figure 18a shows the waveform of the measured signal. Its Fourier spectrum (Figure 18b) shows peaks, corresponding to the periods: (1 h 35 m, 52 m, 10 m, 4 m 20 s)—typical of seiches, 7.7 s—typical of swells, 1.8 s—typical of wind waves, 0.8 s—typical of short wind waves (wavelength is 1 m or less). On the time-frequency spectrogram (Figure 18c), the time dynamic of the frequency composition of the wave is very clearly visible. At the bottom, a horizontal band is visible at low frequencies, corresponding to the swell wave; it is present throughout the entire recording and does not depend on local weather conditions. At the beginning of the spectrogram, a bright area corresponding to intense wind waves is visible for 2 h, obviously caused by a strong northwest wind in Amurskiy Bay. After the third hour of recording, the responses of short waves that occur during south wind begin to prevail in the spectrogram at the top. Earlier, it was repeatedly observed with the help of data from an automatic meteorological station that with a north-westerly wind, wind waves prevail with periods of 2–3 s, going towards the coast of the sea station. With a southerly wind, short waves prevail (with periods of 0.6–0.8 s). This wind blows into the sea from the coast and manages to form only short waves to the place where the wavemeter is installed (100 m from the coast, see Figure 17). Furthermore, relatively short inclined linear tracks are noticeable in the spectrogram—the responses of ship waves from ships passing by the Bay. The slope of the tracks is due to the dispersion of the wave packet coming from a vessel towards the wavemeter; the frequency of the recorded waves gradually increases (the period decreases). By the slope and duration of the tracks, the distance to the ship that caused the ship waves can be roughly estimated.

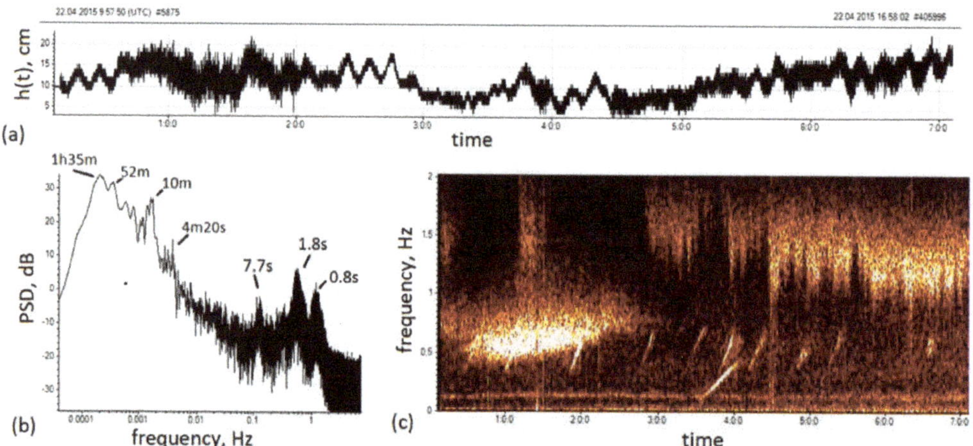

Figure 18. Analysis of the 7 h recording of the wave signal (22 April 2015, 09:57–16:58): (**a**) waveform, (**b**) spectrum, (**c**) spectrogram.

Thus, with the help of videowavemeters, it is possible to organize very high-quality observations of the entire spectrum of oscillatory movements of the sea surface, from short wind waves to seiche and tidal fluctuations.

During the winter months, many bights and bays of the Primorsky Krai freeze over, so the use of marker systems placed in the sea is not possible. However, it is possible to organize observation of the movements of markers installed on the ice surface. Since 2015, winter observations of vertical ice movements have been regularly carried out in Alekseev Bay using a summer videowavemeter camera. Some results of summer and winter observations are given in [21]. The main conclusion: practically all systems of sea-level fluctuations that are observed in the summer months are manifested in the signals of fluctuations of the ice surface, from seiche fluctuations of the bay itself (period 4 min 20 s) to tidal diurnal and semidiurnal fluctuations

It should be noted that, although the measurements of sea-level fluctuations using the videowavemeter are very consistent with the known experimental and model data [20], direct synchronous observations using alternative sea-level meters and surface waves were not done. Such observations will need to be carried out, first of all, to assess the accuracy of measuring the amplitude characteristics of various wave systems by the videowavemeter.

3.3. Synchronous Measurements of Hydrosphere Pressure Variations by a Laser Hydrophone, and of Underwater Currents Variations by Underwater Video Analysis with QAVIS

Three fixed underwater video surveillance cameras located in Alekseev Bay (point 2 in Figure 1) and Vityaz Bay (point 5 in Figure 1) were included in the general system of scientific video monitoring of Peter the Great Bay. The goal was to develop technologies for long-term continuous monitoring of the state of the underwater biodiversity in the bay [22]. In addition to solving biological problems, methods for estimating by video the parameters of underwater currents and surface waves were developed [23]. To assess the correctness of these methods, data from other underwater observations, carried out by the scientific groups of the Institute, were used.

Figure 19 shows the results of 12-day measurements carried out in Vityaz Bay from 15–27 November 2016, using a laser hydrophone and an underwater camera.

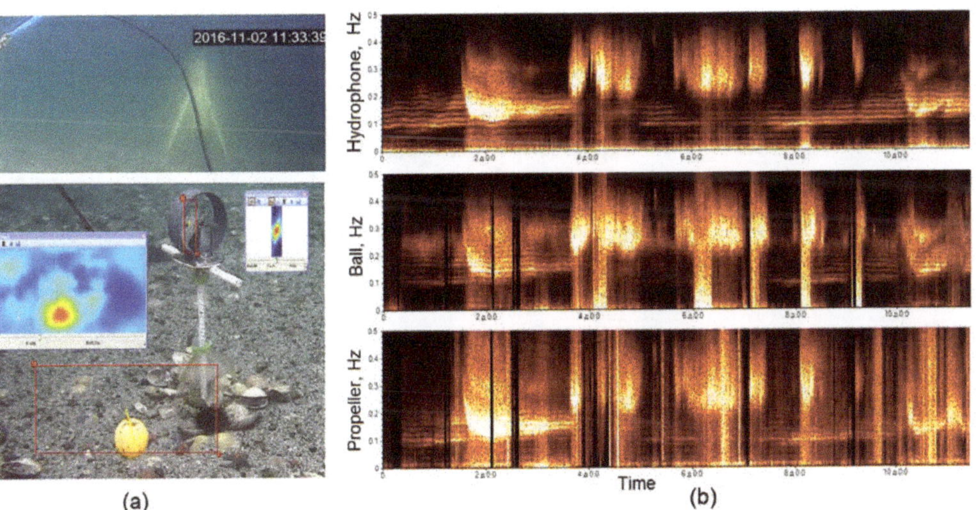

Figure 19. Underwater observations in Vityaz Bay on 15–27 November 2016: (**a**) laser hydrophone and measuring markers Ball and Propeller on the bottom of the bay; (**b**) spectrograms of Hydrophone, Ball and Propeller signals.

A laser hydrophone installed on the seabed measured variations in hydrosphere pressure in the frequency range of up to 1000 Hz. In the low-frequency part of its spectrum (0–0.5 Hz), sea-level fluctuations and wave processes on the sea surface were primarily manifested. The underwater camera was used to measure variations of underwater currents based on QAVIS-tracking of horizontal movements of artificial or natural marker objects. Both instruments were installed at a shallow depth, about 5 m from each other (see Figure 19a). QAVIS correlation tools were used to measure horizontal displacements of a tennis ball with small negative buoyancy (Ball) and rotation of the propeller blades (Propeller) under the influence of water mass movements. The main reason for these movements at such a shallow depth was wave processes on the sea surface, primarily swell and wind waves. Both meters evaluated the one-dimensional projections of the current vector variations. The optical axis of the camera was oriented orthogonally to the coastline towards the opposite coast. Therefore, the Ball signal measured the projection of currents along the coast of the bay. The rotation plane of the propeller blades was initially oriented orthogonal to the optical axis of the camera to estimate the velocity component of currents across the bay. However, at the beginning of observations, the propeller was turned 30–40 degrees, presumably by a spotted seal (Phoca largha). Before that, the largha was seen several times near the propeller on the video. Figure 19b shows the spectrograms of 12 daily recordings of all three signals. The band of analyzed frequencies is 0–0.5 Hz. The signal of variations in hydrosphere pressure will be called below as Hydrophone. The spectrograms are visually similar, all of them well and synchronously track the very complex, non-stationary dynamics of frequency composition of wave processes in the bay. Long horizontal tracks in the low-frequency part of the spectrogram represent swell; bright spots in the high-frequency part represent wind waves.

The purpose of these observations was to test the performance of both optical methods for measuring variations of underwater currents, using a ball and a propeller as markers. The technique with a ball tracks the frequency properties of surface waves better than with a propeller. Both methods are inferior to the hydrophone when estimating the parameters of the seiche vibrations of the bay. In general, the ability to use an underwater camera, in addition to its main biological tasks, for solving the problems of continuous monitoring of wave processes on the sea surface and underwater is very useful. The results of many scientific experiments conducted in Vityaz Bay significantly depend on the hydrodynamic conditions in the water layer thickness. Therefore, information about spectral composition and intensity of variations of underwater currents in different periods may be important for researchers doing such work. This information is also important for marine biologists in the context of accompanying video information about the state of marine biodiversity with information about the external conditions of its existence.

Another way to test the objectivity of underwater QAVIS measurement techniques is to check if there is a significant correlation between the swell and wind wave signal estimates recorded by QAVIS and the hydrophone. The correlation of the Ball and Propeller with each other would prove that both methods work and measure the same processes. Correlation with the hydrophone measurement would prove that processes caused by surface waves are being measured. The second case is more difficult because the hydrophone was located 10 m from the camera. In Vityaz Bay, a complex hydrological regime leads to instability of the directions of motion of wave systems relative to the hydrophone-camera axis. This would lead to the fact that the time delay between the arrival of waves to the first and second meters will be unstable; accordingly, the correlation maximum would be blurred during long-term observation.

The described effect was confirmed when studying the correlations in the above-described 12-day records of the Hydrophone, Ball, and Propeller signals. To analyze the correlation, a fragment of the record was selected, and its spectrum was calculated. The wave systems detected in it—swell waves and wind waves—were filtered. After that, the cross-correlation function was calculated for each wave system. If a visually noticeable sharp peak stood out in it, then the signals were recognized as correlated. The position of the peak determines the delay between the arrival of waves to the meters. Preliminary calculations confirmed that the Ball and Propeller signals are almost always correlated (the correlation coefficient K is between 0.6 and 0.9), and the time delay is close to zero. The correlation function between long (from several hours to several days) QAVIS-records and hydrophone records usually had a diffuse peak. The correlation function between short recordings (from half an hour to two hours) sometimes had sharp correlation peaks for the analyzed wave systems. Figure 20 shows an example of the analysis of correlations for a short recording—1.5 h long. OceanSP was configured to analyze swell waves and wind waves with periods 7 s and 3.5 s, respectively. Figure 20a shows oscillograms of the analyzed signals, and Figure 20b their spectra. The spectra of the filtered swell and wind waves are shown in blue and green. Figure 20d,e shows the correlation functions between signals H3, B3, P3, H7, B7, and P7. The letter in the designation identifies the measured signal, the number, and the main period of the analyzed wave system.

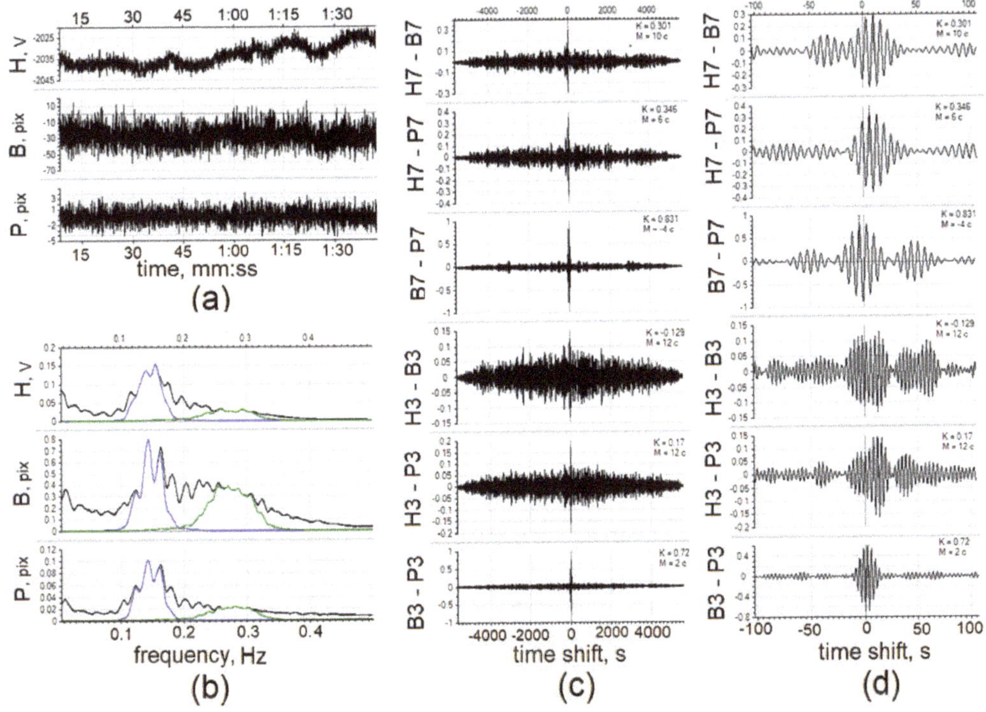

Figure 20. Analysis of correlations in the signals of a laser hydrophone (H) and underwater meters of variations in underwater currents, Ball (B) and Propeller (P): (**a**) oscillograms; (**b**) spectra (the spectra of 7-s swell and 3-s wind waves are shown in blue and green, respectively); (**c**) correlation functions between signals of different meters, separately for swell and wind waves; (**d**) the same correlation functions, but more detailed in the central region of time shifts. Signals B and P are measured in pixels, signal (H) is measured in units of electrical voltage (V) received at the output of the hydrophone. To convert the H signal to a scale of hydrospheric pressure variations, a scale factor of 105.1 Pa/V must be applied.

Figure 20c displays the correlation functions over the entire area of time shifts; such a representation is convenient for an expert to decide about the significance of the correlation between the pair of signals under consideration. If the peak noticeably exceeds the immediate environment, then it can be considered significant. In the considered case, significant correlations between the components of the Ball and Propeller signals are clear. The correlation coefficients at the maximum points are 0.86 for swell and 0.72 for wind waves, even though different projections of currents were measured. This fact proves that both QAVIS methods are workable and can be used to assess one-dimensional projections of underwater currents. The correlation peaks between the QAVIS signals and the hydrophone signal can also be considered significant. Moreover, they are sharp, which means that both wave systems were stable in direction and speed during the observation. Figure 20d shows the central part of the correlation functions. It can be used to visually determine whether the program's estimate of the time delay (M) between signals is unambiguous. In the case under consideration, the decision on the time delay M = 12 s for the H3–B3 pair may be ambiguous, since there is almost the same peak with a shift of M = −2 s on the left. However, the same 12 s delay was obtained for the H3–P3 pair.

In general, the presented results of the analysis of the correlations of underwater QAVIS measurements and laser hydrophone data can be considered another argument in favor of using underwater measuring techniques. These measurements are certainly not very accurate and complete. They estimate only one projection of the speed of the water movement, and the signal of the propeller has few gradations in the rotational speed.

Nevertheless, even in this form, the data obtained with their help correlate with the data of more accurate, but also more expensive devices. It is also clear that there are tasks where high accuracy in measuring the parameters of underwater currents is required. Qavis, in its current form, is not ready for such tasks. However, it could be useful at the stage of developing high-precision measuring systems based on streaming video analysis for organizing long-term continuous observations of underwater currents.

3.4. Observations of Light Marker Buoys, Comparison with Laser Hydrophone Data

In the video monitoring system of Peter the Great Bay, small marker buoys that fell into the view area of CCTV cameras were often used to quickly estimate the wave characteristics. They were usually installed by fishermen or scientists for their own purposes. As a rule, buoys were anchored and moved on the surface within a limited area. In Section 3.1, it was noted that uncontrolled horizontal movements of markers present a problem for QAVIS measurement of water surface level fluctuations $h(t)$ based on tracking the vertical movements of a marker in the observation plane. This problem is especially critical for assessing the characteristics of sea-level fluctuations, like seiches, but less critical for assessing the properties of other wave processes (swell, wind, and ship waves) since, as a rule, their spectra are separated from the spectra of horizontal movements. Nevertheless, the question of how correct it is to apply the results of QAVIS observations of the vertical movements of marker buoys to describe the properties of sea waves requires additional research using alternative measurement techniques.

In the summer of 2018, a small comparative experiment was carried out at Cape Schulz using the laser hydrophone described in the previous section and a stationary TANTOS IP camera. The hydrophone was installed in Vityaz Bay 110 m from the shore at a depth of about 18 m. The camera was installed on the shore on the roof of a boat hangar; installation height approx. 4 m above sea level. A small white buoy, which marked the location of the hydrophone, was visible in the video as an object with a diameter of about 5 pixels (Figure 21a). Its actual diameter is 25 cm. The Digital Zoom mode with 4× magnification was used to improve the visibility of the buoy. The zoom was implemented in TANTOS using bilinear interpolation. Thus, the size of the pixel at the location of the buoy was $25/5/4 = 1.25$ cm. Measurements of vertical movements of the buoy (Buoy) were taken using the QAVIS Correlation tool (Figure 21b) with a sampling rate of 5 Hz. The buoy signal was combined with the concurrent hydrophone measurements for analysis.

Figure 21c–f shows the results of the analysis of one of the recordings with a duration of 5 h. Figure 21c shows the waveforms of the Hydrophone and Buoy signals. In both, an increase in the level associated with the tidal process is noticeable. However, the signals differ significantly. This fact confirms the validity of the recommendation not to use such QAVIS meters for recording sea-level fluctuations. Figure 21d shows spectrograms of both signals. In the spectrogram of the Hydrophone signal in the low-frequency region, two bright horizontal stripes are noticeable, corresponding to the periodicities of 5.6 and 9.5 s. They most likely correspond to two swell systems that arrived at the bay. Above, up to a frequency of 1 Hz, there are no other features; the spectrogram is uniformly black. The spectrogram of the Buoy signal also clearly shows two horizontal bands at the same frequencies. In this case, the entire spectrogram contains noise that is characteristic of roughly quantized signals. In the high-frequency region, the spectrogram shows about ten tracks of waves from vessels passing in the bay.

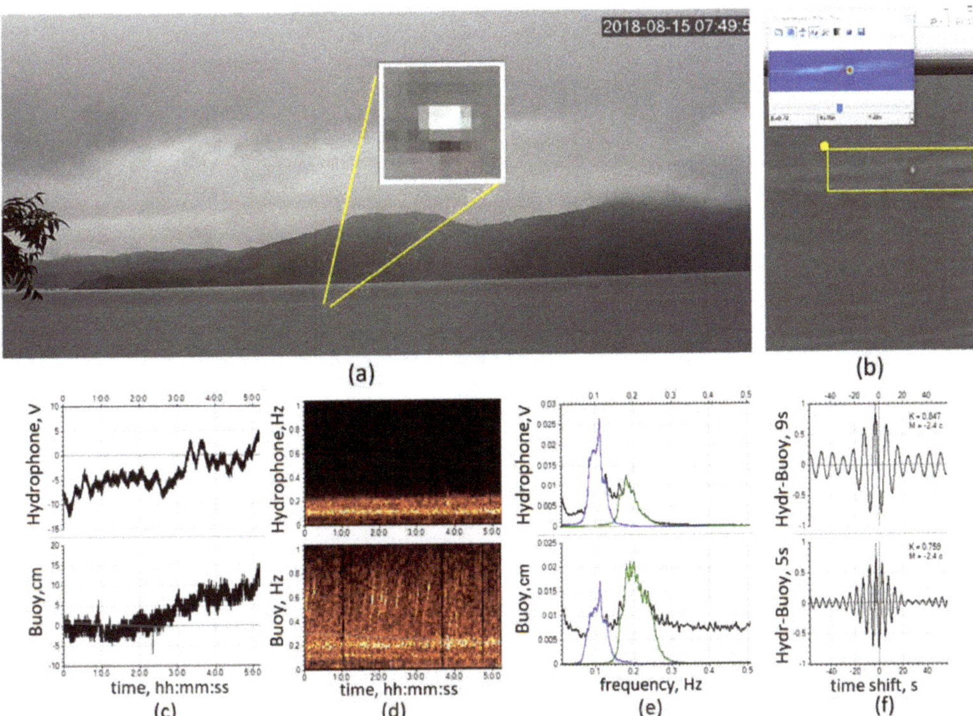

Figure 21. Comparative observations of wave processes in Vityaz Bay (28 August 2018) using a laser hydrophone and QAVIS: (**a**) scene observed by the TANTOS camera; (**b**) measurement of vertical motions of the buoy with the QAVIS Correlation tool; (**c**) oscillograms of 5 h signal records; (**d**) their spectrograms; (**e**) Fourier spectra; (**f**) correlation functions between wave systems detected by both instruments (separately for swell with a period of 9.5 sec and swell with a period of 5.6 s).

The spectra of both signals in Figure 21e contain peaks corresponding to swell periods of 5.6 and 9.5 s. At the same time, no other spectral features appeared in the Buoy spectrum, which could be caused by the buoy design and its weight. Although the spectra are similar, the ratio of powers of the 5 and 9 s swells in them is different. The 5.6 s swell has greater power in the hydrophone signal, while the 9.5 s swell prevails in the buoy signal. In this case, the buoy movements more accurately convey the actual power ratio of the wave process. With the increasing installation depth of the hydrophone, it becomes less sensitive to the effects of short-period waves. In addition to discussing the visual similarity of the spectra, it is important to investigate how correlated the wave systems are that both instruments detect. For this purpose, the swell components with the main periods of 5.6 and 9.5 s have been extracted using a bandpass filter. In Figure 21e, for clarity, their spectra are superimposed on the original spectra (green—swell with a period of 5.6 s, blue—swell with a period of 9.5 s). Then cross-correlation functions were calculated between the Hydrophone and Buoy signal components, separately for each swell system. In both cases, large values of the maxima of the normalized correlations were observed (0.84 and 0.76, respectively). Both components of the swell in the measurement data of both instruments are significantly correlated.

The described example showed the prospects of using the QAVIS methods for measuring the movements of marker buoys for the rapid detection and description of the frequency properties of wave systems present in the water area. At a larger scale of the scene, the wave signals $h(t)$ would be measured with greater accuracy, which is determined by the size of a pixel in the observation plane of the marker buoy. This will reduce the noise level in the spectra and spectrograms. It is worth expanding the research by conducting synchronous observations using several cameras with different optical magnifications, and by placing the hydrophone at a shallower depth, where it will record all wave systems without attenuation. However, the ability to quickly obtain information on the spectral composition of waves with the accuracy shown in Figure 21 can already be useful in research work.

There is, however, one significant limitation. High waves between a buoy and the camera can disrupt the buoy's visibility. Therefore, if the camera is not installed high enough, the described technique will often not work.

4. Conclusions

The article presented a technology developed for the analysis of data of the scientific video monitoring system of Peter the Great Bay (the Sea of Japan/the East Sea). Its key element is the QAVIS computer program. It implements a set of tools for studying the spatial and temporal characteristics of wave processes in coastal waters, based on real-time processing of streaming video from coastal cameras. The article described the QAVIS tools and several measurement techniques, implemented on their basis.

Using the example with a synchronous registration of the arrival of a 9 s swell in the bay on a satellite image and in the records of the seismic station, the capabilities of QAVIS for studying wave structures on satellite images of the sea surface were demonstrated.

The possibilities of methods for measuring wave signals, based on tracking vertical movements of anchored marker objects at distances up to 5 km from a camera, were demonstrated. The limitations of these methods were noted; they are good at registering signals of wind waves and swells, but they are not effective for measuring seiche and tidal fluctuations at the sea level. The reason for this is the horizontal movements of markers on the water surface, which make an uncontrolled contribution to the vertical displacements of the marker on the video frame.

A comparative experiment was carried out, which proved that the preliminary bicubic interpolation of the scene with a marker object allows achieving subpixel resolution in the task of measuring wave signals based on tracking the movements of markers.

The results were presented, proving that the method of measuring wave signals, using stationary videowavemeters, is the most effective. It allows studying the entire spectrum of wave movements, from short wind waves, with characteristic periods of 0.5–0.8 s, to tidal daily and half-a-day fluctuations.

The paper presented the results of 12 days observations in Vityaz Bay using a laser hydrophone and an underwater camera installed at a depth of 5 m. It was shown that the frequency properties of the surface wave signals measured by a hydrophone and the signals of underwater currents measured by video tracking of the Ball and Propeller markers are very similar. In general, both optical markers track the responses of wind waves and swells well, but they are inferior to the hydrophone in tracking seiche fluctuations at the sea level.

In addition to those discussed in this article, there were many more examples of the joint use of QAVIS and other measurement techniques. It should be noted also that the QAVIS-technology is easy to use and can easily be applied to the study of sea processes in any coastal waters of the World Ocean, from which video is streamed to the Internet. For example, in [24], the results of measurements of seiche fluctuations in the port of Crotone (Italy), carried out with the help of QAVIS, are given. Comparison with the GLOSS data of the CR08 station (http://www.ioc-sealevelmonitoring.org/station.php?code=CR08, accessed on 25 September 2021) located 400 m from the CCTV camera confirmed the sufficiently high quality of QAVIS measurements.

Author Contributions: Conceptualization, G.I.D. and V.K.F.; methodology, G.I.D., V.K.F. and P.S.Z.; software, A.A.G. and A.V.G.; validation, P.S.Z., A.E.S., N.A.K. and G.I.D.; resources, G.I.D., A.E.S. and A.V.G.; data curation, P.S.Z., A.E.S., N.A.K. and A.A.G.; writing, V.K.F. and N.A.K. All authors have read and agreed to the published version of the manuscript.

Funding: The work was carried out with the financial support of the project by the Russian Federation represented by the Ministry of Science and Higher Education of the Russian Federation, the Agreement No. 121021500054-3.

Data Availability Statement: The data presented in this study are available on request from the corresponding author.

Acknowledgments: The authors would like to thank: 1—Kamchatka branch of the Unified Geophysical Service of the Russian Academy of Sciences for the installation in 2008 of an excellent broadband seismic station at "Cape Schultz"; 2—European Space Agency and the developers of the Sentinel-Hub EO Browser (Sinergise Ltd., https://apps.sentinel-hub.com/eo-browser/, accessed on 25 September 2021) for the data of the Sentinel satellites and excellent means of access to them; 3—developers of the FFmpeg software package (https://www.ffmpeg.org/, accessed on 25 September 2021) for the very convenient means of broadcasting streaming video from coastal and underwater cameras; and 4—the developers of the best, in our opinion, Fast Fourier Transform library FFTW (https://www.fftw.org/, accessed on 25 September 2021).

Conflicts of Interest: The authors declare no conflict of interest.

References

1. Holman, R.A.; Guza, R.T. Measuring run-up on a natural beach. *Coast. Eng.* **1984**, *8*, 129–140. [CrossRef]
2. Holman, R.A.; Bowen, A.J. Longshore structure of infragravity wave motions. *J. Geophys. Res.* **1984**, *89*, 6446–6452. [CrossRef]
3. Holman, R.; Stanley, J. The history and technical capabilities of Argus. *Coast. Eng.* **2007**, *54*, 477–491. [CrossRef]
4. Stockdon, H.; Holman, R.A. Estimation of wave phase speed and nearshore bathymetry from video imagery. *J. Geophys. Res. Space Phys.* **2000**, *105*, 22015–22033. [CrossRef]
5. Holman, R.A.; Symonds, G.; Thornton, E.B.; Ranasinghe, R. Rip spacing and persistence on an embayed beach. *J. Geophys. Res. Space Phys.* **2006**, *111*, C01006. [CrossRef]
6. Chickadel, C.C.; Holman, R.A.; Freilich, M.H. An optical technique for the measurement of longshore currents. *Geophys. Res.* **2003**, *108*, 3364. [CrossRef]
7. Davidson, M.; Van Koningsveld, M.; de Kruif, A.; Rawson, J.; Holman, R.; Lamberti, A.; Medina, R.; Kroon, A.; Aarninkhof, S. The CoastView project: Developing video-derived Coastal State Indicators in support of coastal zone management. *Coast. Eng.* **2007**, *54*, 463–475. [CrossRef]
8. Fischenko, V.K.; Golik, A.V.; Subote, A.E.; Zatserkovniy, A.V.; Dubina, V.A. Scientific video monitoring system of the Peter the Great Bay (Japan sea). *Geoinformatics* **2011**, *4*, 30–41. (In Russian)
9. Longuet-Higgins, M.S. A theory of the Origin of Microseisms. *Philos. Trans. R. Soc. Lond. Ser. Math. Phys. Sci.* **1950**, *243*, 1–35.
10. Dolgikh, G.I.; Valentin, D.I.; Dolgikh, S.G.; Kovalev, S.N.; Koren, I.A.; Ovcharenko, V.V.; Fishchenko, V.K. Application of horizontally and vertically oriented strainmeters in geophysical studies of transitional zones. *Izv Phys. Solid Earth* **2002**, *38*, 686–689. (In Russian)
11. Dolgikh, G.I.; Budrin, S.S.; Dolgikh, S.G.; Zakurko, A.G.; Kosarev, O.V.; Ovcharenko, V.V.; Plotnikov, A.A.; Chupin, V.A.; Shvets, V.A.; Yakovenko, S.V. Complex spatially dispersed polygon in the far east for earth observations. *Meas. Tech.* **2016**, *59*, 252–255. (In Russian) [CrossRef]
12. Yakovenko, S.A.; Budrin, S.S.; Dolgikh, S.G.; Chupin, V.A.; Shvets, V.A. Hydrophysical laser-interference complex. *St. Petersburg Polytech. Univ. J. Phys. Math.* **2016**, *2*, 294–298. (In Russian)
13. Dolgikh, G.; Budrin, S.; Dolgikh, S.; Plotnikov, A. Supersensitive Detector of Hydrosphere Pressure Variations. *Sensors* **2020**, *20*, 6998. [CrossRef] [PubMed]
14. Addison, P.S. Wavelet transforms and the ECG: A review. *Physiol. Meas.* **2005**, *26*, R155–R199. [CrossRef]
15. Landau, L.D.; Lifshitz, E.M. *Fluid Mechanics, Course of Theoretical Physics*, 2nd ed.; Pergamon Press: Oxford, UK, 1987; Volume 6, ISBN 978-0-08-033932-0.
16. Lendaris, G.G.; Stanley, C.L. Diffraction Pattern Sampling for Automatic Pattern Recognition. *Proc. IEEE* **1970**, *58*, 198–216. [CrossRef]
17. Rabiner, L.R.; Gold, B. *Theory and Application of Digital Signal Processing*; Prentice-Hall: Englewood Cliffs, NJ, USA, 1975.
18. Mangor, K.; Drønen, N.K.; Kærgaard, K.H.; Kristensen, S. *Shoreline Management Guidelines*; DHI: Hørsholm, Denmark, 2016; pp. 37–40.
19. Rodriguez-Padilla, I.; Castelle, B.; Marieu, V.; Morichon, D. A Simple and Efficient Image Stabilization Method for Coastal Monitoring Video Systems. *Remote Sens.* **2019**, *12*, 70. [CrossRef]

20. Smirnov, S.V. On calculation of seiche oscillations of the middle part of Peter the Great Gulf. *Numer. Anal. Appl.* **2014**, *17*, 203–216. (In Russian) [CrossRef]
21. Fishchenko, V.K.; Dolgikh, G.; Zimin, P.S.; Subote, A.E. Some Results of Oceanological Video Monitoring. *Dokl. Earth Sci.* **2018**, *482*, 1244–1247. [CrossRef]
22. Fishchenko, V.K.; Zimin, P.S.; Zatserkovnyy, A.V.; Subote, A.E.; Golik, A.V.; Goncharova, A.A. Stationary systems for underwater video surveillance of coastal water areas. *Underw. Investig. Robot.* **2020**, *1*, 60–71. [CrossRef]
23. Fishchenko, V.K.; Zimin, P.S.; Golik, A.V.; Goncharova, A.A. Utilization of stationary underwater surveillance systems for the estimation of underwater currents and sea disturbance parameters. *Underw. Investig. Robot.* **2020**, *2*, 62–73. [CrossRef]
24. Dolgikh, G.; Fishchenko, V.K.; Goncharova, A.A. Potential for Recording of Waves and Sea Level Fluctuations in the World Ocean Coastal Areas by Internet Video Analysis. *Dokl. Earth Sci.* **2019**, *488*, 1264–1267. [CrossRef]

Journal of
*Marine Science
and Engineering*

Article

Detection and Analysis of the Causes of Intensive Harmful Algal Bloom in Kamchatka Based on Satellite Data

Valery Bondur, Viktor Zamshin *, Olga Chvertkova, Ekaterina Matrosova and Vasilisa Khodaeva

AEROCOSMOS Research Institute for Aerospace Monitoring, 105064 Moscow, Russia;
vgbondur@aerocosmos.info (V.B.); chvertkovaolga@gmail.com (O.C.); ematrosova95@gmail.com (E.M.);
vasilisa.95@inbox.ru (V.K.)
* Correspondence: viktor.v.zamshin@gmail.com

Abstract: In this paper, the causes of the anomalous harmful algal bloom which occurred in the fall of 2020 in Kamchatka have been detected and analyzed using a long-term time series of heterogeneous satellite and simulated data with respect to the sea surface height (HYCOM) and temperature (NOAA OISST), chlorophyll-a concentration (MODIS Ocean Color SMI), slick parameters (SENTINEL-1A/B), and suspended matter characteristics (SENTINEL-2A/B, C2RCC algorithm). It has been found that the harmful algal bloom was preceded by temperature anomalies (reaching 6 °C, exceeding the climatic norm by more than three standard deviation intervals) and intensive ocean level variability followed by the generation of vortices, mixing water masses and providing nutrients to the upper photic layer. The harmful algal bloom itself was manifested in an increase in the concentration of chlorophyll-a, its average monthly value for October 2020 (bloom peak) approached 15 mg/m^3, exceeding the climatic norm almost four-fold for the region of interest (Avacha Gulf). The zones of accumulation of a large amount of biogenic surfactant films registered in radar satellite imagery correlate well with the local regions of the highest chlorophyll-a concentration. The harmful bloom was influenced by river runoff, which intensively brought mineral and biogenic suspensions into the marine environment (the concentration of total suspended matter within the plume of the Nalycheva River reached 10 mg/m^3 and more in 2020), expanding food resources for microalgae.

Keywords: coastal water areas; red tide; harmful algal bloom; sea surface height; sea surface temperature; chlorophyll-a; biogenic slicks; river runoff; Kamchatka

Citation: Bondur, V.; Zamshin, V.; Chvertkova, O.; Matrosova, E.; Khodaeva, V. Detection and Analysis of the Causes of Intensive Harmful Algal Bloom in Kamchatka Based on Satellite Data. *J. Mar. Sci. Eng.* **2021**, *9*, 1092. https://doi.org/10.3390/jmse9101092

Academic Editor: Lev Shemer

Received: 8 September 2021
Accepted: 4 October 2021
Published: 7 October 2021

Publisher's Note: MDPI stays neutral with regard to jurisdictional claims in published maps and institutional affiliations.

1. Introduction

Harmful algal blooms are dangerous natural processes which are difficult to be predicted or prevented. It is necessary to study these processes [1]. Such a task can be effectively solved using satellite remote sensing data [2] through the application of specialized methods of their processing [3–5].

Satellite monitoring methods have a wide range of capabilities to study marine water areas which include wide coverage; efficiency; capability to work in any hard-to-reach areas of seas and oceans; obtaining data of various spatial and temporal resolution in different electromagnetic wave spectral regions; a wide range of registered parameters of water environment; and high reliability of obtained data [6]. The efficiency of these methods is largely increased when combining them with sea truth and modeling results [7,8]. The abovementioned features provide for the high efficiency of marine water area remote sensing when solving various thematic tasks related with coastal water area pollution detection, including to study and monitor microalgae blooming [9,10].

Up-to-date scientific publications on the topic of this paper shows that the problem of satellite data application for monitoring and studying harmful algal blooms (HAB) can have different formulations. On the one hand, specialized methods and algorithms for satellite (mostly, optical multispectral) data processing are being developed and applied that allow for the creation of information products aimed directly at detecting and determining algae

191

type and describing conditions of HABs. The examples of such efforts include a red tide index study [11,12], use of machine learning algorithms for automatic distinction of various types of environmental conditions and HAB detection in satellite optical imagery [3], use of a special algorithm for algae detection for various waters [4], studying the alternative floating algae index (AFAI) [5], et al.

On the other hand, comprehensive studies applying a large amount of heterogeneous data are carried out that are aimed at the understanding of natural processes occurring during HAB preparation and development. Algal blooms are driven by various factors including physical and chemical ones. Physical factors include horizontal inhomogeneity and stratification of water, water illumination and temperature, wind, currents, and tides. Chemical factors include a high level of available mineral and organic nutrients coming from coastal runoffs, and excreta of hydrobionts from both atmospheric precipitation and from bottom sediments during vertical water movements [1,13–15].

Taking this into account, it is possible to study HAB conditions and development features using a time series of well-proven standard information products of satellite oceanography, such as chlorophyll-a concentration in the near-surface layer of the marine environment and sea surface temperature. For example, in [7,8,16], to monitor HAB development and to reveal the causes of its occurrence, various combinations of standard information products are used, formed on the basis of data from satellite spectroradiometers MODIS, SeaWiFS, and MERIS (sea surface temperature, chlorophyll-a etc.), as well as HYCOM and NCEP products describing the ocean level, current, and near-surface wind characteristics. A very important significant water environment parameter registered by satellite methods for HAB study is chlorophyll-a. The effort [17] is focused on a detailed analysis of chlorophyll-a concentration variability for HAB monitoring purposes. This effort, "Detection and analysis of the causes of intensive harmful algal bloom in Kamchatka based on satellite data", carries on the abovementioned comprehensive research. Here we develop an original technique for analyzing large arrays of heterogeneous data from satellite oceanology to study a single large-scale HAB case.

This study was conducted in coastal waters east of Kamchatka. Kamchatka is a peninsula in the northeastern part of Eurasia, surrounded from the west by the Sea of Okhotsk, and from the east by the Bering Sea and the Pacific Ocean.

On the southeastern coast of the peninsula there is the Avacha Gulf, where this study was carried out (see Figure 1, red polygon). The area of the Avacha Gulf is ~2400 km^2. The depths of the Avacha Gulf vary. After the elevations of 150–200 m, there is a sharp slope, turning into the valleys of the Pacific Ocean. The climate of Avacha Bay, due to its exposure, is extremely unstable; the gulf is open to winds and sea waves. The Kamchatka Current passes along the eastern coast of the Kamchatka Peninsula, which includes Avacha Gulf. The water temperature in winter varies from 1 °C to 3 °C, and in summer months the water warms up to an average of 15 °C [18].

From the end of September to the beginning of October 2020, an intensive harmful algal bloom (hereinafter referred to as HAB) was registered in the coastal water areas of the Kamchatka peninsula. This phenomenon caused a massive loss of aquatic organisms, health problems of people who contacted with water, produced plenty of foam on the coast, and created unnatural seawater color [9]. Most of these negative phenomena were observed in the coastal zone of the Avacha Gulf about 40 km long from the Spaseniya Bay to the Listvennichnaya Bay, including the Khalaktyrsky Beach. The map of the region of interest showing the locations of negative HAB effects is given in Figure 1.

Intense HABs are accompanied by a series of negative effects, in particular, toxins released by microalgae enter the water environment, which have a detrimental effect on hydrobionts (up to their death) and the entire ecosystem as a whole [1,13].

This effort is dedicated to the detection and analysis of causes of the intensive Kamchatka 2020 HAB according to satellite data of various types.

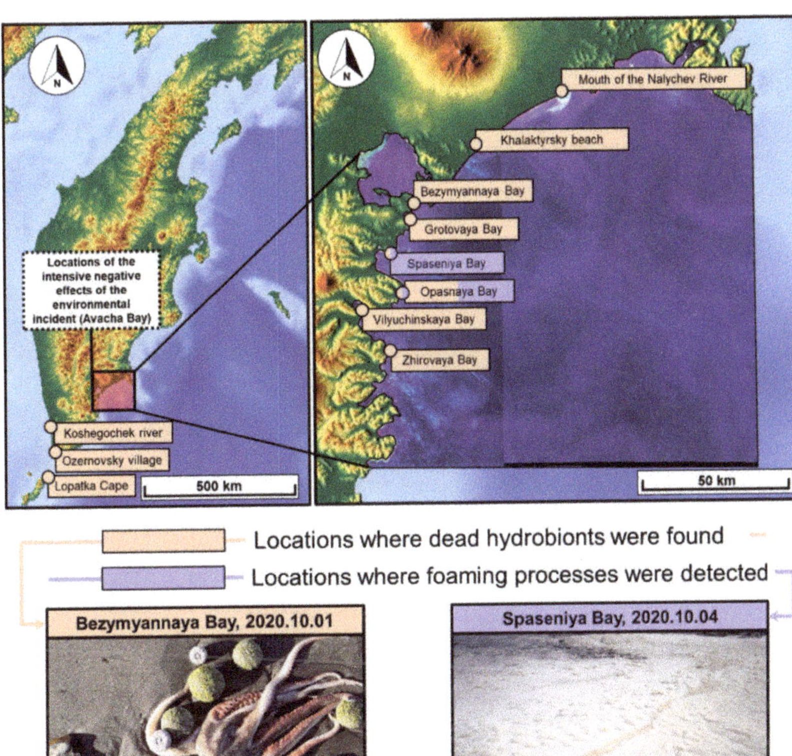

Figure 1. Schematic map of the study area with the locations of registration of negative harmful algal bloom indicators registered near the Kamchatka shore in the fall of 2020.

2. Materials and Methods

2.1. General

Our research is based on the analysis of long-term time series of inhomogeneous source satellite data on the sea surface characteristics and the estimates of variability of key factors causing and affecting the development of a HAB in the studied region. The following source satellite and simulated data were used: HYCOM sea surface height [19], and NOAA OISST sea surface temperature [20] data. LAS information product (Live Access Server): Aviso+ was applied to interpret HYCOM data [21]. Also, satellite optical multispectral and radar imagery obtained by Sentinel-2A/B [22] and Sentinel-1A/B [23] satellites, respectively, were used, as well as sea surface temperature and chlorophyll-a concentrations (MODIS Ocean Color SMI [24] and NOAA VIIRS [25]).

Satellite data processing was carried out in three directions, namely:

- Processing long-term retrospective satellite data series on the ocean level, sea surface temperature, chlorophyll-a concentration to reveal precursors and indicators of HAB development in the fall of 2020.
- Processing radar satellite imagery obtained by Sentinel-1A/B during harmful algal bloom in the fall of 2020 to detect and assess the properties of slicks with their further analysis to reveal the signs of algal blooming in the region of interest.

- Processing Sentinel-2A/B optical multispectral satellite imagery in 2019 and 2020 for the region of the inflow of the Nalycheva River into the water area of interest to assess possible impact of the river runoff on HAB in the fall 2020.

During the research, first of all, the fact was taken into account that HABs with such strongly expressed negative consequences were recorded for the first time off the coast of Kamchatka.

Hence, we believe that the disaster under consideration was preceded by specific conditions of the water environment, which had not been observed previously in the studied water area.

2.2. A Method for the Processing Long-Term Retrospective Series of Satellite Data on the Ocean Level, Sea Surface Temperature, Chlorophyll-a Concentration

Thus, the method for the processing long-term retrospective series of satellite data on the ocean level, sea surface temperature, and chlorophyll-a concentration was based on the comparison of the features of the dynamics of these significant parameters of the water environment in 2020 with the characteristic dynamics of the same parameters recorded in previous years in the study area of interest. The study was carried out within the Avacha Gulf water area covering certain open ocean area adjacent to the gulf (Figure 1).

As it is mentioned above, the following significant water environment parameters were used:

(a) Sea surface height (SSH) influencing the current mode and conditions for upwelling, which in its turn is important for the formation of food resources of microalgae (data are available since 1 January 1993).
(b) Sea surface temperature (SST) (data are available since 1 January 1982).
(c) Chlorophyll-a concentration in the near surface layer of the water environment (data are available since 1 May 2000).

The method of research involved the construction and analysis of graphs showing the intra-annual variability of each studied significant parameter of the water environment within the area of interest. SST and SSH graphs were plotted with one day discreteness, which corresponded to the source temporal resolution of the information products used. The daily value of the studied parameter was averaged over the area:

$$P_{date} = \sum_{i=1}^{n_{date}} p_i / n_{date},\tag{1}$$

where P_{date} is the calculated average value of the water environment parameter for the current date; n_{date} is the number of measurements (nonempty pixels) within the studied area for the target date; and p_i is the value of the studied parameter in the current pixel.

Chlorophyll-a concentration graphs were built with one month discreteness. The computation of P_{month} monthly average values was carried out according to the principle of "cumulative average" for each element of the grid of values within the studied area:

$$P_{month} = \sum_{i=1}^{n_{month}} \overline{P}_i / n_{month},\tag{2}$$

where P_{month} is the calculated average value of the water environment parameter for the current month; n_{month} is the number of pixels within the study area, where there was at least one measurement per month; and $\overline{P}_i$ is the monthly average value of the studied parameter in the current pixel.

The results of within-year dynamics of the studied water environment parameters obtained for 2019 and the previous years were generalized through the calculation of the average $\overline{P}$ value and σ standard deviation:

$$\sigma_{date} = \sqrt{\frac{\Sigma\left((P_{date})_j - \overline{P}_{date}\right)^2}{N}},\tag{3}$$

where N is the number of a sample (the number of years of observations preceding the year of registration of the studied HAB—2020).

$\overline{P}_{date}$ graphs, together with σ_{date} standard deviation corridor, characterize the climatic norm well and make it possible to adequately assess the features of the temporal variability of significant parameters of the water environment recorded in 2020 during the formation of the prerequisites for the development of HABs in the ecosystem of the studied area [9]. As a result of this technique it is possible to determine dates when anomalous changes of each studied parameter of the water were detected in 2020 year relative to averages over the previous years (going strongly beyond the standard deviation corridor). Finally, comparative analysis with additional satellite products for those days when abnormal deviations were detected is performing, as shown on Figure 2.

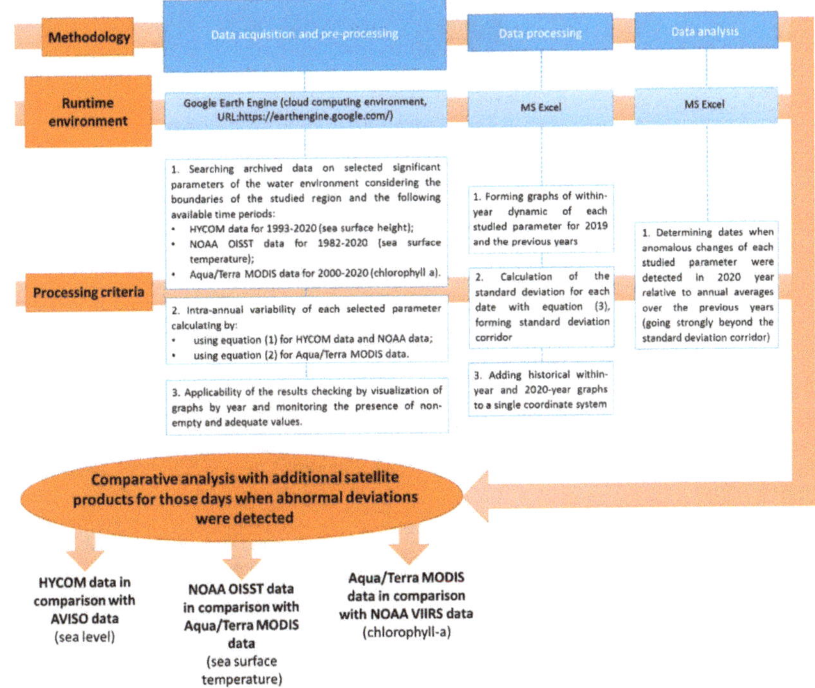

Figure 2. Flowchart of the experiment for processing of long-term retrospective series of satellite data on the ocean level, sea surface temperature, and chlorophyll-a concentration.

Data acquisition and calculations using Equations (1) and (2) were carried out using the Google Earth Engine cloud infrastructure. Calculations by Equation (3) and construction of graphs were carried out in Excel. Comparative analysis with map layouts forming was carried out using the QGIS the ENVI + IDL software.

2.3. Sentinel-1A/B Radar Satellite Imagery Processing Method

In the course of the research, the processing of the Avacha Gulf Sentinel-1A/B satellite radar images was carried out (from 25 August to 5 October 2020). The processing was aimed at the detection and estimation of sea surface slick properties related with biogenic surface active agents films (later on "biogenic SAA").

Using satellite radar images, owing to the resonant mechanism of radio wave scattering by wave-covered water surface, it is possible to monitor slick formations of both natural and anthropogenic origins [6,26]. When interpreting biogenic SAA using radar imagery, it is necessary to provide for their reliable recognition against the background of other slick-forming processes (e.g., calm belts, rain cells, internal wave surface manifestations, atmospheric gravity waves, ship spills etc.) [27].

With intensive blooming of microalgae, biogenic surfactants are formed on the sea surface causing the formation of slicks with specific interpretation features [27]. In satellite radar images, biogenic SAA usually represent a series of long (many kilometers long) often curved or spiral stripes (due to current fields). Such film formations are often observed in the coastal zones or in the upwelling zones. Biogenic surfactants exist in quite a narrow wind speed range (from 2–3 m/s to 5 m/s) [27]. Detection and assessment of the characteristics of such slick formations using satellite radar images can confirm the fact of HAB development, as well as provide detailed information on the spatial and geometric properties and dynamics of biogenic surfactant accumulations.

Flowchart for Sentinel-1A/B dataset processing is shown on Figure 3.

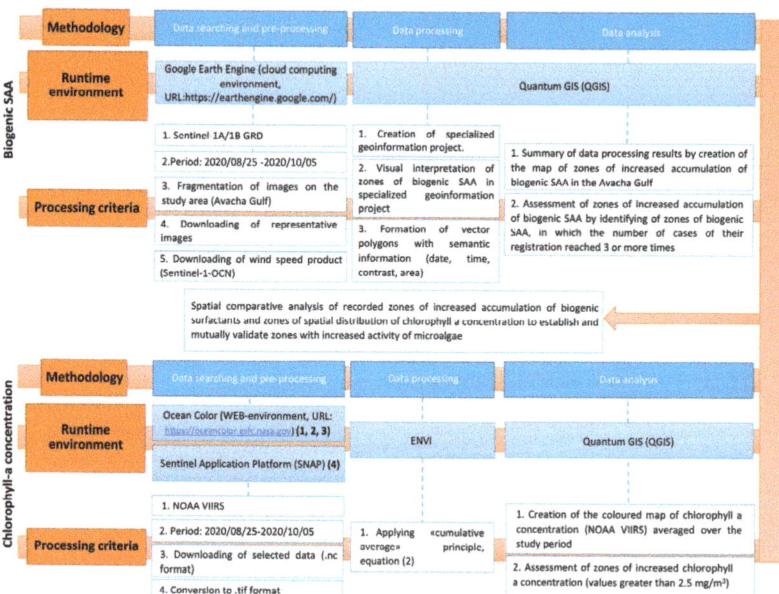

Figure 3. Flowchart of the experiment for processing Sentinel-1A/B radar satellite imagery.

To increase the reliability of biogenic SAA interpretation using satellite radar imagery and to provide for the comprehensive analysis of their spatial and geometric characteristics, a specialized geoinformation project has been developed in this effort that contains source radar satellite imagery and the results of their interpretation as polygons with semantic information (date, time, contrast and area), as well as the geographic features of the studied region, meteorological data (in particular, near-surface wind characteristics), bathymetry, and etc.

Generalized results of detection of slicks related with biogenic surfactants were compared with the data on negative HAB effects (see Figure 1) and spatial chlorophyll-a concentration distributions in the Avacha Gulf obtained according to the "cumulative average" principle (see Equation (2)) for the whole period of radar imagery collection (from 25 August to 5 October 2020).

2.4. Sentinel-2A/B Multispectral Imagery Processing Method

Several rivers, including the rather large Nalycheva river, flow into the Avacha Gulf on the banks of which the most intensive negative consequences of the studied HAB were recorded. Along with the river runoff, nutrients can enter the coastal waters, contributing to the development of microalgae [1,28]. In the present study, the characteristics of the runoff of the Nalycheva River in the interests of studying the features of its influence on the process of preparation and development of intensive HAB, registered in the fall of 2020.

The processing of optical multispectral Sentinel-2A/B imagery was carried out using the automatic neural network Case 2 Regional Coast Colour (C2RCC) algorithm that allows us to assess bio-optical water environment parameters, in particular total suspended matter (TSM) in the surface layer [29,30]. Increased TSM values in marine water areas near river mouths indicate the presence of fine matter of biological and mineral origin [31], which, as a rule, contains nutrients that contribute to the development of algae [32]. TSM calculations were performed using the SNAP software (see Figure 4).

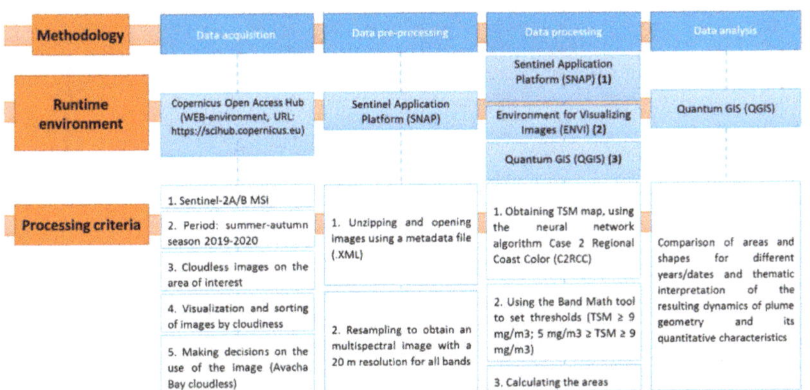

Figure 4. Flowchart of the experiment for processing Sentinel-2A/B multispectral imagery.

The resulting fields of TSM distributions were processed using the threshold method in order to detect a wastewater plume from the Nalycheva River. Two threshold values were used, which ensured the identification of the distribution areas of both the most saturated with suspended matter share of the river discharge (i.e., the plume "core"), and the peripheral areas of the river discharge (i.e., the "transition zone" of the plume). The following condition was used to select the "core":

$$\text{TSM} \geq 9\ \text{mg/m}^3 \tag{4}$$

The selection of the "transition zone" was carried out with another one condition:

$$9\ \text{mg/m}^3 > \text{TSM} \geq 5\ \text{mg/m}^3 \tag{5}$$

Based on the results of identifying the distribution areas of the plume of the Nalycheva River, generalizing schematic maps were formed and analyzed using the geo-information environment.

3. Results and Discussion

3.1. Anomalies in Long-Term Retrospective Aeries of Satellite Data

3.1.1. Ocean Level

Figure 5 shows the graphs of the ocean level dynamics (HYCOM), obtained using the method described above for processing a retrospective data set (since 1993). The insets in Figure 5 show the sea level anomaly (SLA) and the directions of geostrophic currents obtained from the data of the LAS: Aviso + satellite altimetry information product aggregator for some dates in 2020 (when anomalous changes in the sea surface height were recorded).

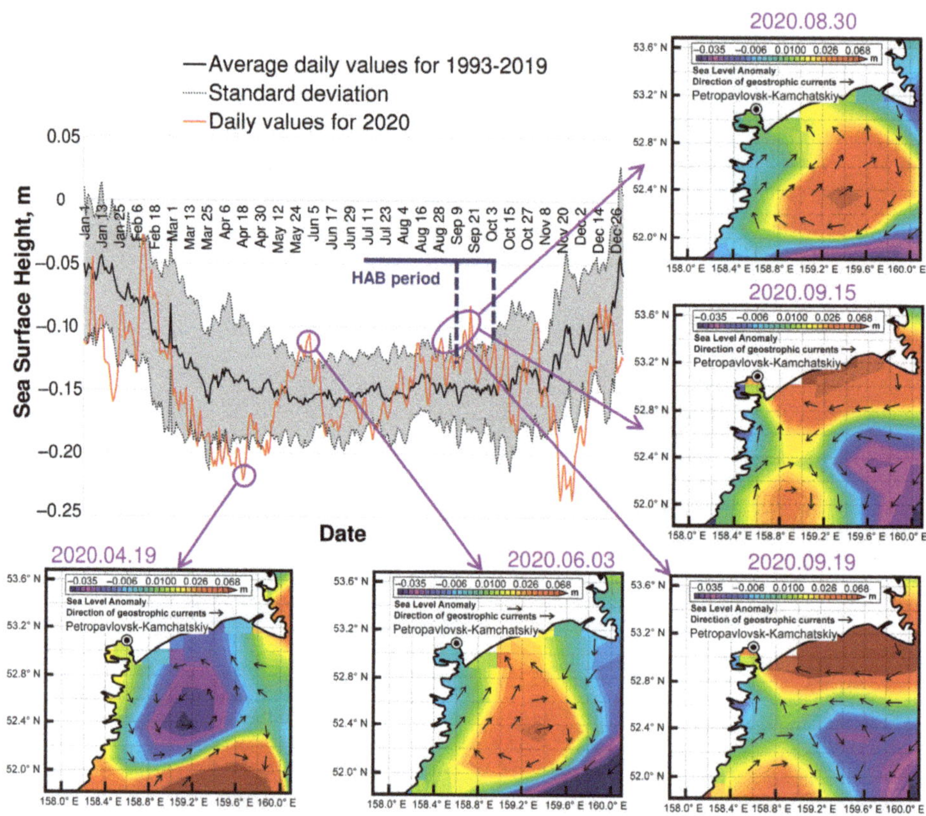

Figure 5. The graphs characterizing the annual variation of daily SSH (sea surface height) values for 2020 (red line) and summarized for the period from 1993 to 2019. SSH values (black line denotes average values, dashed lines denote standard deviations) characterizing the climatic norm (based on HYCOM data processing). The insets show spatial distributions of sea level anomalies and directions of geostrophic currents for the dates of registration of the anomalous sea level: 19 April, 3 Jun, 30 August, 15 September, 19 September (Aviso + data, 2020).

Analysis of the data shown in Figure 5 revealed the following features:

- On 19 April, anomalously low SSH values were recorded relative to the annual average values (HYCOM), which can also be traced in the spatial distribution of sea level anomalies (Aviso +), presented in the inset of Figure 5. These features clearly indicate the presence of a cyclonic oceanic vortex.
- On 3 June 2020, SSH values anomalously exceeding annual average values (HYCOM) were observed, as well as such high values were observed in the spatial distribution of sea level anomalies (Aviso +), shown in the inset of Figure 5. These features clearly indicate the presence of an anticyclonic oceanic vortex in the studied water area.
- On 30 August and the 15 and 19 September 2020, anomalously high SSH values and geostrophic currents were recorded, the configuration of which also corresponded to oceanic vortexes of the anticyclonic type.

Thus, it was found that the intense HAB, which occurred in Kamchatka in the fall of 2020, was preceded by a developed SSH dynamics, characterized by the presence of predominantly positive anomalies and signs of anti-cyclonic vortexes. It should be noted that the cold Vostochno-Kamchatskoe Current passes along the southeastern coast of Kamchatka, which is accompanied by the formation of vortex structures with the subsequent generation of unsteady mesoscale phenomena, including jets and mushroom-like currents. The reason for the formation of such vortices is the influence of the bottom topography and coastline inhomogeneity on the Vostochno-Kamchatskoe Current [33]. These vortices affect the mixing of biologically productive coastal waters and open ocean waters, ensuring the flow of biogenic elements into the upper photic layer and thereby positively affecting the development of phytoplankton [34].

Examples of the positive impact of different types of vortices on the development of phytoplankton are numerous for the regions of the world's oceans, while the mechanisms of the impact of cyclonic and anticyclonic vortices on biota can be different. In most cases, anticyclonic vortices that prevailed in the study area in 2020 are characterized by abnormally high concentrations of chlorophyll-a in the surface water layer along the vortex periphery [35].

Thus, based on the analysis of data on the ocean level dynamics in the study area, it has been established that during 2020, the situation has repeatedly evolved, contributing to the processes of preparation and development of HABs.

3.1.2. Sea Surface Temperature

According to the abovementioned method, the temperature of the sea surface in the studied water area was investigated using the daily information products of NOAA OISST for the period from 1981 to 2020. The obtained NOAA values of the surface temperature for 1981–2019 were averaged and compared with daily values for 2020, while detailed spatial distributions for critical dates were considered according to MODIS data (Figure 6).

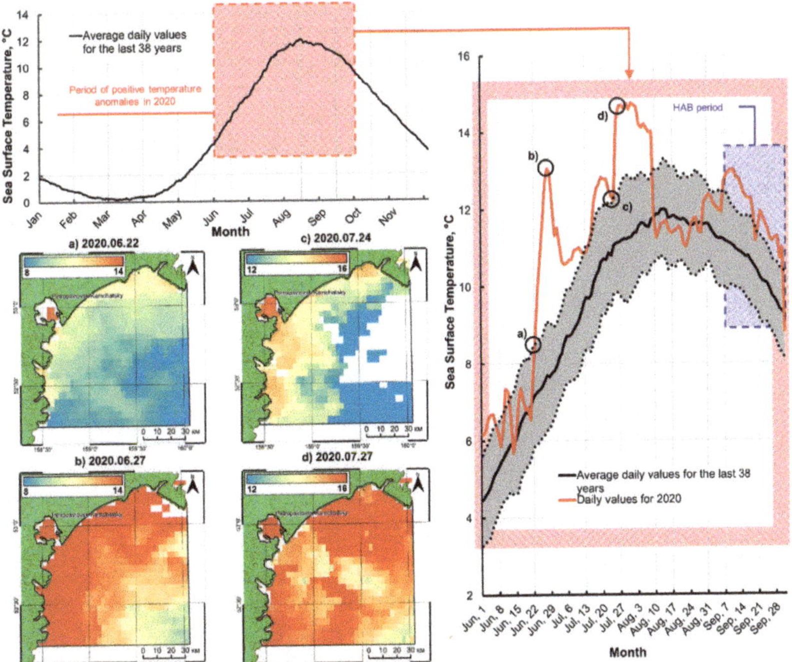

Figure 6. Graphs characterizing the annual variation of daily SST (sea surface temperature) values for 2020 (red line) and summarized for the period from 1981 to 2019 SST values (black line denotes average values, dashed lines denote standard deviations) based on NOAA OISST data processing. The insets (**a**–**d**) based on MODIS data show spatial distributions of SST on 22, 27 June and 24, 27 July (when the most intensive anomalies had been formed).

From the analysis of the data shown in Figure 6, it follows that from 22 June to 12 July 2020 and from 15 July to 8 August 2020 in the study area, anomalous exceeding in sea surface temperature (reaching 6 °C) were recorded compared to average long-term values. The overshoot of sea surface temperature values recorded on 27 June 2020 exceeded 3 intervals of standard deviation. On 27 July 2020, such an overshoot exceeded 2 intervals of the standard deviation (see Figure 6). The spatial distributions of the sea surface temperature show that warm anomalies are most intensive near the coastline, in the area of registration of negative HAB manifestations. It is known (see, for example, [14]) that such positive SST anomalies have a significant effect on the biogenic mode of the upper photic layer of the seawater column and can lead to the intensive development of microalgae, causing red tides.

As follows from the analysis of Figure 6, in 2020 in the study area, warm anomalies were especially strong. In this regard, it is fair to assume that the temperature conditions in 2020 greatly contributed to the preparation and development of the HAB off the coast of Kamchatka, and the temperature factor played a key role in the development of the environmental disaster off the coast of Kamchatka.

3.1.3. Chlorophyll-a Concentration

The chlorophyll-a concentration was studied using the monthly average information products MODIS Ocean Color SMI [24], obtained according to the method described above (see Figure 2) as a result of processing of a retrospective data set from 2000 to 2020 (in the periods from May to October). The resulting monthly average values for 2000–2019 were averaged and compared with the monthly average values for 2020 (see Figure 7, bottom).

Figure 7 (top) shows the spatial distributions of chlorophyll-a concentration obtained from NOAA VIIRS data [25] at the time of intense HAB development for some dates in 2020 (when the transition from the background to anomalous conditions was recorded).

An analysis of the data obtained, presented in Figure 7 (top), made it possible to reveal a sharp increase in the concentration of chlorophyll-a, which occurred in the third decade of September 2020, practically over the entire area of the Avacha Gulf, compared with the previous days. A high level of chlorophyll-a concentration in this water area was also observed in the following days, with a maximum on 1 October 2020. Compared with background values, for example, on 4 September 2020, the chlorophyll-a concentration measured on 1 October 2020 increased by approximately an order of magnitude from 0.1 to 1.0 conv. units and more (see Figure 7, top).

From the analysis of the graphs shown in Figure 7 (lower part) it follows that the values of the average monthly concentrations of chlorophyll-a in September–October 2020 significantly (more than 3.5-fold) exceeded the climatic norm; values of chlorophyll-a concentrations in the same months averaged over 2000–2019. It should be noted that satellite information products on the concentration of chlorophyll-a, calculated based on global algorithms [24,25], must be analyzed taking into account, inter alia, the regional characteristics of the studied water area and atmospheric conditions. However, the unprecedentedly high measured values of chlorophyll-a obtained in this study for October 2020 indicate significant changes in the optical properties of the near-surface layer of the marine environment caused by biological factors. The possible influence of atmospheric phenomena and the regional specificity of the optical characteristics of waters on the accuracy of the data obtained is largely compensated by the use of the accumulated averaged monthly values of chlorophyll-a concentrations for 2020 and their comparison with the monthly values of chlorophyll-a concentrations in the water area of interest, averaged over the 20-year observation period (2000–2019).

3.2. Biogenic Surfactant Films (Sentinel-1A/B)

According to the previously described method, satellite radar images obtained from the Sentinel-1A/B satellites were studied during the period from 25 August to 5 October

2020. In this case, information products of the first and second processing levels were used, including data on the characteristics of the near-surface wind (Sentinel-1-OCN).

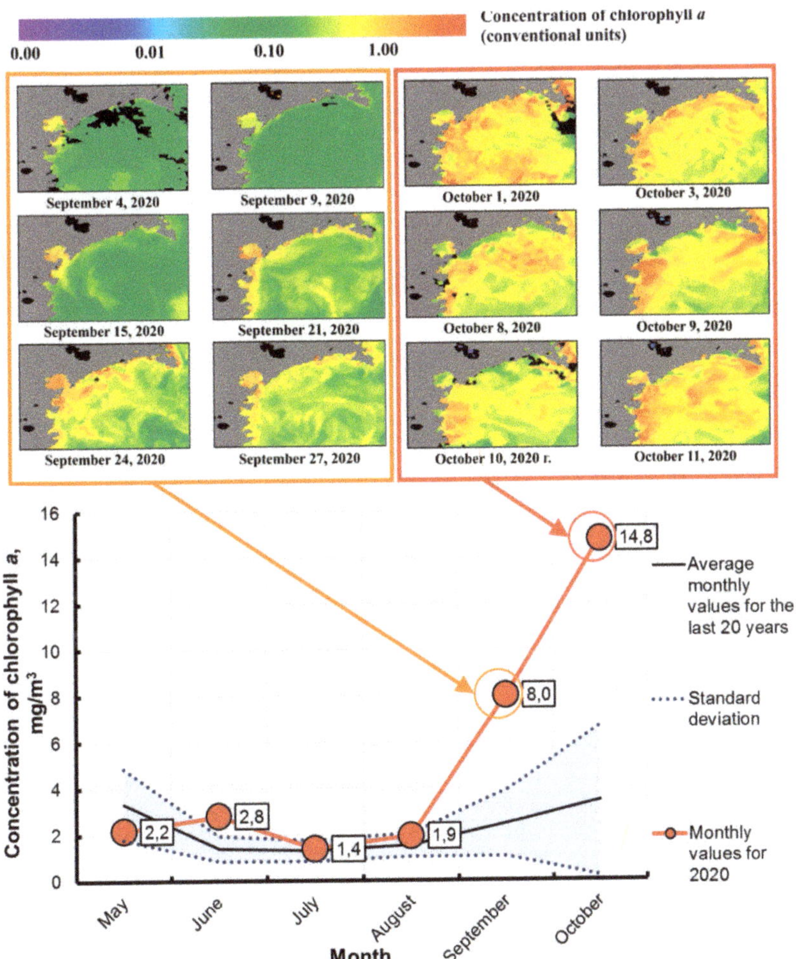

Figure 7. Information products formed during the assessment of phytoplankton bloom intensity in the Avacha Gulf: daily spatial distributions of chlorophyll-a concentration (in conv. units) in September–October 2020, obtained from the processing of VIIRS data (NOAA satellites) (above); the graphs characterizing the monthly average values of chlorophyll-a concentrations from May to October 2020 (red line), as well as from May to October averaged over the time period from 2000 to 2019. (blue line denotes average values, dashed lines denote standard deviation) characterizing the climatic norm (obtained from MODIS data processing (AQUA and TERRA satellites).

Numerous biogenic surfactants were detected on space radar images from 25, 26, 30 August, 7, 19, 23 and 30 September, as well as 1 and 5 October 2020. Based on the results of space radar image processing, a map of zones of increased accumulation of biogenic surfactants was created, in which the number of cases of their registration reached three or more times (see Figure 8). The analysis of these data has shown that biogenic surfactants were distributed unevenly over the studied water area. In different parts of the Avacha Gulf, the number of cases of registration of biogenic surfactants ranged from 0 to 5 (over the observation period). The measured radar contrasts of the detected slicks were ~3.6 to ~5.5 dB.

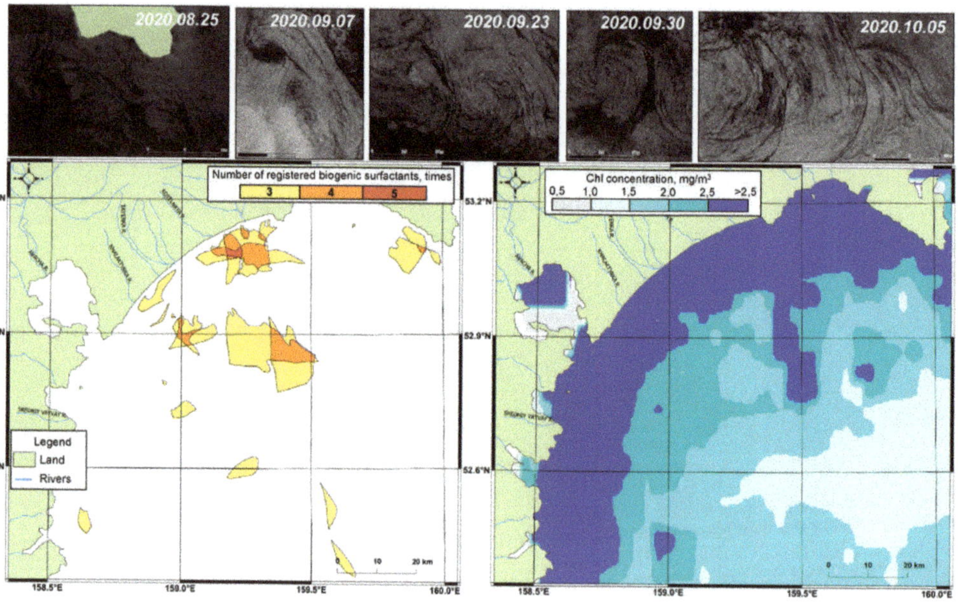

Figure 8. Radar satellite image interpretation result (Sentinel-1A/B) (the images were obtained between 25 August and 5 October 2020). The map of zones of increased accumulation of biogenic surfactants in the Avacha Gulf (left). Right, the map of chlorophyll-a concentration distribution (NOAA VIIRS) averaged over the period from 25 August to 5 October 2020.

Analysis of Figure 8 shows that zones of increased accumulation of biogenic SAA were most often concentrated near the coast of the Avacha Gulf. Zones of increased accumulation of biogenic surfactants, in which the number of cases of their registration was three times, were located chaotically throughout the Avacha Gulf at a distance of ~1–70 km from the coast. Zones of concentration of biogenic SAA with the number of cases of their registration 4 times were observed in the northern and northeastern parts of the Avacha Gulf at a distance of ~1–28 km from the coast. The zone most exposed to biogenic surfactants with the number of cases of their registration 5 times was located in the coastal part of the Avacha Gulf near the mouths of the Nalycheva and Kotelnaya rivers at a distance of ~4 km from the coast (coordinates of the zone are 53.1° N, 159.2° E). The total area of zones of increased accumulation of biogenic SAA, identified by the results of processing space radar images, was ~680 km^2.

The results of registration of zones of increased accumulation of biogenic surfactants in space radar imagery were compared with the map of the spatial distribution of chlorophyll-a concentration in the Avacha Gulf, formed according to NOAA VIIRS data (see Figure 8). As a result of the comparison, it was confirmed that high activity of microalgae was observed in the coastal strip ~25 km wide, as well as in the central part of the gulf near the point with coordinates 52.8° N and 159.5° E.

As a result of the analysis of satellite radar images, it was found that the study area during the HAB in 2020 was subjected to intense natural impacts. Attention is drawn to the large-scale nature of biogenic surfactants registered in the Avacha Gulf.

3.3. River Runoff (Sentinel-2A/B)

Optical multispectral satellite images obtained from the Sentinel-2A/B satellites on cloudless days of the summer-fall season of 2019 and 2020 near the mouth of the Nalycheva

River were recalculated into TSM values in accordance with the method described above, and then processed using two thresholds. Threshold 1 assumed the selection of all pixels for which the TSM value exceeds 9 mg/m^3, and made it possible to determine the distribution area of the "plume core" of waters carried by the Nalycheva River into the bay. Threshold 2 assumed the selection of all pixels for which the TSM value exceeded 5 mg/m^3, but did not exceed the value of threshold 1, which made it possible to determine the boundaries of the "transition zone" of the plume [9,31]. The results of identifying the areas of distribution of plumes of the Nalycheva River runoffs in the Avacha Gulf are shown in Figure 9.

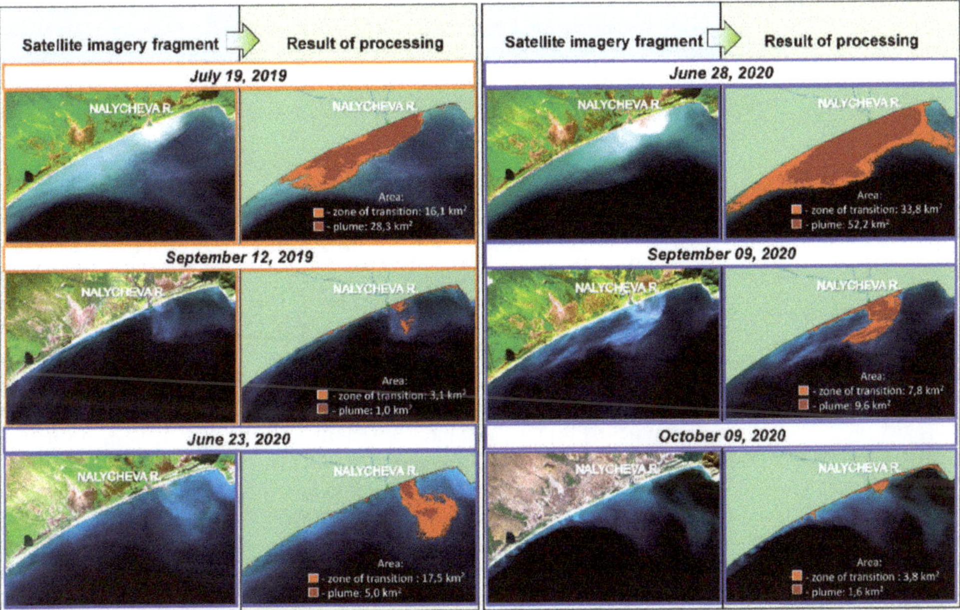

Figure 9. Maps of the Nalycheva river plume propagation areal in 2019 and 2020 (Sentinel-2A/B, TSM concentration spatial distribution thresholding, the plume's "core" and "transition zone" are identified).

The analysis of Figure 9 has allowed us to note the following:

1. The Nalycheva River runoff plume in 2019–2020 had a significant impact on the optical characteristics of the waters of the Avacha Gulf near the mouth of this river, which is reflected in an increase in the TSM concentration values determined by the C2RCC algorithm. The influence of runoff can be traced both in 2019 and in 2020.

2. In 2019, the Nalycheva River runoff plume was reliably recorded two times (Figure 9 (images with an orange frame)), and in 2020 four times (Figure 9 (images with a blue frame)). In this case, the largest plume area (the combination of the "core" and the "transition zone") in 2019 was ~44 km^2, while in 2020, the plume was recorded almost twice as large, amounting to ~86 km^2.

3. Optical properties of waters entering the Avacha Gulf from the Nalycheva River, testified to the bringing suspensions containing nutrients of biological and mineral origins. Attention is drawn to the fact that this process in 2020 run much more intensively than in 2019.

Based on the data obtained and analyzed, it can be assumed that in 2020 the Nalycheva River runoff mode created conditions favorable to the development of HABs in the Avacha Gulf. It is noteworthy that the largest area of the Nalycheva River plume was recorded on 28 June 2020 (see Figure 9), with the strongest temperature anomaly recorded on

27 June 2020 (see Figure 6). An increase in temperature under normal light intensity and the sufficiency of nutrients entering the marine environment, including with river runoff, has a positive effect on the growth rate of algae and their photosynthesis. Thus, the release of a large amount of nutrients into the coastal water area and an anomalous increase in the temperature of the water environment that almost coincided with this event was a powerful factor in the accelerated development of microalgae. An analysis of Figure 8, illustrating the results of processing space radar images, also indicated that HAB developed most intensively directly in the area of the Nalycheva River mouth.

3.4. Validity of the Obtained Results

Let's address the limitations of use of satellite oceanography data. In our research, first of all, we took into account the characteristics of information product accuracy declared by the developers, as well as the impact of weather (particularly, atmospheric) phenomena. For example, it is known that the accuracies of absolute chlorophyll-a concentration value retrieval using global algorithms according to MODIS spectroradiometer (Terra and Aqua satellites) data largely depend on the region of research (including on the water type and atmospheric aerosol types). Thus, the paper analyzes not individual absolute values of recorded parameters of the water environment but the entire available array of data on three types of significant parameters for the selected study area (SSH, 1993/2020; SST, 1982/2020; chlorophyll-a, 2000/2020).

This approach has allowed us to assess the dynamics of the measured significant parameters of the water area at the time of HAB preparation and development in Kamchatka in 2020 compared to the results of previous measurements. The comparative analysis addressed the mean values and standard deviations of the measured parameters, which are characteristic of the study area, recorded in a long historical retrospective (with the same instruments). At the same time, the features of the natural seasonal variation of the studied values were taken into account.

The validity of the outcome obtained is evidenced by the mutual correspondence of the results of detecting anomalies in the sea surface level, temperature, and concentration of chlorophyll-a obtained using different-type observation and data processing systems. For example, revealing the event of sharp increase of chlorophyll-a concentration was carried out using data from both VIIRS and MODIS spectroradiometers. At the same time, temperature anomalies registered based on long-term NOAA OISST data series processing are confirmed by MODIS data. The SSH HYCOM data processing results are supported with SLA AVISO data.

Control of possible errors of algorithms used to obtain the results is also to be noted. The graphs given in Figures 5–7 were obtained using the technique programmed in Google Earth Engine and MS Excel environments and described in Section 2.2. To control possible errors in the development of the algorithm, verification calculations of the variation of the measured parameters (SST/NOAA, chlorophyll-a/MODIS) were carried out manually for several dates having the key importance for our effort. When compared with the results of automatic processing of the entire data set, no errors were found.

Figure 10 shows an essential graphs of annual variability of two registered key water environment parameters for the representative dates (annual time intervals) revealed in Sections 3.1.2 and 3.1.3 (27 June SST; and September, chlorophyll-a) obtained as a result of verification calculations.

An analysis of the graphs given in Figure 10 allows us to see the deviations of the values of sea surface temperature (Figure 10, top) and chlorophyll-a (Figure 10, bottom) in 2020 relative to the set of measurement results performed in the historical retrospective. We can point out that the graphs shown in Figure 10 describe cases of registration of historical maxima of SST and chlorophyll-a. At the same time, the values of 2020 go quite far beyond the variability of the values of previous years, which corresponds to the results obtained in Sections 3.2 and 3.3. The additional "Data Availability Statement" section contains links to the program codes of the Google Earth Engine, allowing us to obtain the data shown in Figure 10.

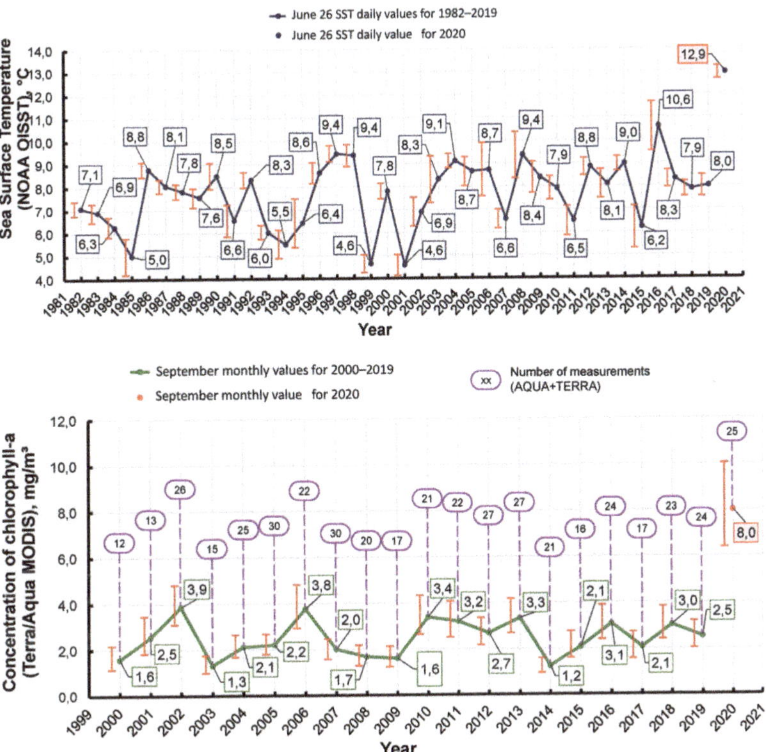

Figure 10. Annual variability of parameters averaged over the area of the study region: top, 27 June SST daily values for 1982–2020; bottom, September chlorophyll-a monthly values for 2000–2020. Red segments denote confidence intervals (95%).

The accuracy, Δ, assessment was carried out for the selected examples. Confidence intervals, $\pm\Delta$, were calculated (see red segments in Figure 10). Let's briefly ground the given accuracy values.

The NOAA OISST package includes a raster layer that characterizes the assessed SST error (standard deviation) for each pixel of daily information products [20]. The accuracy with a 95% confidence level was determined on account of the assumption of normal distribution:

$$\Delta_{SST} \sim \overline{\sigma} \cdot z_{0.95} \sim \overline{\sigma} \cdot 1.96 \tag{6}$$

where $\overline{\sigma}$ is the mean standard deviation of SST measurements within the studied site, $z_{0,95}$ is the quantile of a normal distribution for 95% confidence level.

C_{CHLOR_A} chlorophyll-a concentrations were measured repeatedly and then were averaged. In this case, the accuracy of the estimates obtained can be determined based on a priory known RMS of the algorithm that retrieves chlorophyll-a concentrations based on the measured spectral reflectance of the water surface. In the information product used in this effort, the values of chlorophyll-a concentration were obtained using the OC3M regression algorithm, while it is known that the RMS for Log10 (C_{CHLOR_A}) is 0.255 [36].

Taking into account the above, the accuracy of chlorophyll-a assessment on a logarithmic scale was determined as follows:

$$\Delta_{LOG-CHLOR\_A} \sim \frac{RMS}{\sqrt{n}} \cdot z_{0.95} \sim \frac{0.255}{\sqrt{\frac{n_{month-no-cloud}}{n_{site}}}} \cdot 1.96 \tag{7}$$

where n is the number of observations, defined as the ratio of the number of $n_{month-no-cloud}$ cloudless pixels used in the calculation (equipment of the TERRA and AQUA satellites) to the area of the studied water area (n_{site}) expressed in a number of pixels. The limits of the $\pm\Delta_{LOG-CHLOR\_A}$ confidence interval calculated on a logarithmic scale were converted to the $\pm\Delta_{CHLOR\_A}$ linear representation and plotted. In addition, n «floor» values are shown in this graph (see Figure 10).

In the given example, the accuracy of the obtained SST values in percent varied between 2.1% and 16.1%, and the accuracy mean for the whole observation period was 7.1%. The accuracy of the obtained chlorophyll-a data in percent varied between 18.7% to 37.8%, and the accuracy mean for the whole observation period was 25.2%. Shown in Figure 10 confidence intervals indicate that the abnormally high values of SST and chlorophyll-a concentrations registered in 2020 exceed the values of possible errors of the analyzed information products.

In conclusion, we note that the fact of HABs in Kamchatka in September–October 2020 was confirmed by the results of field studies, which is recorded in the HEADAT international list of HABs [37]. The samples revealed a high concentration of toxic substances related to the poison causing DSP (diarrhetic shellfish poisoning). The main causative agents of such toxic substances are dinoflagellate algae of the Karenia species. The concentration of harmful algae, according to the HAEDAT information sheets, increased from 152,000 cells/L (4 October 2020) to 482,208 cells/L (12 October 2020) and up to 622,000 cells/L (13 October 2020).

4. Conclusions

The processing results for long-term retrospective series of daily satellite data on the sea surface height (since 1993) and temperature (since 1982) have allowed us to observe the preconditions of intensive harmful algal blooms in Kamchatka in the fall 2020. Registered ocean level variability in the 2020 summer months affected the current mode (predominantly anticyclonic vortexes were observed) and microalgae food resource formation. The strongest temperature anomalies of the water environment on record (reaching 6 °C in comparison with the climatic norm in June–July), combined with the specific conditions of the circulation of coastal waters, provided extremely favorable conditions for the seasonal development of phytoplankton, which ultimately led to an intense red tide and had a negative impact on the coastal waters of the Kamchatka Peninsula (up to the mass death of hydrobionts).

The harmful algal bloom that occurred in Kamchatka in the fall of 2020 was most clearly recorded in the form of an abnormal increase in the chlorophyll-a concentration in the coastal waters of the Kamchatka Peninsula compared to previous years (2000–2019). The monthly average value of this parameter in October 2020 (blooming peak) approached 15 mg/m^3 what exceeds the climatic norm almost four-fold. At the same time, based on the results of space radar image interpretation, a large number of surface films of biogenic origin were recorded localized mainly near the coast, in the areas where rivers flow into the water area of the Avacha Gulf. During the study period (2019–2020), river runoff intensively introduced mineral and biogenic suspensions into the marine environment, expanding the food supply for microalgae (the values of total suspended matter within the Nalycheva River plume reached 10 mg/m^3 or more in 2020).

The main novelty of this effort is the study and presentation of a specific, previously unused experimental setup for a comprehensive study of long-term retrospective series of heterogeneous information products of satellite oceanography to explain the causes and features of a single large-scale ecological disaster (i.e., the harmful algal bloom in

Kamchatka which occurred in 2020). We can conclude that the final results are sufficiently valid taking into account the limitations of the data, methods, and approaches involved in the study, as well as the mutual correspondence of the results obtained on the basis of processing and analyzing information from various sources.

Other known events of intensive harmful algal blooms that had negative impacts on the ecosystems of marine areas can be studied using the suggested method, which allows one to process long-term retrospective series of satellite data on sea surface height and temperature, and chlorophyll-a concentration, as well as operative radar and optical imagery.

Author Contributions: Conceptualization, formal analysis, and validation, V.B. and V.Z.; methodology and software, V.Z.; SSH, SST, and chlorophyll-a investigation, O.C.; biogenic slicks investigation, E.M.; river plume investigation, V.K.; data curation, V.Z., O.C., E.M. and V.K.; writing—original draft preparation, O.C., E.M. and V.K.; writing—review and editing, V.B. and V.Z.; visualization, O.C., E.M. and V.K.; supervision, V.B.; project administration, V.Z.; funding acquisition, V.B. All authors have read and agreed to the published version of the manuscript.

Funding: The work was carried out with the financial support of the Ministry of Science and Higher Education of the Russian Federation, the Agreement No. 075-15-2020-776.

Institutional Review Board Statement: Not applicable.

Informed Consent Statement: Not applicable.

Data Availability Statement: Links to the Google Earth Engine codes for obtaining test data: chlorophyll-a key anomaly [38], SST key anomaly [39].

Acknowledgments: The authors are grateful to the Google Earth Engine service, the QGIS platform, and the SNAP software for providing the functionality that allowed us to conduct the research.

Conflicts of Interest: The authors declare no conflict of interest. The funders had no role in the design of the study; in the collection, analyses, or interpretation of data; in the writing of the manuscript; or in the decision to publish the results.

Abbreviations

AFAI	Alternative Floating Algae Index;
AVISO	Archiving, Validation and Interpretation of Satellite Oceanographic data;
C2RCC	Case 2 Regional Coast Colour;
ENVI	ENvironment for Visualizing Images;
GRD	Ground Range Detected;
HAB	Harmful Algal Bloom;
HAEDAT	Harmful Algae Event Database;
HYCOM	HYbrid Coordinate Ocean Model;
IDL	Interactive Data Language;
LAS	Live Access Server;
MERIS	MEdium Resolution Imaging Spectrometer;
MODIS	Moderate–resolution Imaging Spectroradiometer;
MSI	Multispectral Instrument;
NCEP	National Centers for Environmental Prediction;
NOAA	National Oceanic and Atmospheric Administration;
OISST	Optimum Interpolation Sea Surface Temperature;
QGIS	Quantum GIS (QGIS) is a user friendly Open Source Geographic Information System (GIS);
RMS	Root-mean-square deviation;
SAA	Surface Acting Agent;
SAR	Synthetic Aperture Radar;
SeaWiFS	Sea-viewing Wide Field-of-view Sensor;
SLA	Sea Level Anomaly;
SMI	Standard Mapped Image;

SNAP   SentiNel Application Platform;
SSH   Sea Surface Height;
SST   Sea Surface Temperature;
TSM   Total Suspended Matter;
VIIRS   Visible Infrared Imaging Radiometer Suite.

References

1. Anderson, D.M.; Glibert, P.M.; Burkholder, J.M. Harmful algal blooms and eutrophication: Nutrient sources, composition, and consequences. *Estuaries* **2002**, *25*, 704–726. [CrossRef]
2. Stumpf, R.P.; Tomlinson, M.C. *Remote Sensing of Harmful Algal Blooms: Remote Sensing of Coastal Aquatic Environments*; Springer: Dordrecht, The Netherlands, 2008; pp. 277–296. ISBN 978-1-4020-3099-4.
3. Cheng, W.; Hall, L.; Goldgof, D.; Soto, I.; Hu, C. Automatic Red Tide Detection from MODIS Satellite Images. In Proceedings of the IEEE International Conference on Systems, Man and Cybernetics, San Antonio, TX, USA, 11–14 October 2009; pp. 1864–1868.
4. Hu, C. A novel ocean color index to detect floating algae in the global oceans. *Remote Sens. Environ.* **2009**, *113*, 2118–2129. [CrossRef]
5. Wang, M.; Hu, C. Mapping and quantifying Sargassum distribution and coverage in the Central West Atlantic using MODIS observations. *Remote Sens. Environ.* **2016**, *183*, 350–367. [CrossRef]
6. Bondur, V.G. Satellite Monitoring and Mathematical Modelling of Deep Runoff Turbulent Jets in Coastal Water Areas. In *Waste Water-Evaluation and Management*; InTech: Rijeka, Croatia, 2011; pp. 155–180. ISBN 978-953-307-233-3.
7. Bakhtiar, M.; Rezaee Mazyak, A.; Khosravi, M. Ocean Circulation to Blame for Red Tide Outbreak in the Persian Gulf and the Sea of Oman. *IJMT* **2020**, *13*, 31–39.
8. Zhao, J.; Ghedira, H. Monitoring red tide with satellite imagery and numerical models: A case study in the Arabian Gulf. *Mar. Pollut. Bull.* **2014**, *79*, 305–313. [CrossRef]
9. Bondur, V.G.; Zamshin, V.V.; Chvertkova, O.I. Space Study of a Red Tide-Related Ecological Event near Kamchatka Peninsula in September–October 2020. *Dokl. Earth Sci.* **2021**, *497*, 255–260. [CrossRef]
10. Binding, C.E.; Greenberg, T.A.; McCullough, G.; Watson, S.B.; Page, E. An Analysis of Satellite-Derived Chlorophyll and Algal Bloom Indices on Lake Winnipeg. *J. Great Lakes Res.* **2018**, *44*, 436–446. [CrossRef]
11. Ahn, Y.H.; Shanmugam, P. Detecting the red tide algal blooms from satellite ocean color observations in optically complex Northeast-Asia Coastal waters. *Remote Sens. Environ.* **2006**, *103*, 419–437. [CrossRef]
12. Lou, X.; Hu, C. Diurnal changes of a harmful algal bloom in the East China Sea: Observations from GOCI. *Remote Sens. Environ.* **2014**, *140*, 562–572. [CrossRef]
13. Sakamoto, S.; Lim, W.A.; Lu, D.; Dai, X.; Orlova, T.; Iwataki, M. Harmful algal blooms and associated fisheries damage in East Asia: Current status and trends in China, Japan, Korea and Russia. *Harmful Algae* **2021**, *102*, 101787. [CrossRef] [PubMed]
14. Sukhanova, I.N.; Flint, M.V. Anomalous blooming of coccolithophorids over the eastern Bering Sea shelf. *Oceanology* **1998**, *38*, 502–505.
15. Bondur, V.G.; Grebenyuk, Y.u.V.; Sabinin, K.D. Peculiarities of internal tidal wave generation near Oahu Island (Hawaii). *Oceanology* **2009**, *49*, 299–309. [CrossRef]
16. Shi, W.; Wang, M. Observations of a Hurricane Katrina-induced phytoplankton bloom in the Gulf of Mexico. *Geophys. Res. Lett.* **2007**, *34*, L11607. [CrossRef]
17. Wang, M.; Lide, J.; Xiaoming, L.; Karlis, M. Satellite-derived global chlorophyll-a anomaly products. *Int. J. Appl. Earth Obs. Geoinf.* **2021**, *97*, 102288. [CrossRef]
18. Luchin, V.A.; Kruts, A.A. Properties of cores of the water masses in the Okhotsk Sea. *Izv. TINRO* **2016**, *184*, 204–218. [CrossRef]
19. Chassignet, E.; Hurlburt, H.; Metzger, E.; Smedstad, O.; Cummings, J.; Halliwell, G.; Bleck, R.; Baraille, R.; Wallcraft, A.; Lozano, C.; et al. Global Ocean Prediction with the Hybrid Coordinate Ocean Model (HYCOM). *Oceanography* **2009**, *22*, 64–75. [CrossRef]
20. Reynolds, R.W.; Banzon, V.F.; NOAA CDR Program. NOAA Optimum Interpolation 1/4 Degree Daily Sea Surface Temperature (OISST) Analysis, Version 2. *NOAA Natl. Cent. Environ. Inf.* **2008**. [CrossRef]
21. LAS (Live Access Server): Aviso+. Available online: https://www.aviso.altimetry.fr/en/data/data-access/las-live-access-server.html (accessed on 1 February 2021).
22. ESA European Space Agency—Missions—Sentinel-2. Available online: https://sentinel.esa.int/web/sentinel/missions/sentinel-2 (accessed on 13 May 2021).
23. ESA European Space Agency—Missions—Sentinel-1. Available online: https://sentinels.copernicus.eu/web/sentinel/missions/sentinel-1 (accessed on 13 May 2021).
24. NASA Goddard Space Flight Center, Ocean Ecology Laboratory, Ocean Biology Processing Group. Moderate-Resolution Imaging Spectroradiometer (MODIS). Available online: https://modis.gsfc.nasa.gov/ (accessed on 13 May 2021).
25. NASA Goddard Space Flight Center, Ocean Ecology Laboratory, Ocean Biology Processing Group. Visible and Infrared Imager/Radiometer Suite (VIIRS). Available online: https://www.nasa.gov/mission_pages/NPP/mission_overview/index.html (accessed on 13 May 2021).

26. Bondur, V.G.; Grebenyuk, Y.V.; Sabynin, K.D. The spectral characteristics and kinematics of short-period internal waves on the Hawaiian shelf. *Izv. Atmos. Ocean. Phys.* **2009**, *45*, 598–607. [CrossRef]
27. Ivanov, A.Y. Slicks and Oil Films Signatures on Syntetic Aperture Radar Images. *Issled. Zemli Iz Kosm.* **2007**, *3*, 73–96.
28. Erokhin, V.E.; Gordienko, A.P. Influence of organic pollutants on the growth of dinophytic microalgae. *Issues Mod. Algol.* **2019**, *21*, 48–55. [CrossRef]
29. Brockmann, C.; Doerffer, R.; Peters, M.; Kerstin, S.; Embacher, S.; Ruescas, A. Evolution of the C2RCC Neural Network for Sentinel 2 and 3 for the Retrieval of Ocean Colour Products in Normal and Extreme Optically Complex Waters. In Proceedings of the Living Planet Symposium (ESA SP-740, August 2016), Prague, Czech Republic, 9–13 May 2016; p. 54.
30. Pugach, S.P.; Pipko, I.I.; Shakhova, N.E.; Shirshin, E.A.; Perminova, I.V.; Gustafsson, O.; Bondur, V.G.; Ruban, A.S.; Semiletov, I.P. Dissolved organic matter and its optical characteristics in the Laptev and East Siberian seas: Spatial distribution and interannual variability (2003–2011). *Ocean. Sci.* **2018**, *14*, 87–103. [CrossRef]
31. Bondur, V.G.; Vorobjev, V.E.; Grebenjuk, Y.V.; Sabinin, K.D.; Serebryany, A.N. Study of fields of currents and pollution of the coastal waters on the Gelendzhik Shelf of the Black Sea with space data. *Izv. Atmos. Ocean. Phys.* **2013**, *49*, 886–896. [CrossRef]
32. Filimonov, V.S.; Aponasenko, A.D. Seasonal dynamics of suspended matter in waters of lake Khanka. *Atmos. Ocean. Opt.* **2013**, *26*, 524–531. [CrossRef]
33. Vakulskaya, N.M.; Dubina, V.A.; Plotnikov, V.V. Eddy structure of the East Kamchatka current according to satellite observations. *Phys. Geosph.* **2019**, *1*, 73–81. [CrossRef]
34. Andreev, A.G. Water circulation in the north-western Bering sea studied by satellite data. *Issled. Zemli Iz Kosm.* **2019**, *4*, 40–47. [CrossRef]
35. Mikaelyan, A.S.; Zatsepin, A.G.; Kubryakov, A.A. Effect of Mesoscale Eddy Dynamics on Bioproductivity of the Marine Ecosystems (Review). *Phys. Oceanogr.* **2020**, *27*, 590–618. [CrossRef]
36. Chlorophyll-a (Chlor-a) ATBD. Available online: https://oceancolor.gsfc.nasa.gov/atbd/chlor_a/ (accessed on 29 September 2021).
37. Harmful Algal Event Database. Available online: http://haedat.iode.org/ (accessed on 20 January 2021).
38. Google Earth Engine Code (Chlorophyll-a Key Anomaly). Available online: https://code.earthengine.google.com/0401cac5c4a3 cfcffed3501a08c98a36 (accessed on 20 August 2021).
39. Google Earth Engine Code (SST Key Anomaly). Available online: https://code.earthengine.google.com/40ab28733e0347cd56674 612f1c8369d (accessed on 20 August 2021).

MDPI
St. Alban-Anlage 66
4052 Basel
Switzerland
Tel. +41 61 683 77 34
Fax +41 61 302 89 18
www.mdpi.com

Journal of Marine Science and Engineering Editorial Office
E-mail: jmse@mdpi.com
www.mdpi.com/journal/jmse